D1486219

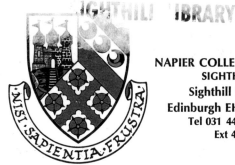

Alexander Holley and the Makers of Steel

Jeanne McHugh enjoyed a long association with the American Iron and Steel Institute. During that affiliation, she was responsible for organizing and establishing a technical and research library for the institute and also served as the assistant to the vice president of research and technology. She is a member of the American Society for Metals and the Society for the History of Technology.

JOHNS HOPKINS STUDIES IN THE HISTORY OF TECHNOLOGY

general editor: Thomas P. Hughes
advisory editors: Leslie Hannah and Merritt Roe Smith

The Mechanical Engineer in America, 1830–1910: Professional Cultures in Conflict, by Monte Calvert

American Locomotives: An Engineering History, 1830–1880, by John H. White, Jr. (out of print)

Elmer Sperry: Inventor and Engineer, by Thomas Parke Hughes (Dexter Prize, 1972)

Philadelphia's Philosopher Mechanics: A History of the Franklin Institute, 1824–1865, by Bruce Sinclair (Dexter Prize, 1975)

Images and Enterprise: Technology and the American Photographic Industry, 1839–1925, by Reese V. Jenkins

The Various and Ingenious Machines of Agostino Ramelli, edited by Eugene S. Ferguson, translated by Martha Teach Gnudi

The American Railroad Passenger Car, New Series, no. 1, by John H. White, Jr.

Neptune's Gift: A History of Common Salt, New Series, no. 2, by Robert P. Multhauf

Electricity before Nationalisation: A Study of the Development of the Electricity Supply Industry in Britain to 1948, New Series, no. 3, by Leslie Hannah

Alexander Holley and the Makers of Steel, New Series, no. 4, by Jeanne McHugh

The Origins of the Turbojet Revolution, New Series, no. 5, by Edward W. Constant II

The locomotive Whistler, painted by Alexander Holley in 1851, while at Brown University. Fir engine to run between Providence, Rhode Island, and Stonington, Connecticut, on the N York, Providence, and Boston Railroad. (Courtesy of Brown University.)

Jeanne McHugh

Alexander Holley
and the Makers of Steel

THE JOHNS HOPKINS UNIVERSITY PRESS
Baltimore and London

Printed in the United States of America

The Johns Hopkins University Press, Baltimore, Maryland 21218
The Johns Hopkins Press Ltd., London

Library of Congress Cataloging in Publication Data

McHugh, Jeanne.
    Alexander Holley and the makers of steel.

    (Johns Hopkins studies in the history of
technology; new ser. v. 4)
    Includes index.
    1. Holley, Alexander Lyman, 1832-1882.
    2. Businessmen — United States — Biography.
    3. Steel industry and trade — United States.
I. Title. II. Series.
HD9520.H58M3   338.4'7669142'0924 [B]   79-27414
ISBN 0-8018-2329-3

For Harrison

Contents

Illustrations

Acknowledgments

he writer of a biography travels a long road that has many unforeseen turns and bypaths. Because my subject, Alexander Holley, traveled widely, information about him is particularly scattered; thus, I owe a deep debt of gratitude to the helpful staffs of many libraries in several countries. The Connecticut Historical Society was an important source. In New York City, the Engineering Societies Library, the New York Public Library, the New-York Historical Society, and of course, the American Iron and Steel Institute Library, with which I was associated for many years, provided much material.

Also helpful were the Library of Congress; the Smithsonian Institution, particularly Philip Bishop, then its Curator of History and Technology; the Cleveland Public Library; and the Western Reserve Historical Society. The Bass, De Golyer, Western History Collections in the University of Oklahoma Library also were of great assistance, as was the university's Engineering Library.

I am especially grateful for the facilities provided me by the Iron and Steel Institute (London), the British Museum, and the Technische Hochschule in Vienna.

Miss Jean Wesner, Chief Librarian of Bethlehem Steel Company, has remained interested and helpful during the many years it has taken to write this book.

To the late Mrs. H. P. Bellinger, daughter of Alexander Holley's best friend, Rossiter Raymond, I am grateful for personal reminiscences. I am deeply indebted to Mr. Maurice Firuski of Lakeville, Connecticut, who so willingly helped me in my search for members of the Holley family.

I wish to express my sincere appreciation to the late Allan Nevins, who started me on my researches, and to the late Walter S. Tower and Charles M. Parker, who knew the path would be difficult but urged me on. I regret that neither lived to read the manuscript in print. I am grateful, as well, to Dr. Thomas Hughes, who understood the problems I would encounter and whose suggestions and comments eased my task.

To the Huntington Hartford Foundation I offer very special appreciation for the fellowship that made it possible for me to devote uninterrupted time to writing.

Not every biographer has the wholehearted support and even affection of the subject's descendants. I feel most fortunate in having had the dedicated interest of Holley's family: Mr. Francis Olmsted, Mr. and Mrs. Paul K. Randall, Mrs. Charles Rudd, Mr. John K. Rudd, Mrs. Ashbel Wall, and Mr. John W. Wall.

The book is dedicated to my husband, who lived with the successes and failures of Holley for many years. Although active in another field, he found great interest in visiting steel plants both in the United States and Europe with me. His encouragement meant a great deal to me.

I · In the Beginning

For centuries a glowing night sky has signified the presence of iron or steelmaking furnaces in operation. In modern practice, additional indications are present. Flaming coke tumbles from the long narrow ovens it has entered as coal. Skip cars scurry endlessly up and down the towering blast furnaces carrying iron ore, flux, and coke to fill the huge structure. These materials, which constitute the substance and strength of iron, work their way down the furnace stack, gradually absorbing the heated air piped in from the stoves nearby. By the time they reach the hearth, the raw materials will have been transformed into a molten incandescent mass.

The molten iron, heavier than the slag that floats on its surface, lies in the bottom of the hearth. It is removed through the iron notch, an opening that is plugged with clay except when the furnace is casting. In this operation the molten metal is channeled through a series of runners to refractory-lined ladles that take it either to the open hearth furnace or to the furnace used in the basic oxygen process. A small percentage of the iron goes to the pig-casting machine to be formed into solid blocks for use in the foundry industry. That part of the metal that is taken to a furnace will, with added ingredients and further processing, become steel.

Huge iron and steel plants, scattered from north to south and from ocean to ocean, are the mainstay of American industry. The American economy depends upon their production. In prosperous times, the molten stream can pour from steelmaking furnaces at the rate of more than 10 million tons per month.

There are many hypotheses about the discovery of iron. One early theory holds that prehistoric man discovered outcroppings that had been fused by raging forest fires or bolts of lightning. Some scholars believe that the first iron was meteoric. Support for their opinion can be found in the Egyptian word for iron, *ba-en-pet,* which is translated as "metal from heaven." Students of the works of ancient Romans like Pliny and archeologists who have excavated in Greece or translated hieroglyphics in Egyptian tombs agree that early civilizations were familiar with the working of iron and perhaps even of steel. A small piece of iron was found in the deepest recesses of the Great Pyramid,

Khufu, which was built about 2900 B.C. There it was protected from the damaging effects of oxygen. Exposed to air, iron usually rusts and disintegrates, leaving stains in the earth as the only traces of its existence.[1]

For centuries before the first settlement of America, iron was produced in Europe, as well as in Asia and even in Africa. From the time of Sir Walter Raleigh, iron ore was known to be present in the Colonies, but there is no record that the Indians ever made use of it. Its existence, however, was one inducement for colonizing the New World.[2]

Charcoal, produced from wood, was as important as ore in the early manufacture of iron. In 1558 the first of many such English laws was passed prohibiting the operation of furnaces or the building of new ones in areas where timber was scarce or likely to become so.[3] To produce two tons of iron a small furnace typically required the charcoal yield of an entire acre of woodland. So it is small wonder that the vast forests across the Atlantic attracted the ironmasters.[4]

What is thought to have been the first furnace in America was built in 1622 in Virginia on the banks of Falling Creek, not far from the present city of Richmond. Erected by a group of Englishmen under a patent granted by the Virginia Company of London, the furnace may not have produced even a pound of iron. The ironworks was demolished in an Indian uprising on 22 March 1622. The manager, John Berkely, and the workmen were massacred, and their tools were either destroyed or thrown into the river.[5]

The first significant production of iron in the Colonies took place in Saugus, Massachusetts, a quarter of a century later. For some years, successive managers of the furnace, which operated between 1646 and 1670, were in litigation with the owners. The Massachusetts court records are a source of detailed information about this first commercial furnace. All of the raw materials needed for its operation were nearby. Deposits of usable bog ore and rock mine were plentiful along the coast of Massachusetts. A cleverly proportioned mixture of rock mine and bog ores made the combination self-fluxing. Charcoal was produced from the trees of surrounding forests.[6]

As colonization moved toward the west and the south, more small furnaces and forges made their appearance. If the settlers were fortunate enough to discover a large deposit of iron ore and if the necessary water power also were available, the community flourished. Water power was needed to turn the huge water wheels. They provided the motive power for operating the large leather bellows that, in turn, supplied the air blast required for the furnaces and forges.

Such a combination was found in Litchfield County, located in the northwestern corner of Connecticut where Massachusetts forms its northern boundary and New York its western edge. Water power was abundant and the iron deposit was large, given prerevolutionary levels of production. The deposit extended down from Vermont through the corner of Connecticut and into New York and New Jersey. Because the composition of the ore was almost uniform throughout the vein, it became known as Salisbury ore, after the town in Connecticut where it was discovered.[7]

Most of the ore came from Old Hill (also called Ore Hill) in Salisbury, two miles west of Lakeville. Originally Lakeville was called Furnace Village and, before 1745, Salisbury bore only the name Town M. From about 1732, this area, coupled with the Davis and Chatfield mines a short distance away, and the lake bearing the impressive Indian name of Wonoscopomuc, formed a vital part of the Connecticut mining industry.

Old Hill, the largest of the deposits, was primarily responsible for establishing the reputation of the Salisbury area as a source of high quality iron. Lacking, as yet, any knowledge of metallurgical chemistry, the users of Salisbury iron ore could not know that the reason for the high quality of the iron produced in the district was the ore's remarkable freedom from phosphorus and sulphur.[8] The ore lay very close to the surface and the workmen mined it by the open pit method or by working horizontal drifts about thirty-five feet below the surface. The miners called the openings woodchuck holes, which describes their appearance.[9]

The fame of the Salisbury iron increased when Thomas Lamb, an energetic promoter, in 1734 built a forge at nearby Lime Rock that soon was producing from 500 to 700 pounds of iron a day.[10] A forge built in Lakeville in 1748 was operated with some success. In 1762 it passed into the ownership of a group that included Ethan Allen of Ticonderoga fame. This group added a blast furnace reputed to be the first real furnace to smelt ore in the valley.[11]

Like most early furnaces, it was built into the side of a hill with the top bridged to the nearby road so that the iron ore, limestone, and charcoal could be charged into the furnace without using a hoist. A continuous procession of carts loaded with charcoal and of horses carrying ore in leather saddle bags made its way over the planks to dump the materials into the open top of the furnace. The flame and smoke that belched forth produced a dramatic scene, particularly at night, but the workmen learned that the evil gases lurking in the escaping smoke could spell death, and they approached the spectacular display with caution.[12]

The furnace came into the ownership of Richard Smith, a Tory, in 1768. When the Revolution broke out, Smith considered it safer to return to England. The state of Connecticut seized the furnace and Colonel Joshua Porter was appointed as overseer to operate it for the benefit of the cause. The furnace remained in blast throughout the war, supplying cannon for the navy as well as large quantities of shot and shell. After the war, since the reputation of the iron had become so well established, the production of guns was continued. Toward the end of the century, the frigates *Constellation* and the celebrated *Constitution,* better known as *Old Ironsides,* were fitted with Salisbury cannon.[13]

Seven generations of Holleys had lived in Connecticut prior to the birth of Alexander Lyman Holley on 10 July, 1832. On his arrival from England in 1644, the first Holley had settled near Stamford. Others of the family later chose Stratford, a small town near the mouth of the Housatonic River. As early as 1740 the name of a member of the family appears in the town records of Sharon as being one of the original purchasers of the town. Other Holleys settled in the towns later to be known as Lakeville and Salisbury. This area, within a range of twenty miles, continued to be the home of succeeding generations, many of whom died in the houses of their birth.[14]

Until the time of Luther Holley, who was born in 1751 in Sharon, family members were primarily farmers. When he injured his leg so severely that he no longer could work on his farm, Luther Holley turned to other pursuits. After first teaching school he entered the mercantile business. Although virtually without schooling, he had a wide knowledge of literature. His admiration for John Milton was so great that he memorized all of *Paradise Lost* and named the first of his sons to live beyond infancy, John Milton. It was Luther Holley who first turned the interest of the family toward the iron mines and iron production of the district, but it was his son John Milton who would establish the Holley name in the industry.[15]

In 1777, the year in which John Milton Holley was born, Furnace Village was a hive of activity as the furnace worked day and night turning out shot and shell for the Revolution. Only a few weeks before the farmers at Bennington, Vermont, had stood their ground, as the village feared attack by either Tory partisans or General Clinton's troops which were bivouacked not far away.[16]

The effects of war were felt in the colony for some years thereafter. Growing up in this period, John Milton Holley felt unsettled about his future. His father, although still primarily engaged in agriculture, no longer worked his farm himself. Uninterested in a life as a farmer, the son cast about for some other occupation.

Young Holley was aware that he possessed a certain flair for mathematics and, therefore, began the study of surveying with Samuel Moore in Salisbury. In 1796, when he was nineteen, he joined a surveying team then being organized under the leadership of Moses Cleaveland to plot the Western Reserve.[17] The trip was rugged and hazardous. Details of

John Milton Holley house, now called the Holley-Williams house. Bought in 1808 by John Milton Holley, it was the home of Alexander Hamilton Holley until his marriage. Now a museum. (Courtesy of Harrison Kerr.)

the party's daily life are set forth in Holley's *Journal*.[18] Storms whipped up on shallow Lake Erie made sailing difficult and dangerous. Swamps and thickets made the overland treks slow and wearisome even for the tall and sturdy Holley. Danger threatened when, at Buffalo, Cleaveland and his party were met by unfriendly representatives of the Mohawk and Seneca tribes headed by the famous Red Jacket and Joseph Brant.

The six-month trip must have satisfied the youth's craving for adventure. The word *home,* printed in large capitals in a *Journal* entry as he left the Western Reserve testifies that his enthusiasm for exploring the wilderness had run its course. Shortly after his return, Holley married Sally Porter, daughter of the colonel who had taken charge of the furnace after its seizure from the Tory, Richard Smith.

Now there was plenty of work in Lakeville to keep him occupied. The mines were producing more ore than ever, and the furnaces, which had contributed their share to the Revolution, were still active. A year or two before 1800, Luther Holley purchased the furnace in Lakeville, the water rights, and a large share of Old Hill.[19] Young Holley joined his father in the enterprise. Inasmuch as there were no schools to teach him metallurgy, he served an apprenticeship in which he gained his knowledge from the men at the furnace.

Sometime thereafter John Milton Holley entered into partnership with John Coffing, a member of another Lakeville family that long had been connected with the iron industry. The new firm of Holley and Coffing increased its holdings by purchasing a half-finished furnace nearby on the steep slopes of rugged and desolate Mount Riga.[20]

Local interest in national political leaders was intense in the newly created state of Connecticut. John Holley long had admired Alexander Hamilton; and when a son was born in 1804, shortly after Hamilton's death at the hand of Aaron Burr, the child was given the dead states-man's name.[21]

As soon as he was old enough, Alexander Hamilton Holley, was sent to school in Sheffield with an older brother John. He was not happy either at this school or at several others that he attended. In later life he often commented that he had learned more outside these schools than in them.[22]

When his older brother entered Yale, it was assumed that Alexander would follow. However, the severe headaches and bad health that had plagued his youth, and his desire to enter the business world changed these plans. In 1819, at the age of sixteen, he began his career as a clerk in his father's counting room and store.[23]

Business activity in Lakeville in the early part of the nineteenth century depended almost entirely upon the output of the iron furnaces; the Lakeville-Salisbury district now was at the peak of prosperity. Ore was plentiful and lay near the surface; mines dotted the countryside and appeared even in some backyards. Now there were five furnaces in Salisbury alone, as well as many others scattered about the neighbor-hood, some as far away as the Massachusetts and the New York borders.

Charcoal piles as large as forty feet in diameter smoldered on the slopes of Mount Riga.[24] One of young Holley's jobs was to search for wood to keep the vital piles smoldering. Along with the skill, courage, and physical stamina he developed on the lonely rides, during summer and winter, through the mountain ravines in search of suitable timber came a woodsman's knowledge of trees.[25] Most valuable was oak, because of its tight grain; in the market it brought nearly twice the price of white pine. Young wood was desirable because the saplings produced a strong, hard charcoal. Old wood, if sound, was usable, but dead and decayed trees were useless.[26]

As the demand for charcoal increased, Alexander Holley's travels took him farther and farther afield. The trips were not without their hazards, for the charcoal burners with whom he had to deal were a wild and rough lot who led an untrammeled existence and paid little attention to the accepted ways of the community.

At other times, Holley worked in the store, where he gained a knowl-edge of human nature that served as an education. The different groups of workers, which generally kept their distances, had separate characteristics: the charcoal burners were notorious, lawless rascals; the miners were tough and, for the most part, ignorant; the men who produced the iron were, on the whole, more law abiding and better educated than the others. Young Holley soon learned that substantial amounts of tact and even caution were required to serve these motley groups.[27]

The varied scene proved of constant interest to the maturing boy. Communication with the rest of the country and the world was increasing steadily. The forge on Mount Riga was making anchors for the Greek frigates then sailing against Turkey. Potash kettles were carried by sleigh or wagon to the new counties in New York. Musket iron was made for the United States armories at Springfield, Massachusetts, and Harpers Ferry, Virginia, as well as for various private armories. Holley's brother George expressed the belief that life in the small New England town was at its most pleasant period when he wrote, "The intelligence, enterprise and sterling integrity of the people and the high religious and social tone of society made life worthwhile."[28]

Soon after he entered his father's business, Holley began to recognize his own inadequacies. His frequent contacts with uneducated and uncultured men presented an object lesson that he could not ignore. He now resolved to arise early each morning and to devote the time to study. With spartan will Holley stuck to his resolution. To the end of his life, he believed that while others slept, he achieved the best work of the day.

In 1822, young Holley's health began to decline. His father, perhaps sensing the young man's need for freedom as he reached the age of eighteen, or recognizing a propensity to wander that appears in the several generations of Holleys, granted him a year's furlough. Making his first trip to western New York state, Alexander visited the newly opened sections of the Erie Canal, then went to Niagara Falls and across New York to Boston. It was a satisfactory interlude, and the young man then was content to return to his duties.[29]

Although he persevered in his early morning studies, Holley was not unaware of the gayer aspects of life. His diary mentions frequent parties at which there was "agreeable conversation, musick on the piano, singing and dancing," as well as weekly dancing classes, sleighing parties, chess, and frequent drives in the company of young ladies of the neighborhood.[30]

In 1828, because of his continuing indifferent health, Holley left the long hours of the store. Instead, he supervised the operation of his father's farm and traveled on business for the company. The compensations of the latter are evident from an entry made in his diary a few weeks later: "Pittsfield with a load of iron. Passed time on Piazza of 3rd story of Russell's Hotel where we smoked a few segars, moistened our throats with a little mint julep and dried them again with a few dry jokes and some very agreeable conversation under the influence of a bright moon which made the jokes and drinking after all a mere matter of moonshine for that evening."[31]

For some months, Holley had been making frequent trips to Goshen to visit the attractive daughters of Erastus Lyman. Soon after his return from Pittsfield, he drove the fifteen miles to show off a new carriage. The next morning he noted in his diary, "Rode to Goshen at 9. Called at Capt. Lymans and found Mr. R. and the ladies all in fine spirits. After chatting an hour or so we invited the ladies to ride with us to town hill instead of riding to Salisbury as it was intended. It is often thus when there are ladies in the case and a man should be pardoned if all his expectations in business are not realized and he finds himself so pleasantly studied. I did not find myself exactly in a mood for sleeping—

House in Lakeville, Connecticut, where Alexander Lyman Holley was born. In an article that appeared in the *Lakeville Journal* for 18 March 1899, the Holley family historian, Malcolm D. Rudd, writes that John Milton Holley bought what now is known as the Holley-Williams house in 1808, when Alexander Hamilton Holley was four years old. Young Holley lived here until his marriage in 1831, when he moved into the house across the garden in which Luther Holley had lived from about 1782 until 1828. In that year the building was enlarged by adding a second story and garret, a lean-to eighteen feet wide that ran across the back of the house, and a piazza across the front. Alexander Lyman Holley was born in this house in 1832 and the child's mother died here. Alexander Hamilton Holley lived here until the house near Lake Wononscopomus was completed in 1853. The photograph was made from a glass plate negative and was taken on 6 March 1899, shortly before the house was demolished. (Courtesy of John Rudd.)

the retrospect was so agreeable and the occurrence of the day so strongly impressed on my mind."[32]

On New Year's Day of 1831, Holley took stock in his diary of his strengths and weaknesses. He believed his morals to have improved somewhat and his judgment of men and things corrected to a certain degree but, he wrote, "I am not entirely free from that odious and hateful practice of occasional profanity not because I delight in sin but force of habit and influence of many with whom I am necessarily in business and intercourse of life."[33]

By February, 1831 he was confiding in his diary, "My ride today with Miss Jane the younger Lyman daughter was decidedly more interesting than any I had ever before taken in the society of a young lady and has led me to believe seriously that pleasure of ones existence could be greatly increased by constant society of such a companion."[34]

Two weeks later, at an appropriate moment during a carriage ride, Holley proposed. Miss Jane kept him waiting three weeks for her answer, which was not forthcoming until long past the middle of March.[35]

The couple's marriage in October of 1831 was followed by a wedding trip by carriage to visit Holley's relatives in New York and Massachusetts so that they could meet his bride. Their house across the garden from that of the elder Holley was ready upon their return, and the young couple settled down in what is described as idyllic happiness.[36]

NOTES

1. J. Newton Friend, *Iron in Antiquity* (London: Griffin & Co., 1926), pp. 10, 30, 39, 92.
2. E. N. Hartley, *Ironworks on the Saugus* (Norman: University of Oklahoma Press, 1957), p. 28.
3. James M. Swank, *History of the Manufacture of Iron In All Ages*, 2d ed. (Philadelphia: American Iron and Steel Association, 1892), p. 50.
4. Arthur Cecil Bining, *Pennsylvania Iron Manufacture in the Eighteenth Century* (Harrisburg: Pennsylvania Historical Commission, 1938), p. 75.
5. Hartley, *Ironworks on the Saugus*, p. 37.
6. Ibid., p. 167. Rock mine first appeared in the company accounts very early and is mentioned in the *Winthrop Papers*, Volume 5, ed. Allyn B. Forbes (Boston: Massachusetts Historical Society, 1944), p. 239. It was a dense, igneous rock quarried on the nearby peninsula of Nahant.
7. Charles Rufus Harte and Herbert C. Keith, "The Early Iron Industry in Connecticut" (Paper presented at the Fifty-first Annual Meeting of the Connecticut Society of Civil Engineers, New Haven, Conn., 20 February 1935), p. 11.
8. Ibid., p. 12.
9. Harte and Keith, "Early Iron Industry in Connecticut," p. 12.
10. Alexander Lyman Holley, "Notes on the Salisbury, Conn., Iron Mines and Works," *Transactions of the American Institute of Mining Engineers* 6 (1878): 222.
11. Swank, *Iron in All Ages*, p. 129.

12. Chard Powers Smith, *The Housatonic* (New York: Rinehart & Co., 1946), pp. 259-60.
13. Ibid., p. 264.
14. Ronnie D. Judd. *The Educational Contributions of Horace Holley* (Nashville, Tenn.: George Peabody College for Teachers, 1936), pp. 1-3, passim. Judd obtained information about the early members of the Holley family from an interview with Malcolm Rudd, the Holley family historian, who died soon thereafter. Many of the early Holley papers in his possession were dispersed. Before 1790 the Holley name was spelled Holly.
15. *Alexander Hamilton Holley Memorial Volume* (Privately circulated, 1888).
16. Smith, *The Housatonic*, P. 261.
17. James Harrison Kennedy, *A History of the City of Cleveland* (Cleveland: Imperial Press, 1896), pp. 29, 31.
18. Ibid., p. 65.
19. Holley, "Notes on the Salisbury, Conn., Iron Mines and Works," p. 222.
20. *Alexander Hamilton Holley Memorial Volume.*
21. Ibid.
22. Ibid.
23. Ibid.
24. Smith, *The Housatonic,* p. 264.
25. *Alexander Hamilton Holley Memorial Volume.* The word *pit* is misleading. It meant the structure as a whole and was not a hole in the ground.
26. Frederick Overman, *The Manufacture of Iron, in All Its Various Branches,* 2d ed. (Philadelphia: Henry C. Baird, 1851), pp. 83-4.
27. *Alexander Hamilton Holley Memorial Volume.*
28. Ibid.
29. Ibid.
30. Diary of Alexander Hamilton Holley, 13 January 1829, Holley Family Papers.
31. Ibid., 14 August 1829.
32. Ibid., 14 September 1829.
33. Ibid., 1 January 1831.
34. Ibid., 10 February 1831.
35. Ibid., 23 March 1831.
36. Ibid., 3 October 1831.

II · Alexander Lyman Holley

On 20 July 1832, the year following their marriage, a son was born to Alexander and Jane Holley. The mother died within eight weeks of the child's birth, leaving him the heritage of her New England character and her maiden name. Alexander Lyman Holley's father reacted to the tragedy by closing the house and sending the infant to his maternal grandparents in Goshen, Connecticut. He soon realized, however, that his action did not solve his problem and that his sorrow must not be visited upon the child. During the following year, Holley reopened the house, engaged a nurse, and brought his son back to Lakeville.[1]

Three years later, on 10 September, 1835, he married Marcia Coffing, a childhood acquaintance who was the daughter of his father's former

Alexander Lyman Holley and his half-sister Maria at an early age. Date and painter unknown. (Courtesy of John Rudd.)

11

partner. [2] The romance had been slow in developing. Miss Coffing was deeply religious and doubted the possibility of becoming interested in someone who did not share her devotion to the Bible. In an extensive exchange of letters her suitor asserted his interest in religion and endeavored to convince the skeptical young woman of his devoutness. Miss Coffing, in turn, frequently expressed doubt about his sincerity, but the differences finally were resolved. She became the only mother that young Alexander Holley was to know, and although five children subsequently were born to this second marriage, the relationship between the two always was close and affectionate. [3]

After two or three years at the district school, young Holley was transferred to the academy in Salisbury. [4] As he matured it became evident that he had inherited his father's good looks: the generous mouth, the blue-gray eyes, and the abundant, wavy, light brown hair with even an unruly lock that fell over his rather high forehead.

Young Holley soon exhibited an exuberant and enthusiastic approach to life that his father never could condone entirely. He might climb to the head of the Old Man of the Mountain during a family vacation in the White Mountains of New Hampshire or disappear from the group if a railroad or locomotive were in the vicinity. One escapade fortunately never reached his father's ears. Early in his stay at the Salisbury school, his classmates dared one another to climb to the belfry of the school building. With great effort several boys did so, but Holley was not content with that feat. He climbed to the golden ball that surmounted the belfry tower and, to the shivering delight of the boys below, stood on the gilded sphere. [5]

Holley's interest in engineering became evident before he reached his teens. One of his first engineering projects was a miniature rustic bridge, which he constructed over the brook near his house. Proud of his accomplishment, he persuaded the entire family to attend the opening ceremony. As they walked solemnly across the structure, Holley, costumed as a troubadour, sat nearby playing a guitar. [6]

When young Holley was eleven or twelve, his father, noting that all pocket knives used in the United States were imported from Sheffield, England, razed the ancient Lakeville furnace and built a factory on the site for the manufacture of cutlery. The boy spent much time there watching this unfamiliar industry. Soon he was studying the machinery and brashly making suggestions for its improvement. Although some of the ideas were crude and impractical, to his father's amused astonishment, others were good enough for adoption. [7]

Having left the academy at Salisbury, Holley spent the next four years in various schools preparing for entrance to Yale. [8] During a stay at the academy in Farmington his interest in machinery in general and steam engines in particular developed from pastime to near obsession. His class was taken on an excursion to an old copper mine nearby. The event was important to Holley and he wrote excitedly to his father, "The steam engine attracted considerable of my attention, of course. It was splendidly made and fitted and went so still that one would hardly know that it was in the room. Mr. Hart [his teacher] told the gentleman that showed us around that he would have me draw a plan of the engine from memory which I have done, and which Mr. Hart is much pleased with. He says that he is going to send it to the aforesaid gentleman at Bristol."[9]

Although his son's enthusiasm for steam engines and railroads must have proved trying at times, the father chose to treat the youthful exuberance with care and discretion. If some of his exploits seemed to indicate the contrary, the boy did endeavor to please his father. His letters frequently displayed his eagerness to "get on the right side of virtue and hard work." In one letter he writes, "It seems as if I should dive and dig and plow, and if that did not succeed, back out and plow again, in my studies, as faithfully as the locomotive, old Connecticut did this morning in the drifts, with seven long cars, all alone."[10]

The young enthusiast read current papers and magazines with care, searching for information about locomotives and machinery. Letters to his father often consisted mainly of detailed descriptions of a new invention or idea. In one of them he noted, "I have seen, in a newspaper, an account of a man in England who makes steel that will cut iron or any hard substance without dulling it. I should like to hire out to that man for a year or so. ... I wish that I could learn the art of making steel."[11]

At Farmington Holley also became interested in literature and composition. His early efforts at composition were less advanced than his knowledge of machinery. In 1847 he joined Societas Literarum, a debating society. The society's constitution and bylaws are in Holley's handwriting, indicating that he may have been both founder and member.[12]

Early in February of the same year the fourteen-year-old railroad enthusiast wrote zestfully from Farmington that a new locomotive was establishing records in its run between Plainville and Southington by negotiating the distance of five miles in six and one-half minutes. Then, he added, possibly for his stepmother's pleasure, "Before I came here I thought I should be satisfied if I could have a watch. I had one and I thought that if I had a melodeon I would then never ask for anything else. But after I had one I was not satisfied and if I should have the whole world I would not have enough and I mean to try to get along now with what I have. There is, at present, a revival of religion in this place and several are hoping that they have become pious."[13]

Although he was a capable student in the areas of his interests, Holley was not always an obedient one. It was difficult for him to accept the school restrictions, and his frequent excursions to nearby towns to visit machine shops or railroad yards were an incessant source of trouble. Characteristically, instead of lying about his indiscretions, when he was in difficulties, his letters home became less and less frequent. The older Holley, sensing the situation, would as a consequence write more frequently, impatiently urging his reluctant son to do likewise.

Since both assumed that the boy would enter Yale, the elder Holley decided to send his son to the Cottage School in Bridgeport for the necessary tutoring in classics. Bridgeport, across the state from Lakeville, offered the boy a large industrial territory in which to roam. This, in combination with his social instincts, created difficulties almost immediately.

Toward the end of that year, his stepmother learned that Holley had been drinking spirits. How much he had consumed is not known; the fact that he had even tasted them was sufficient to horrify the family. His distraught stepmother did her best to make the boy understand the evil of such an action and young Holley, in turn, was properly repentant since he was well aware of the convictions of his parents.[14]

The reports that young Holley's father received from the Cottage School were discouraging. As the year wore on, the boy was doing less and less well. He seemed unable to settle down to the study of Greek, Latin, and the other subjects that were required for his entrance to Yale. He did continue his interest in composition. The kind of writing he was doing is indicated by a list that he sent his father:

1st. A Dream, personifying wisdom (a judge), History, Chemistry, Mathematics and Philosophy, who speak of their respective merits. Philosophy wins the prize. 6 pages of foolscap.

2nd. A Fragment (revised from a former) containing a few thoughts like those in Gray's Elegy. 2 pages.

3rd. History of Audax Promptusque. A novel. 26 pages.

4th. War (revised from a former). 4¼ pages.

5th. Pride (rather disconnected). 2 pages.

6th. A Picture. 6¼ pages. On this I have spent in all at least twelve hour's hard study, and it is the best composition I ever wrote. I would like to have you read it.

7th. A Dream and Consequent Wish. The best I ever wrote except No. 6.

8th. A composition in blank verse. Unfinished.[15]

The enthusiasm for locomotives still burned at fever pitch. The day in December 1848, when he succeeded in actually riding one from Bridgeport to Fairfield and was permitted to visit the engine house there was

so exciting that he could not resist writing his father about the achievement. To his father, this was nearly the last straw, and the harried
man replied acidly, "What about spending this time on lessons. To drag
along at heel of the class or school is not creditable to you and and will
only enable you to drag along at heel of your class in college which will
be still more disreputable. You must remember you have both a name
and a reputation to sustain. No one else can sustain it for you. It must
be done by your own efforts." [16]

When Holley, with six friends, arrived home from Bridgeport for the
summer vacation, a report from the principal, the Reverend Henry
Jones, followed him. It carried the news of such disobedience and poor
accomplishment that his father decided to send his son to Williams
Academy at Stockbridge in the fall of 1849. [17]

At Williams the head of the school, E.W.B. Canning, understood the
boy and his driving interests. Canning in later years remembered the
erring student as a fair, fresh-cheeked, blue-eyed wide awake boy of
sixteen when he arrived at the academy to pursue studies preparatory
to a college course.

> He had a geniality, generosity and overflowing good humor which made him
> popular with his schoolmates and gave him the lead in activities in and out of
> school. He excelled in every branch of study; but his interest centered in
> Natural Philosophy and mechanics. He was a prominent member of a literary
> society of the Institution called the 'Philologian,' in which he manifested
> talent unusual for his years for debate and free discussion; while his fun-
> loving propensities found vent in conducting mock trials and in humorous
> essays and declamations. He established and mainly conducted a periodical
> entitled *The Gun Cotton*, issued fortnightly on a large sheet in manuscript,
> which was read by myself from the desk, and afforded great interest and
> amusement by the variety and spice of its contents. This he edited during his
> stay at the Academy.
>
> Though excelling in all the branches of study required of him, his penchant
> for mechanics and invention developed itself markedly when he attacked
> Natural Philosophy and Physics. Dissatisfied with the meager description in
> textbooks of the steam engine, with which, he seemed to be better acquainted
> than the author of the treatise, he, at my request, made drawings in detail
> of a stationary engine and of a locomotive, with an accuracy and skill that
> would have done credit to a professional engineer or draftsman. These I used
> in demonstrating, in preference to the imperfect model among the school
> apparatus.
>
> During one of his vacations, he came up from his Salisbury home expressly
> to show me a miniature engine of his own building. It was complete in all
> respects, and of skillful workmanship; and on being fired up, ran with admira-
> ble success. Thus he foreshadowed the devotion to the mechanical arts which
> so eminently characterized his manhood. One number of his *Gun Cotton* I
> remember, was devoted to the description of an aerial voyage by a machine

of his own devising, whose practical workings were related with as much interest as the details of the wonders of sightseeing it enabled its inventor to describe. This, he prophesied, would be substantially the vehicle of locomotion in A.D. 1950. [18]

Holley wrote to his father about the drawings for the engine that Canning mentions. He had devoted his entire leisure for nearly two weeks to making sectional and perspective views of the internal works, machinery, and steam works of the most improved locomotive engines. By means of seventeen drawings and ten explanatory pages, he showed how the steam was produced, applied and cut off at half-stroke and how the engine functioned. He said that Mr. Canning was so pleased that he was going to have them framed and hung in the classroom. "I have explained them in such a manner that anyone can understand them, and I really hope that people will look at them, for there is more ignorance among scientific and educated men on this point than on any other. People who pretend to 'know the ropes' cannot explain the simple form of the steam engine, even with a model. Of mechanics and chemistry, I intend to get the most thorough knowledge if I have the opportunity (and, in fact, I intend to get it anyway) both practical and theoretical. These are the studies I have always liked and I am bound to investigate and become master of them." [19]

Canning must have believed that idle hands make mischief because he permitted Holley a limited amount of free time. There is a detailed description of his activities in a letter he wrote to his father as the end of his stay at Williams approached.

> I write you in great haste, to say, that at the close of the term we're to have an exhibition, I was first chosen to write and speak a valedictory oration. Then I have to write *The Gun Cotton* and a continuation of a treatise on the manufacture of pocket cutlery (which I have been at for several weeks) for the *Experimenter* all of this for the day of the Exhibition. Then the forenoon of the day is to be occupied with a Lawsuit. The Philologian Society have, for the occasion, resolved themselves into a 'high court of inquiry,' and I am chosen for 'State's Attorney,' to prosecute a man for assault and battery. This will be public. Besides this, I have to write a continuation of my treatise on cutlery for the *Experimenter* next week, and to write two numbers of another paper, I publish, called the *Locomotive*. Then I have to prepare also for a public trial next week in the Philologian, and also to learn (probably) a part in a dialogue for Exhibition, and all this in addition to Latin and Greek. My request is this, that I may omit my music lessons till next term, as I have my hands full, and shall burn some irons, if I don't look out.
>
> I shall also want a new coat to expatiate in on the great day. Please send as soon as convenient. [20]

Whether young Holley actually was engaged in all of the activities he listed or merely wished to avoid his music lessons remains a question.

The mention of music lessons does appear to be the culminating point in the recital of his participation in school activities.

In 1849, as the time neared for his admission to Yale, Holley became increasingly depressed as he visualized four years of study in the classics stretching before him; four years tied to Latin and Greek, which he dreaded and detested; four years away from locomotives and machinery. Tortured with the thoughts of such a period, he wrote a frantic letter to his father.

> I have had everything done for me that could be done, everything that kind parents could possibly do for a son. I have made three trials, and each one has proved unsuccessful—three trials to get an education and now having spent so much of your money and trouble, I wish to make one trial at work; and I know I can succeed. ... It is a waste, sir, I think to send me to school. It is not in my nature to deprive myself of an education if I was able to get it: but I see the folly of it. I have tried as hard as I could to learn out of books, but it is folly in the extreme. If I have any talent at all, it is for writing. When I lay myself out, and spend many hours on a composition, I can, by copying all the good parts, make out quite a piece, but this is all the talent I have got, anyway. Now I have got hands and can work. ... There is John [his younger half brother] a natural scholar, who could not bear hard work, and who will make by his mind alone, a great man; but I can work, and do nothing else to any advantage. ... I am very anxious, of course, to hear your decision, and will come if you say so, with a right good-will; but if not, if all looks dark ahead, I do not know what I shall do. ... P.S. Please write tomorrow, as I am very anxious to hear.[21]

His father, recognizing some of his own symptoms that thirty-one years earlier, had led him to enter business instead of college, was understanding but firm. After a time he answered.

> I am very sorry to see you yielding to so desponding a tone of spirit. You are not quite the dunce you would attempt to make me believe you are, nor have all your experiments to secure an education failed, as you would intimate. You certainly know more than you did, when you commenced your career abroad, about books, and about men and things. I have certainly had two or three reports to that effect. Why then do you despond? I have not complained of your scholarship in writing to you. ... This complaint has been on your part and not on mine. ... No, my son, you must not despond. If dark days now appear, brighter ones will follow. ... Have you often heard me say how grateful I always felt to my father, that he did not indulge me in my propensity for change? I think I should have been a ruined man if I had been permitted to hop from this thing to that, and hither and yon, just as fancy for the time might dictate. He said no, and I felt hard about it; but I have often since thanked him for that no. Besides, suppose you were to leave school and engage in business systematically (which you never have done yet) and were subject to the drill which you must be, when that time does come, would not that after a while, be as irksome as anything else? ... You are in the right way now, my

son, and your future happiness, usefulness, and respectability depend upon a continuance in this. Of this I assure you, as a father most earnestly desiring your happiness as well as usefulness.

If you are not ready to enter college at the next commencement, then go on till you are ready. There are more years to follow. Only make good use of your time, and we shall all be satisfied. . . . You say you think you have one talent at any rate. I think you have more; but cultivate that at all events and do all you can toward cultivating others. The cultivation of this very talent may yet be essential to your success in life. [22]

With little hope of an immediate change in the situation, the son prepared to accept the ultimatum and to go to Yale. He took his father's advice to concentrate on writing. His treatise on cutlery was expanded and dispatched to the *Railroad Journal*, which was edited and owned by Henry Varnum Poor. [23] The lengthy article was published in nine successive issues of the magazine. [24]

His father had been more deeply concerned than he had admitted by his son's expressed desire to avoid Yale. He had come to the realization that Yale and a classical course might be suitable for one son but certainly not for the other. In casting about for a scientific school, the elder Holley learned that an expanded curriculum that would emphasize science and its applications was being planned for the fall of 1850 at Brown University in Providence. In a letter to Canning, Holley suggested that his son now be prepared for Brown rather than for Yale.

Young Holley received the news with great excitement and relief, and all his enthusiasm exploded in a letter to his father.

I understand a new school is to be started in Providence which you want me to enter. Mr. Canning thinks it would be better for me to study Philosophy and Chemistry which I always liked and stand some chance of getting a name and fortune than to spend four years in pursuing studies which are a bore and which I cannot learn to enjoy. I can draw all day and half the night in perfect contentment and happiness, or build a machine or perform experiments in chemistry, but to study Greek for an hour, I feel fidgety, blue and nervous. I believe in doing what I like to do and what is naturally pleasant and then I can do it to advantage. In the present age of the world, the chemist and philosopher are looked to from all quarters for improvement in traveling, lighting up and firing up our world and in performing many useful operations they stand good chance to make a great name and a great fortune for themselves and to benefit the world by their discoveries. The names of Paine and Morse and Davy will never be forgotten and their discoveries will always benefit the world and be a blessing to mankind.

One more point. To study hard and diligently on the classics which are almost useless in after life are said to strengthen the mind and why will not intense and persevering study on that which will prove of infinite value in after life strengthen the mind full as much? I feel a new vigor and a new incentive to study in anticipation of going to a scientific school and of avoiding the dull and dreary life of college. . . . Shall I relinquish Greek?[25]

Holley entered Brown University in the fall of 1850 and soon was following a familiar pattern. Almost at once, Alexis Caswell, the professor of mathematics, was attracted by his student's interest in mechanics. Soon Holley was painting pictures of locomotives and constructing working models to be used in the teacher's classes.

A month after he had entered Brown, Holley outlined his intensive schedule in a letter home. "I have succeeded by working nights and in odd hours, in making a painting of the fastest locomotive in the city which I showed to the Professor of Mathematics. He was pleased with it, and asked me to go with him and look at the locomotive, and explain to him a new invention in locomotive engineering, called 'link motion,' which I did satisfactorily. I considered this quite an honor, as he has explained the steam engine to classes for a dozen years or more. He then took me into several factories where I could not have gone without him."[26]

Early in the semester, Holley's handling of money became a matter of discussion between father and son. The young man was on his own, for the first time, and his allowance never seemed to go as far as his father thought it should. In one long letter he explained why it was important that he buy an expensive ticket to hear Jenny Lind sing. Young Holley believed that the concert was one of those "extraordinaries which seldom occur."[27] His father was in no position to question this particular expense since he himself had paid four dollars a ticket to hear the Swedish singer.[28]

Although he wrote long letters to his father, the son seldom wrote to his stepmother. This appears to have distressed the elder Holley, for he wrote his son that she was hurt at not having heard directly from her stepson for some time. "You seem my son to be rushing through life too hurriedly. There seems to be hardly deliberation and system in your movements. I say this because your movements are rapid and sometimes not well considered." His father then expressed the belief that his son was not reading the letters from home with any care since he rarely referred to their contents except to acknowledge the receipt of checks.[29]

The reason why his stepmother received so few letters from him was understandable. When he was unwise enough to tell his stepmother that he had been to the theater, she wrote that she was pleased that he had told her so but continued, "Theater appeals to the passions, whilst the preaching of the gospel appeals to the nobler emotions of man. The theater presents life in an unreal and exaggerated aspect—the gospel presents a more sober and truthful view of it." Having established this premise, his stepmother devoted six additional closely written pages to an expansion of the theme.[30]

In this belief she was joined by the president of the university, Dr. Francis Wayland. In the spring of 1850, he had said in his report to the Executive Board of the university, "Allurements to vice and dissipation have increased to a painful degree in our city. Two theaters are open every night, concerts, lectures, billiard rooms are exerting all their solicitations." The committee appointed to review the problem reported, "While the Professors adhered to celibacy, and slept in the Colleges, they were able to exercise a kindly supervision of the Undergraduate. But when they exchanged the always anxious and not seldom vexing duties inseparable from the government of a hundred young men, for the light cares of domestic life the students were left in the exclusive possession of one of the Colleges, and in the other there was no one invested with authority except the Steward."[31]

Because there were no funds to pay for proctors, the committee suggested that the students live with private families. This move, which would have meant losing the rents from the students, was rejected. Instead, stricter supervision by the faculty was recommended.[32]

During the Christmas holidays, Holley found a shop where he could satisfy his deep yearning to operate machinery. Tea and dinner were forgotten in his anxiety to give a satisfactory performance. The work consisted of turning and filing. When he learned that the lathe that he had been entrusted to operate cost $500 he reported the fact to his father with great pride. The apprentice was convinced that if he turned to the occupation of machinist he could earn $3.00 per day and that, even if everything else failed, he could make $1,000 per year running a locomotive.[33]

With respect to studies, Holley excelled in mathematics and rhetoric, but his French was not satisfactory. His health was a constant source of discomfort during his first winter in Providence. He complained of toothaches and dysentery. In 1851 he began the year with measles and in March his frightened parents made a hasty trip to visit him when they learned that he was seriously ill with pneumonia. The period of convalescence was of some duration and, so that he might receive proper care, they took him back to Lakeville for a week.[34]

Upon his return to Brown, Holley buckled down to work in an effort to make up for lost time. In July, however, a letter from President Wayland concerning the son's behavior brought the elder Holley hurriedly to Providence to investigate. His son was not at his boarding house when his father arrived. After a considerable search, he discovered the young man working in a machine shop. The two spent the morning discussing the situation. Later, the father had an hour's visit with the president; as a result, young Holley both returned to the college and took up residence there. The remainder of the afternoon was spent in buying fur-

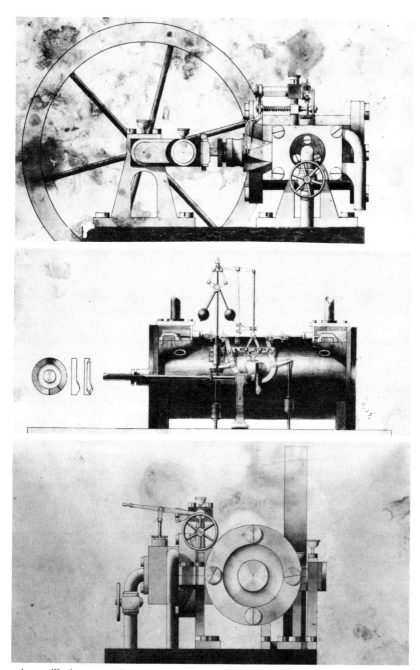

An oscillating, reversing, expansion, high-pressure steam engine. Designed and drawn by A. L. Holley in 1851. (Courtesy of Mrs. Paul Randall. Smudges are due to smoke damage.)

nishings for the room after which, much relieved, his father returned to Lakeville. [35]

Meanwhile, Holley had been turning his thoughts toward rectifying the known defects in locomotives. One of the most difficult problems in the development of the locomotive was the efficient use of steam. Intrigued by the possibilities of the future of steam, Holley invented an improvement that, he assured his father, would be patented if he had to sell the coat off his back to pay for it. By January 1852, he succeeded in putting his idea on paper. The invention provided for a cut-off of the steam admitted to the cylinder when the piston had traveled part of the stroke, allowing the steam to work expansively during the remainder of the stroke. [36]

With great enthusiasm he wrote his father.

> Professor Caswell says that it will work as well as Corliss's celebrated cut-off. Professor Norton says it is very ingeniously contrived and is better than Corliss's. Today I took it to Mr. Greene (of the firm of Thurston & Greene), the most extensive engineers of the city and he said that it would work as well as any cut-off he ever saw. He said that I could not get it patented, for this reason: A man some years ago got up a cut-off, and in his patent it was provided that he should own all improvements in the cutting off of steam by the regulator, wherein the valve-motion was obtained by a connection with the working part of the engine. My cut-off is entirely different in principle and form from his, and from the others; yet Corliss himself, whose cut-off is different from his and mine, is today involved in a lawsuit with the original inventor. . . . I feel satisfied at all events, that my plan is entirely original, and as good as anybody's. . . . I can have a picture and description of Holley's Improved Cut-Off circulated in *Appleton's Engineers Journal* free of expense, and shall immediately do so. You advised me when I was home, to consult Mr. Adams, C. E. of New York, about my oscillating engine. I did consult him, and found that my fundamental improvement, on which the rest depended, was patented in England by a celebrated engineer, some years ago. I thought it all out myself, and never heard of it until I had put it on paper myself. [37]

As time passed, Holley's relations with President Wayland worsened. In June, Mrs. Holley wrote her stepson that she had received a letter from the president saying that Holley could not try for a prize because he had not pursued his assigned studies as thoroughly as he should have and that his rhetoric was satisfactory as far as recitation was concerned, but that he had neglected his writing almost entirely. Furthermore, her son had manifested a spirit of insubordination in more instances than one and also was in the habit of using profane language. [38]

Holley was working so intensely on his mechanical projects that it was quite likely that his other studies suffered. In September 1852 his father asked, hopefully, if he thought that this might be the last year of his educational career. In light of the occurrences of previous years, Holley senior had reason to ask. Although his son optimistically assured his father that he would graduate, the final year presented difficulties. [39]

In February 1853, perhaps to escape the watchful eyes of the college authorities, Holley and two close friends left their quarters at Brown and moved to Mansion House. President Wayland must have written to Mr. Holley immediately concerning the move expressing dissatisfaction with young Holley's behavior, because on 7 March the father replied as follows:

Rev. and Dear Sir: Returning from New York Saturday evening I found your kind note of the 2 inst. I am much obliged by your Notice and equally distressed at its purport. When my son was last at home, I intimated to him my entire dislike of his former mode of boarding and expressed the wish that he should go into some respectable family—indeed insisted upon his leaving his eating room.

Soon after his return to Prov. he informed me that he had the consent of the Faculty to leave the College building for a room—that the Mansion House was a very respectable boarding house—that officers and many respectable students were boarding there and that transient boarders were not received. I felt quite pleased with the idea of his coming again under family influences, in a large family of as respectable boarders as those of the Mansion House were said to be and felt too that it was all agreeable to the Faculty. Your Note does not say indeed that it is not, but your responsibilities in regard to him are strictly modified. Of course, if there is anything in this change contrary to your wishes, or if this place is an unsuitable one for him to continue in, I am anxious to know it and shall be equally anxious to correct any improper course that he may have pursued.

I had become quite satisfied from some personal investigation (if I have not been greatly deceived) as well as from his improved state of health, that his habits were especially improved and was happy under the impression that he was making exertions to acquit himself like a man in his closing year. . . .

We have enjoyed his last vacation more than usual, and have put more confidence in his efforts and promises than we have done before in some time—and we shall feel greatly grieved if we find that we have been deceived.

I cannot but entertain a father's hopes and a father's anxieties that, if he has made efforts to correct his course, that he should be encouraged. If he is worthy of it, and is really making an effort to reform, I wish you would still not only bear with him, but also lend him a helping hand.

If he is not making such efforts as occasion requires, I should be pleased to hear from you again. [40]

President Wayland's previously expressed fears were ill founded and, probably to his father's relief, Holley graduated in September 1853, with the degree of Bachelor of Philosophy. His oration, "The Natural Motors,"

Alexander Hamilton Holley. (Courtesy of John Rudd.)

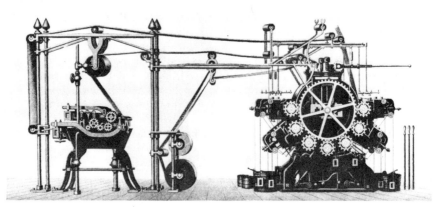

Drawing by A. L. Holley, Providence, 1853. Machine used for cleansing rubber blankets used by calico and De Laine printers.

Alexander Hamilton Holley house, built in 1852. (Courtesy of Harrison Kerr.)

was a typical commencement address with an added eulogy to the object of his deepest interest.

"Stand by yonder Railway," he declaimed, "Far off in the distance we catch low rumblings which draw nearer and increase in volume. On a sudden, with the speed and fury of the hurricane, a hissing hot, impetuous monster breathing through its dingy nostrils smoke and coals comes clattering down the iron way. The earth shakes under its ponderous tread, and the rent air closes behind it in a whirlwind. The ear is deafened by the hideous tumult of its whirling limbs, the eye bewildered by its lightening speed. Yet, we wonder at its presence, — we hear it rumbling in the distance."[41]

Two months after Holley's graduation the house known as Ion Hill, which his father had begun building in 1852, was completed. Located on the edge of Lakeville, it stood on a small knoll in a grove of fir trees some distance away from the road. Not far away was Lake Wononsco-pomuc. While the interior of the house with its large and sunny rooms was gracefully designed, the brown stucco exterior in which was embedded gravel dredged from the bottom of the nearby lake, represented a less attractive period of New England architecture.

Young Holley's stepmother had been ill since the preceding May and entered the new home as an invalid. She never was able to leave the first floor, where one of the sitting rooms had been converted to a bedroom. With aid of a chair on rollers, she could be taken from her room to the sitting room, where she spent long hours watching the lake. She died in 1854, the year following her stepson's graduation from Brown University.

NOTES

1. *Alexander Hamilton Holley Memorial Volume* (Privately circulated, 1888).
2. Ibid.
3. Ibid.
4. *Memorial of Alexander Lyman Holley* (New York: American Institute of Mining Engineers, 1884), p. 100.
5. Notebook of Maria Holley, Holley Papers, Connecticut Historical Society (hereafter cited as CHS).
6. *Memorial of Alexander Lyman Holley*, p. 101.
7. Ibid.
8. Rossiter W. Raymond, Holley's close friend in later years, states that he was at Williams Academy, Stockbridge in 1846 and 1847 and again in 1848 and the Cottage School in Bridgeport in 1849. The principal at Williams claims that Holley entered his school in 1848 but says nothing of 1846 or 1850. Letters and report cards indicate that he was certainly at Farmington during the winter term of 1847-48, at Williams in the beginning of 1849, that he went to Bridgeport later in that year and then back to Williams in January of 1850.
9. *Memorial of Alexander Lyman Holley*, pp. 101-2.
10. Ibid., p. 102.
11. Ibid.
12. The bylaws read as follows:
    1. It shall be the duty of each member to try to sustain the society.
    2. Each member shall pledge his word of honor to keep all the proceedings a secret.
    3. The business of this society shall be debating and composition reading.
    4. This society shall support a semi-monthly paper, called the _____ [name omitted].
    Bylaws.
    1. The officers of the Society shall be a Secretary and President.
    2. There shall be as many as eight members of the society.
    3. It shall be unlawful to use vulgar or profane language (Scrapbook, Author's Collection).
13. Alexander Lyman Holley (hereafter cited as ALH) to Maria Holley, 18 February 1847, CHS.
14. A temperance society had been formed in Salisbury a year before Holley was born and from the initial twenty members the membership increased to three hundred seventy in the first two years. When two members were so unfortunate as to break the pledge, a committee was organized to confer with the backsliders. The Salisbury group was proud of the fact that four blast furnaces and all the forges were managed without the use of spirits. This was indeed an amazing victory in an industry in

which the workers customarily received spirits in addition to wages and in which interference with the whiskey rations was considered a legitimate reason for a strike. The Salisbury statistics listed the five stores that sold spirits and the six that did not. Ninety farms were cultivated without their use, but sixty-four did include rations as part of the wages. The town admitted to the presence of fifty-six drunkards among its populace. (See Holley Papers, CHS.) At Maria Holley's behest, her husband had joined the temperance movement. The general sentiment of good society in Lakeville is revealed in a letter that his stepmother wrote to her stepson at about the time when he had been caught drinking.

The feeling of the public is much excited in relation to a shocking act committed by young Timothy Chittendon .... He took over the hearse to .... Edmund's grandfather to carry his remains to the grave. On his return he called at Mr. Bradley's, Falls Village and deposited in the hearse a barrel of rum for Mrs. Holcomb's establishment .... Steadying the load with the straps used in confining the coffins that contain the loved remains of our departed friends. There is a feeling of indignation felt on the subject. Many have resolved never to make use of the desecrated hearse again. There is a move on to buy one for the town (Maria Holley to ALH, 21 December 1848, CHS).

15. *Memorial of Alexander Lyman Holley*, p. 105.
16. Father to ALH, 12 December 1848, CHS.
17. Diary of Alexander Hamilton Holley, 1 May 1849, Holley Family Papers (hereafter cited as HF).
18. *Memorial of Alexander Lyman Holley*, p. 104.
19. Ibid., pp. 102–3.
20. Ibid., p. 106.
21. Ibid., pp. 107–8.
22. Ibid., pp. 108–9.
23. Poor was the grandfather of the painter of the same name.
24. There was considerable delay in payment, however, and having received no reply to numerous letters he wrote to Poor Holley took matters into his own hands and made a surreptitious journey from Stockbridge to New York. He walked before sunrise to West Stockbridge where he caught the train for Hudson and from there went down the river to New York on the steamboat *Alida*. Poor, no doubt surprised by the appearance in his office of the young man whose letters he had ignored, had the grace to apologize and to strike a bargain with the author. Holley was to receive $25.00 in cash and 27 copies of each issue of *Railroad Journal* in which a chapter of the article appeared—243 issues of the magazine. Records do not disclose whether or not Holley tried to carry the copies back with him. Later, the son admitted the trip to his father in a letter with an account of the return trip on a train that had two locomotives and eleven cars (*Memorial of Alexander Lyman Holley*, p. 107).
25. ALH to father, 15 June 1850, CHS.
26. *Memorial of Alexander Lyman Holley*, p. 109.
27. ALH to father, 30 October 1850, CHS.
28. Diary of Alexander Hamilton Holley, 5 November 1850, HF.
29. Father to ALH, 3 December 1850, CHS.
30. Maria Holley to ALH, 23 December 1850, CHS.
31. Walter C. Bronson, *The History of Brown University* (Providence: Brown University, 1914), pp. 294, 295.
32. Ibid.
33. ALH to father, 7 December 1850, CHS.
34. Diary of Alexander Hamilton Holley, 13 March 1851, HF.
35. Ibid., 14 and 19 July 1851, HF. Young Holley was not alone in his dissatisfaction with

the conditions as they existed at Brown. Wayland's efforts to enforce stricter discipline, as well as his arbitrary handling of several requests made by the students, brought the student body to the point of open revolt in November. Walter Bronson described the contributing causes for the revolt when he wrote, "There followed the so-called rebellion of 1851, during which the students used the 'most abusive epithets toward the President and Faculty,' and made disturbances in the chapel" (see Bronson, *History of Brown University*, p. 296). One of the requests to President Wayland was presented by the literary societies that wished to conduct their meetings in the evening rather than in the afternoon. Wayland refused permission, and the members rebelled in such a way that most members of the group were threatened with expulsion. Holley was as angry as the rest but, in spite of his deep dislike for Wayland, he did not join the insurrection for fear of jeopardizing his already uncertain standing. He wrote his stepmother that he believed the issues were not of enough importance to warrant his joining the malcontents but in his letter he worked off considerable temper as he narrated the story of the troubles at Brown (*Memorial of Alexander Lyman Holley*, pp. 110-11).

36. *Memorial of Alexander Lyman Holley*, p. 110.
37. ALH to father, 14 January 1852, CHS. Holley was in error about circulating the invention through *Appleton's*. The cut-off was illustrated and described in the July 1852 issue of *Mechanical Magazine*.
38. Maria Holley to ALH, 18 June 1852, CHS.
39. Father to ALH, 21 September 1852, CHS.
40. Father to President Wayland, 7 March 1853, Brown University Archives.
41. "Commencement Address," Scrapbook, Author's Collection.

III · Holley and the "Old Jigger"

O f the outstanding machine shops in the United States in 1854, two were in Providence: Thurston Greene and Company and Corliss, Nightingale and Company. The second, headed by George Corliss, was conducted in the tradition of the famous English inventor-machine shop combinations.

Corliss already was famous for the steam engine that bore his name; it was considered one of the most important advances since the time of James Watt. Now obsolete, this engine type was used when mechanical rather than electrical transmission of power predominated. Today, it is known mainly because students of engineering may be asked during the course of their studies to describe the construction and operation of the engine.

No doubt the rumor that had circulated in Providence that Corliss was working on a new idea for a locomotive had attracted Holley to the inventor. Holley's knowledge of and enthusiasm for locomotives was impressive. Convinced of the qualifications of the eager but untried engineer, the austere Corliss engaged him to work in his shop. Perhaps Corliss also saw his own reflection in Holley's intense interest in machinery.

Born in 1817 in Easton, New York, Corliss was the son of a surgeon but had received only a common school education. His first job, as a storekeeper, was one that he often neglected in favor of his interest in mechanics. Entirely self-taught, he invented an intricate and ingenious sewing machine for stitching harness leather that was patented three years before Howe's better-known sewing machine.[1]

Corliss traveled to Providence in 1844 to seek backing for his machine. Although unsuccessful in obtaining backing, he did secure a position as a draftsman with the firm of Fairbanks, Bancroft and Company. Within three years the company belonged to Corliss, and two years thereafter, in 1849, he invented the engine that would carry his name throughout the world.[2]

Holley's career with Corliss, Nightingale and Company began late in the fall of 1853. His concern over this first major venture that he could consider permanent made small problems seem disproportionately large. When his belongings failed to arrive from home, he was left for a week or more with only his best suit and four shirts. He was very conscious of his appearance, and was reluctant to appear for work in dress clothes. Much annoyed, he wrote his father a scolding letter insisting that he communicate immediately about the delay with the president of the railroad. His father must have ignored his son's plea inasmuch as Holley began work in his Sunday suit.[3]

Boarding in the home of Mr. Clark, second in charge at the Corliss shop, the new job holder considered that he had found a bargain. For three and one-half dollars a week, Clark provided him with a bedroom and sitting room as well as his board, fuel, and light. Corliss frequently joined the Clarks for lunch, giving the new employee an opportunity to become better acquainted with the man and his ideas. In addition, as he walked to and from the shop with Clark, he eagerly gathered knowledge from a man similarly dedicated to engineering.[4]

For the first few days of Holley's employment, both Corliss and Clark were out of town. Without an assignment, yet eager to prove his worth, Holley took it upon himself to repair the company's drawing tools. Impressed by the youth's initiative, Corliss suggested that sometime he would like to have Holley do a rendering of the engine then being constructed in the plant. To young Holley, "sometime" became "at once," and he set to work. The finished drawing, skillfully done, exhibited still another facet of Holley's talents.[5]

Corliss was a man who commanded from the employees feelings of admiration and respect mixed with awe. One described him as dignified, tall, and ruggedly handsome. His piercing black eyes, overhanging brows, and two slightly protruding upper side teeth resembled Thomas Nast's conception of Uncle Sam. Although a tall man, Corliss wore a high, pearl-gray stovepipe hat that increased his stature to such an extent that he could be seen by the workers from anywhere in the shop. He spent much time wandering about, watching, inspecting, correcting, and experimenting.[6]

Day after day, Holley listened as Corliss held forth on the subject nearest to his heart. The quality of the engines constructed in his shop was of great importance to the inventor. It was his theory that because of the pounding they received on the cobblestone pavements, horseshoes were the best source of iron. Huge piles of discarded horseshoes were kept in the yard adjoining the plant; they were heated in a special furnace and forged into bars to be used wherever iron of exceptional strength was needed.[7]

Corliss began his experiments with locomotives in 1851 and had been working on his theory of valve gear arrangements for two years when Holley came to him in 1853. Most of this time had been spent in working out changes in locomotive design. The actual construction of the engine had been started only a short time before Holley's arrival. He could see the locomotive being built but had not been asked to do any work on it. Repairing drafting instruments and making drawings for the various Corliss machines kept the young man busy but not necessarily contented. His hands fairly itched to touch the mass of metal, so that when Corliss asked him to prepare drawings of it and at last Holley could measure and work close to it, he felt deeply gratified.

"I can hardly realize that I am here and that you are on Ion Hill," he wrote his father. "However, I am already quite at home and I believe that the past three weeks have been the happiest three weeks with me, that I can recollect."[8]

Holley watched the engine's first unsatisfactory preliminary test runs on the Worcester Railroad with the critical eye of a professional. He was disgusted. As long as he was dallying with the measurements that kept him close to the mass of boiler plate, he was reasonably content, but after she was taken on her next unsuccessful run, Holley no longer could contain himself. He talked at length with Corliss about her.

In reporting the discussion to his father, he wrote, "Finally for some reason Corliss gave the machine into my hands and told me to have her finished and painted as I pleased and then to run her and test her capacity and at the same time to gain accurate knowledge of locomotive machinery in the best possible way. Then he wants me to make working drawings, and if his machine works to the satisfaction of all, to superintend the building of locomotives. There are a thousand things about all railroad machinery that no one can understand without seeing personally all the workings. Hence I am going into everything which can give me information on the subject." About the middle of January the locomotive was hauled up for the alterations that the young engineer had suggested.[9] The letters continued to arrive detailing the progress of the Corliss locomotive. Holley's son wrote about *our* locomotive although *my* would have described his feeling for the engine more appropriately.[10]

On 28 May 1854 the engine was finished and the Lakeville household received the final report. "She looks splendidly—more showy than any engine I ever saw and I doubt not that she will do the work triumphantly."[11]

The locomotive, christened the Advance, became the bosom friend, confidant, and instructor of Holley. In his estimation she was a fine engine. Others who had taken her on test runs, he believed, were prejudiced and did not handle her with the sympathy and understanding that she so badly needed at this stage of her career. Holley admired her principles and thought her arrangement excellent. Her few defects he would remedy after he knew her better. He was positive that he could "set this so far abortive principle on its legs and put seven-league boots on those legs." He also believed that the company out of gratitude and confidence would better his salary and that inasmuch as men under him received $500, he would be entitled to at least that amount.[12]

Holley at last was permitted to take the Advance on the tracks of the old Boston Railroad. He assured his father that there was little chance of a collision since there was only one train a day. The track was straight and laid out for fast running. Congratulating himself for having been given this opportunity, he went on to say that this job was the luckiest thing that ever had happened to him. Corliss had suggested that after the engine had been tested thoroughly, she might be run as an express train on the Harlem or the Hudson River Railroad to establish her reputation. As reflected in a later letter, Holley let his imagination run along the track ahead of him. He dreamed of taking the Advance out over the Cleveland and Pittsburgh road and, after that, taking charge of building locomotives, perhaps turning out one a week. He confessed to his father that Corliss had suggested only part of this but asserted that all of it might be accomplished and that he was going to endeavor to make it all come true.[13]

Entrusting responsibility to another, particularly to so young a man, was in direct contradiction to Corliss's usual behavior. Heretofore, he had insisted on being the sole head of the company and had hesitated to delegate authority. He took pride in his demonstrated self-sufficiency. Equipment was never bought if it could be produced in the shop, and he did not read technical publications for fear of being exposed to ideas other than his own. Busy with his engineering plans, he was impatient with financial details. As a result, the affairs of the company suffered.[14]

Corliss had a number of pronounced theories. Orders often were refused if they did not conform to his own design ideas. Because of his religious views, tests on engines never were conducted on Sundays. He also had been known to allow a distraught customer to wait weeks for a badly needed boiler because a robin had nested in a wheel of the only wagon large enough to transport the boiler.[15]

During the summer of 1854, Holley was either too busy to take a vacation or unwilling to leave the locomotive that Corliss had entrusted to his care.[16] During August the Advance was taken on frequent trial runs over the Stonington Railroad. She proved to be a difficult and unpredictable engine and Holley's patience was sorely tried. The frequent breakdowns made it increasingly difficult for him to defend her in his reports to the disgusted Corliss. Of one trip he wrote his father:

> The Advance was got out, tried and worked when she held together, to the perfect satisfaction of all. But she only made a few trips, when it was discovered, as I prophesied, that her machinery (valve work) was not strong enough. Twice it broke down- on the road. I stopped 'disconnected' and

brought in the train with one side (i.e. one of her steam engines, the other being taken apart) the first time, and the second, being near Stonington we disconnected, sent on the train, and loafed home with one engine at our leisure. We were cautioned to run slow, and we spent five hours in going fifty miles. Everytime we came to a good patch of blackberries we stopped, lay off, and ate at our leisure. This is the first time I ever went blackberrying on a locomotive. The Advance is now in the shop, probably for the last time. We are satisfied that she will save wood, and all that remains is to simplify the machinery to some extent, and make it stronger.[17]

The Advance made many such trips. Inevitably she limped home, urged on by the man who rapidly was becoming the only one who still believed in her. In later years, when time had eased the pain of the early frustrations, Holley could speak with some amusement of a typical trial run.

The idea began to obtain that science should be pursued not in books, but in things; and I commenced the pursuit of science in and on and under one of the awfullest things this world ever saw. It was Corliss' original locomotive, euphoniously called the 'Old Jigger.' This locomotive was of a certain inborn cussedness which could hardly be the attribute of a mere machine—her spiritual nature was a sort of a Mephistophelian cross with a Colorado mule; and as to her constitution and membership, a cotton-factory 'mule' was simple in comparison. The Old Jigger had, as nearly as I can remember, 365 valves, one to break down every day of the year. And as to valve motion, well, nobody ever counted the number of its pieces. They were as the sands of the sea-shore. Most of them used to jar off, the first few trips of the week, after which all the men in the shop could comparatively keep track of the rest of them. I will say for the Old Jigger that she made the best indicator-card I ever saw from a locomotive; clean cut-off, almost a theoretical expansion curve, and an exhaust as if she had knocked out a cylinder-head. Well, once in a while, after she had been jackassing over the road about four hours behind time, and we had pinch-barred her into the roundhouse, we used to pull out these indicator-cards of hers, and talk them over right before her, and we would look at her and ask one another why in thunder an engine that could make a card like that would act as if the very old—chief engineer was in her. And next morning she would rouse up and pull the biggest train that had ever been over the road—ahead of time.

But she was an inconstant old girl—and lazy, too—used to prefer to work with one side, and always made some plausible excuse for breaking down the other. I remember, one March morning, when nobody was looking, she kicked off about two dozen pieces of her starboard valve gear, and brought up all standing, over a culvert about ten feet wide and full of ice water. As I was standing in this culvert up to my middle, disconnecting her eccentric straps, a college professor came along, and rammed his umbrella into me, and asked me to explain to him the difference between this locomotive and any other locomotive. I then delivered my first scientific lecture; and I am now of the opinion that its diction would have been modified by a divinity student.[18]

Toward the end of September, Holley wrote a curious letter to his father:

> In private life my affairs are changed. I have always been desirous of doing well and being moral but never succeeded at all. My principles were not right. But now I am entirely disgusted with the old way of life and on new principles and for the strongest reasons have commenced a more reputable course. Once in my life I have fully made up my mind and the keeping of my resolutions comes easy, comparatively. Business prospects, characters, reputation and various circumstances respecting these have led me into the change. Perhaps I was more loose in my morals than my friends supposed. Now, I *have* given up every variety of intemperance as far as I can discover it. As for ever being under the influence of liquor I had rather be hanged and I hope I shall be when so seen. I have given up smoking, which I practiced to excess. I have left off tea and coff. I go to church regularly and read my Bible daily because it is proper to do so. I do not yet discover any other motive.[19]

In reply to this effusion, his father reminded his son that he had made resolutions before and asked wryly whether this letter indicated a recent reform. Realizing his father's skepticism, young Holley wrote in reply that he was arising at six in the morning in order to read an hour before breakfast and he threatened to rise even earlier if he believed it best to do so. He was setting aside three hours a day for the betterment of his mind: one hour for reading, one for writing, and one for painting. At the office he was busy making drawings of the Advance for the patent office. Swearing utmost devotion to this mechanism, he vowed that he would rather see his own arm broken than permit any harm to visit the locomotive. He laid such stress on his hermit-like existence that it might have appeared obvious to his father that something was wrong.[20]

Indeed, something was radically wrong, but it was not until 15 December 1854 that his position became so precarious that he was forced to confess his difficulties at home. Letters exchanged between father and son indicate that he had incurred a sizeable debt and had no means to pay it.

Holley had inherited eleven hundred dollars from his grandfather Lyman in July 1854. Whether the money was invested by his father or had been given to the young man, the results were the same. The knowledge that he possessed this amount of money must have had the effect of enlarging the opening of an already gaping purse.[21]

The older man, who patiently had extricated his son from numerous escapades over the years, was overwhelmed by the magnitude of his son's present trouble. The confession was a bitter blow to his father, who had hoped that his son had settled down, was working diligently, and had left behind his weakness for being what his father termed "a good fellow."

"Your letter of December 15 has been before me for two hours," his father replied. "I am utterly confounded and can scarcely permit myself to utter a word upon so little reflection. . . . I prefer to aid you in your good motivations at this moment rather than to argue abstract principles with you and to spare you of the fact I enclose my check to you for $354 the entire yes as you say the entire amount of your indebtedness. . . . It starts the perspiration on a gray and wrinkled brow to think how I am to replace this money in these times when men are suffering everything but crucifixion to obtain money for honest purposes. . . ."[22]

The father was hurt grievously that his son should have permitted himself to be placed in such a position. When young Holley wrote acknowledging the check and commented that, after months of sleepless nights and low spirits, finally he was at peace, the proud older man scolded him for thinking of his own comfort rather than the principle involved. When the repentant son, seeking some excuse for his behavior, said he could not be penurious, the indignant parent replied, "No man is penurious, who is economical to this last degree while he is in debt. No man is penurious who spends but one cent per annum, when his income is but two. There is no more dependent, degrading and mortifying condition to be in than that of being in debt. The money in your pocket, the coat on your back, the hat on your head is not your own while you owe for them."[23]

Young Holley had been frightened badly. He promised his father.

If I am not now able and ready to live right I never shall be and no resolutions or promises can make the case any more. . . . My situation and prospects are fine and my principles in general as far as I know, my judgement tells me, are right and if I am blessed with moral strength I can and will be somebody. . . . I have never written you a letter under these circumstances. To be delivered from the fear of exposure and to assure myself of confidence (which though always extended has never been met) constitutes one of the happiest eras in my history. But not until necessity drove me to it did I allow myself to take the great step. It is very interesting that a man should be such a fool as to withhold for years from his best friend the honest sentiments of his soul. That folly I intend to practice no more.[24]

His father was not impressed by extravagant expressions of gratitude. On New Year's Eve, the repentant son assured his father, "Your comments upon my principle, my judgement tells me are correct. It is not proper to spend the money until the money is at hand. But it is very difficult to believe and act in accordance with this principle. The extravagance of the times . . . has probably caused in a great measure the present money panic . . . [It] forces young men of no courage to spend in prospect of coming wealth . . . renders elegance fashionable and in a great measure substitutes show for value and dress for worth. . . ."[25]

The complicated situation in which his son had found himself remained so shocking to the thrifty, steady New Englander that his answer contained several acid comments regarding his son's future. The interchange was terminated in January 1855 with the following effusion. From the listing of the pleasures to be forsworn, the reasons for the young man's debt can be deduced.

Dear Father, I am mortified and surprised that I have not succeeded in satisfying you on the points about which you question and I am still more surprised that you think that the principles which I set forth will lead to indefinite results. You seem to attribute to me the using of the very spirit of the times which I deprecate and deplore as a grinding principle, while you make no mention of the other sentiments which are right if there is a god and a conscience.

Lest there should be mistake again, I would respectfully state once more, that my principles are these. 1st. Divine faith and worship. 2nd. The dominion of the mind. 3rd. Social excellence. 4th. Temporal prosperity as far as possible in accord with the above. The way I live up to these principles is very unworthy of the principles because I am weak, human and eminently prone to be wicked in every way. But I approximate as nearly as I can by, 1st. Attending church regularly, reading my Bible and avoiding a guilty conscience, by avoiding the cause of moral phenomenon as often as possible. 2nd. Devoting my leisure time assiduously to intellectual pursuits. 3rd. Cultivating social qualities, with which I am naturally endowed but which are sadly degenerate and inadequate to almost every occasion. 4th. By attending to business as earnestly as I can. I do not drink, not even a glass of wine with a lady and have lost some popularity (apparently tho not really thereby) I am absolutely temperate on the liquor subject. I do not go to houses of ill fame, I do not gamble nor play billiards nor any other game where money or time is lost. I have not been to the theater in three months and have had no social suppers nor rides nor reunions which have cost anything in six months if I recollect right. But I do take considerable pains with my dress, I do smoke, I am sorry to say and do not attend to my professional nor private duties as faithfully as I should but more faithfully than I have before. Altho the fact may not be as satisfactory I doubt not the statement is.[26]

Early in 1855, despite Holley's efforts, the performance of the Old Jigger at last convinced Corliss that his ideas for improving locomotive design were ineffectual. The Advance, which was the dignified name painted on her side, was superior in many ways to existing engines. Her economical use of steam was an excellent feature, but the detached, variable cut-off was too delicate to withstand the tough wear and the jolting caused by the rough track.

Even though this failure caused Corliss to abandon locomotive design, he hoped Holley would stay and work on the development of other engines. But Holley was steadfast in his devotion to locomotives, and decided to leave. On 27 March, 1855, he set out in search of a job where he could work with railroad engines. His letter of recommendation from Corliss was a good one: "We should be glad to avail ourselves of his services in our regular business of manufacturing Stationary Engines. But his mind is on Locomotives and therefor into that branch of Mechanics will he carry that Spirit and aim that will ensure success.

"As we do not propose to pursue the Locomotive business Mr. Holley leaves us and carries with him our best wishes for his success. [Signed] Corliss & Nightingale."[27]

With this and several other such letters, Holley began his search. At that time there were many locomotive shops scattered around the United States, most of them small. The railroads, still in their infancy, had less than 20,000 miles of track. The industry was still comparatively small; most people were not yet convinced of its importance and continued to regard canals and rivers as more desirable means of transportation.[28]

The condition of the railroad industry, Holley learned, was far from robust. The young man's search included every locomotive shop east of the Mississippi River, except one in Milwaukee that he already knew had nothing for him. Little locomotive building took place in Detroit, and no draftsmen were wanted. Cleveland had no position available, and there the manager assured Holley that if there was anything in the world he detested, it was "patented improvements" on engines. One firm in Cincinnati had a first class draftsman, another called in one for short periods only when needed.[29]

His disappointment and misery poured out in a letter written from St. Louis.

> It is strange, that when I have taken such pains to represent myself modestly, and am willing to do anything and work and persevere to the utmost extent, that when I have got such strong letters, and when I know that I can build a better locomotive than all the rest of them together — even in the face of all this, I cannot in the whole Western country get a place to earn my daily bread. The idea is damning to a man's spirits; though sometimes it looks so ridiculous that I cannot but laugh at the world generally. One thing is certain I will build locomotives; and if my life is spared, I will ultimately place myself in a position where I can look down on every man who has neglected me, and laugh with a good will at the bad luck which the improvements they scorn have brought upon them. It may be a long time before I shall get my head out of water, and I sometimes fear that my spirit will fail me before I see the daylight. If I sink, poor and unknown, I honestly believe I shall drag down with me some ideas of which, however humble they may be, the world is not worthy. ... I want to visit Philadelphia, Baltimore and New Jersey works, and see if I can convince myself and my friends that I am good for something. If I fail in doing this, I am ready to sink; for if there is anything

certain in this world, it is, that I will NEVER do anything permanently for a living, but just this one thing, namely, build locomotives.[30]

When Holley's hopes appear to have reached a low ebb, he was hired in September 1855 by the New York Locomotive Works in nearby Jersey City.[31] The superintendent, a Mr. Gould, admitted that the reason for his not engaging the young engineer sooner was that he had not taken him seriously. Holley came from an old family and showed the manners and bearing of a cultivated person. Gould had learned from experience that such young men invariably became disgusted with the dirt and grime of the locomotive shop and soon found excuses to leave.

Holley convinced Gould that he wanted to know every detail of the machinist's trade. He assured the reluctant superintendent that he desired to be in a position to tell any man who eventually might work for him how to do anything connected with the design and construction of locomotives. Holley believed that the most essential practice was to take part in the "rough and tumble of setting up locomotives."[32]

In addition to working with locomotives, Holley also served as drafts-man. He had developed a technique that was sufficiently unusual to attract the attention of Zerah Colburn, editor of the *Railroad Advocate*. Colburn wrote in the *Advocate* that the richness of color and brilliancy of the artistic finish of Holley's drawings was superior to anything he had seen. It was Holley's practice to lay on the watercolor thickly. Instead of producing a ragged or harsh effect, he succeeded in achieving a soft but, at the same time, brilliant tone. In Colburn's opinion, Holley was introducing a new school of coloring and shading in his drawings of locomotives.[33]

Late in 1855 or early in the following year, Holley met Mary Slade, the daughter of a well-known New York merchant. He was carried away by the first intense love of his life. "Petite," as he called her, supplanted the locomotive in his affections, and he wanted to marry her at once. But when approached by Holley for his permission, Mary's father stren-uously opposed the marriage. She was only sixteen and her father could not bring himself to believe that she was mature enough for such a step. Possibly Slade, who was acquainted with the Holley family, also may have thought that young Holley needed to settle down before taking on the added responsibility of a wife.

Having failed to gain permission to marry, the couple eloped. For some weeks the marriage was kept secret. Holley still hoped to persuade the inflexible Slade to consent to the marriage that already had taken place. At last, on June second, having abandoned that hope, he con-fessed the action in letters to both his father and father-in-law, and departed with his young wife for Albany.[34]

The letter to his father came as a thunderbolt, taking him completely

Alexander Lyman Holley and Mary Holley shortly after their marriage. (Courtesy of Mrs. Ashbel Wall.)

by surprise. He left immediately for New York, where he met John Slade at the National Hotel. Although still furious with the couple, Slade, assured Holley that his only reason for objecting to the marriage was his daughter's immaturity. After spending the evening discussing the matter with the calmer and more reasonable Holley, he accepted the accomplished fact and the two fathers sent a message recalling their children from their journey.[35]

An entry in the diary of the elder Holley describes his subsequent trip to nearby Lime Rock, where he told friends and business associates of the marriage. The same entry describes the young couple's June 7 return to Lakeville, where they spent the afternoon sailing the small boat *Maria* on peaceful Lake Wononscopomus.[36]

Holley's stepbrother John, then at Yale, wrote with delight to their sister Maria when he heard the news. "What do you think—Alex has been and gone and done it."[37] His letter probably did not surprise her, for Maria had received earlier a letter from Alex extolling the virtues and beauty of "Petite." Later her father told Maria of the details of the marriage and added, "It is not necessary for you to enter into explanation with anyone about this matter, any further than to say, that they were married some weeks since at the Metropolitan in New York with only a few friends present." Her father pointedly added, "I hope no other child of mine will ever marry without making me their first confident."[38]

For the remainder of the summer, the couple traveled between Lakeville and New York. As the days grew warm, the young bride occasionally remained in Lakeville, while her husband worked weekdays in New Jersey and spent weekends at home. In September they took a honeymoon trip to Niagara Falls and returned thereafter to New York to begin married life. Soon Holley's associates at the locomotive works grew accustomed to seeing the young wife trudging along with her husband as he conducted his experiments with the new coal burning locomotives.[39]

NOTES

1. J. A. Hall and G. W. Richardson, "George H. Corliss," *Mechanical Engineering* 55 (July 1933): 403.
2. Ibid., p. 405.
3. ALH to father, 23 November, 1853, Connecticut Historical Society (hereafter cited as CHS).
4. Ibid.
5. Ibid.
6. Hall and Richardson, "George H. Corliss," p. 408. The men, covered with oil and grease at the end of the day, never understood how Corliss's clothes could lead such a charmed life. At the end of a day in the shop they were spotless. No matter how much the men noticed his clothes, when Corliss made his appearance at the far end of the shop, it was the tall head piece that attracted their undivided attention. Tipped far back, they recognized it as a sign that all was well with their superior and the men stayed at their machines speaking to Corliss as he came past. But if the hat was drawn down over his eyes and they could see his lips were parted showing those two protruding teeth, the machinists hastily found urgent errands in another part of the shop.
7. The same practice took place in China up to World War II. Horse and mule shoes were shipped in burlap bags across the Pacific, then distributed to local blacksmiths. The shoes were heated in a crude forge, straightened out, and hammered into implements such as chisels, rakes, and hoes (Edwin C. Barringer, *The Story of Scrap* [Washington, D.C.: Institute of Scrap Iron and Steel, 1947], p. 188).
8. ALH to father, 3 December, 1853, CHS. Ion Hill later became known as Holleywood, the name it still bears.
9. ALH to father, 3 December 1853, CHS.
10. ALH to father, 12 January 1854, CHS. Because of his exaggerated aversion to the painting or addition of any fancy work on his engines, Corliss never had permitted such practice in his shop. But soon Holley was painting the locomotive and her tender with an elaborate design of his own invention and in colors to his own taste.
11. ALH to father, 24 June 1854, CHS.
12. ALH to father, 12 January 1854, CHS. The financial arrangements that Holley made when he began to work for Corliss are not clear. He commented in a letter written at this time that nothing had been said about salary. Since he already had worked as

painter, draftsman, designer and engineer, he was concerned that the subject of wages had not been discussed.

13. ALH to father, 6 August 1854, CHS.
14. Hall and Richardson, "George H. Corliss," p. 408. At one period, when repair bills had not been sent out for seven years, Corliss reluctantly delegated the task. Statements amounting to almost six hundred thousand dollars were sent out. It was exceedingly fortunate that almost the entire amount was collected. These idiosyncrasies, along with his inability to train others, brought about the eventual decline of the organization after Corliss died in 1888.
15. Ibid., p. 409.
16. ALH to father, 11 August 1854, CHS. He took one day off to sail in a friend's yacht from Providence to see the races off Newport. This was the first yacht race he ever had witnessed, and to be a part of it was a rewarding and exciting experience. The famous *Maria*, designed by Robert Stevens and without question the fastest yacht afloat, was sailed by John Cox Stevens, Commodore of the New York Yacht Club, and won the cup. To Holley's delight, he was invited aboard to meet the famous yachtsman and to inspect the ship.
17. ALH to father, 11 August 1854, CHS.
18. *Memorial of Alexander Lyman Holley* (New York: American Institute of Mining Engineers, 1884), pp. 112-13.
19. ALH to father, 28 September 1854, CHS.
20. ALH to father, 24 October 1854, CHS.
21. Father to ALH, 10 July 1854, CHS.
22. Father to ALH, 19 December 1854, CHS.
23. Father to ALH, 22 December 1854, CHS.
24. ALH to father, 27 December 1854, CHS.
25. ALH to father, 31 December 1854, CHS.
26. ALH to father, January (no day given) 1855, CHS. Holley's predicament evidently was known at Brown, since Professor Caswell wrote to the young man's father late in August 1855 that he was pleased to know that his son had been relieved from the unpleasant position in which he was placed by the suspicions of his associates. The older Holley was equally relieved, since he recently had been elected Lieutenant Governor of Connecticut and hardly could suffer the possibility of gossip concerning his erring son. (Caswell to Alexander Hamilton Holley, 23 August 1855).
27. Corliss & Nightingale to To Whom It May Concern, Scrapbook, Author's Collection.
28. "Railroad Progress in the United States," *Proceedings of the American Iron and Steel Association* (Philadelphia: AISA, 1873), p. 35.
29. *Memorial of Alexander Lyman Holley*, P. 114.
30. Ibid., pp. 114-15.
31. Ibid., p. 115.
32. ALH to father, 7 September 1855, CHS.
33. *Railroad Advocate*, 2 February 1855. Two paintings of locomotives survived Holley's lifetime. One of them, the Whistler, was given to Brown University in Providence, Rhode Island, by Holley's granddaughter, Mrs. Ashbell Wall. The other, a companion to the Whistler, along with much other Holley primary source material, was lost in a fire in 1973.
34. Diary of Alexander Hamilton Holley, 2 June 1856, Holley Family Papers (hereafter cited as HF).
35. Ibid., 4 June 1856.
36. Ibid., 7 June 1856.
37. John Holley to Maria Holley, 8 June 1856, CHS.
38. Father to Maria Holley, 13 June 1856, CHS.

39. Years later, when the daughter of Holley's oldest friend, Rossiter Raymond, was in the final years of her life, she wrote of the couple.

I was only a child, but I remember her vividly. Mrs. Holley was at our house quite often. . . . She was sparkling, the most magnetic woman . . . that I have ever known. Her face was bright with interest and sympathy, her hair was prematurely white and wayward — not untidy — and she had a bubbling flow of talk. . . . I think 'blithe and bonnie' would express my feeling about her delightful personality.

All that I could say of Mr. Holley was that he glowed while his wife sparkled. . . . His blond mustache helped the quiet smiling look, and his hair, not exactly curly was tumbled about his forehead. Of these two I'd like to add that their consideration for children was something sufficiently unusual for me to remember all these years (Mrs. H. P. Bellinger to Jeanne McHugh, 26 November 1951).

IV · The Great Eastern

hile working with Corliss, Holley had written numerous articles about locomotives for several periodicals, one of which was the *Railroad Advocate,* edited by Zerah Colburn. When Holley and Colburn met, they discovered that they shared a passion for locomotives and they soon established a close relationship. The two had been born in the same year, but into families of different means. Colburn's had been able to provide him only a few months' attendance at a district school.

Colburn's flair for engineering had developed so early that, at fifteen, he already had written the *Monthly Mechanical Tracts,* published in Lowell, Massachusetts. His book *The Locomotive Engine* appeared in 1850 before he was twenty. At twenty he became superintendent of the New Jersey Locomotive Works at Paterson, and in 1853 he was made editor of the mechanical department of the *American Railroad Journal.* Restless in this post, Colburn left it after a violent clash with the editor, Henry Varnum Poor; the following year he started the *Railroad Advocate.*[1]

Colburn was Holley's superior in the range and maturity of his professional knowledge, and this was to Holley's great advantage during the short period of their companionship. Soon after his arrival in New York, Holley began to write more frequently for Colburn's journal. When the restless Colburn, having bought Iowa land warrants, decided to build a steam saw mill on the property, Holley seized the opportunity to purchase the *Advocate.*[2] Samuel Cozzens, a friend and fellow student at Brown who was then practicing law in New York, helped in the financing by taking a one-half share.

Holley abandoned his position with the locomotive works and, with the issue of 19 April 1856, became the part-owner and editor of the paper. In August the name of the periodical was changed to *Holley's Railroad Advocate.* The little journal was badly printed on flimsy paper but, since technical journalism was still uncommon, it attracted favorable attention. Holley expended great amounts of time and energy in trying to make the venture a success, both editorially and financially. Colburn was still contributing a few articles, but the major portion of the journal's contents was written by Holley.

Zerah Colburn. From *Cassier's Magazine* 42 (September 1912): 194.

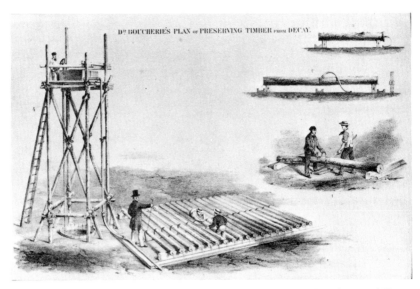

Example of drawing done by Zerah Colburn. Taken from *American and European Railway Practice*, 1861.

Design used on letterhead of the *Railroad Advocate*.

In his efforts to establish the paper, Holley traveled throughout the East and as far south as Richmond in a desperate search for subscriptions and advertising.[3] His willingness to assume any commission that could bring in money is evident from a letter to his father. He said that he had been authorized to sell the Tredegar Iron Works in Richmond and had received two hundred dollars as a retainer with a promise of two thousand in event of a sale, but, considering the temper of the times, he doubted that he would be successful. In this surmise he was correct. The commercial crisis of 1857 was already brewing and it soon burst with a violence that swept the country.[4]

By mid-1857, the *Railroad Advocate* was headed for bankruptcy. Colburn, who quickly had tired of pioneer life in Iowa, had returned to New York and rejoined Holley by buying back his former one-half share in the floundering paper. With the issue of 4 July 1857, they brought the *Railroad Advocate* to an end and the following week issued the first number of the *American Engineer*. As part of their efforts to arouse interest in the new journal, they made preliminary attempts to organize an association to be called the American Association for the Improvement of Railroad Machinery.[5] If the two young editors believed that such an organization could aid in the support of their newly named *American Engineer*, they erred in their estimate of the business situation. The journal ceased publication with the issue of 19 September.

Holley's father had been in no position to help the floundering papers even had he wished to. By 13 October, nineteen banks in New York had suspended operations and the elder Holley fully expected a run on the Iron Bank in Lakeville, in which he had a substantial financial interest and of which he was an officer. On the morning of the thirteenth, fearing the run would occur as the doors opened, the worried man, who had gathered every dollar he could collect, went to the bank and waited for the expected rush. Because of Holley's good name and integrity, there was no run, and the bank stayed open. But in a period of such financial uncertainty he did not dare sponsor a failing railroad journal.[6]

Although the two papers were complete financial losses, they did serve as a proving ground for young Holley. He gained valuable experience in the railroad field and made friendships and formed associations that proved helpful in future years. From this experience, although indirectly, stemmed the first great achievement of Holley's life.

At that time, the construction and management of railroads in the United States still lagged behind the practice in Europe. Holley and Colburn were commissioned to study European railway practices and to report on those features that might be of importance to American management. The presidents of seven American railroads helped sponsor the trip.[7]

The two established headquarters in London, and Colburn took the lead in showing his colleague the sights. London filled Holley with admiration and wonder. "Everything is astonishing, entertaining, instructive and in some cases overwhelming. The Tower fairly reeks with interest," he reported to his sister Maria.[8]

As their first step, the two engineers sought out Daniel Kinnear Clark, whose book on railway machinery, written a few years before, had given him standing both in England and in the United States. The two were impressed that Clark's office was located in Great George Street. Just as a Harley Street address long has been associated with successful men of medicine, so Great George Street had similar connotations for engineering. The quarters of the influential Institution of Civil Engineers were located there. The nearby Houses of Parliament and Westminster Abbey added to the venerable atmosphere of the area.[9]

The first person to greet Holley and Colburn when they arrived at Clark's was James Dredge, who was the son of an eminent engineer and later would become the distinguished editor of *Engineering*. At that time he was serving his apprenticeship as a clerk. To the inexperienced Dredge, the sight of these strangers from across the Atlantic was unforgettable. They appeared as sudden apparitions in the dark and gloomy office. Their exuberance and vitality dazzled the sober young man. In his efforts to compare the two, he concluded that the effervescent Holley seemed like the spirit of light, while the dark and almost sinister Colburn was like the spirit of darkness.[10]

For three months, the two engineers worked with great concentration as they prowled England collecting a vast amount of material covering all phases of railroad activity. But the trip was not all work. Christmas night of 1857 found the two ensconced in their quarters in the Exeter Hotel celebrating the holiday. It was a gay and festive occasion.[11] As Dredge commented in later years when he could look back on Holley and Colburn from the vantage point of maturity, "They were of an age when they could afford to burn candles at both ends and they did under forced draught."[12]

On New Year's Day, 1858, Holley sailed for home on the *Ariel*. In a handwritten account, Holley described the tempestuous voyage. For six days the waves grew higher, and the wind howled with a force that the inexperienced sea traveler could hardly believe possible. On the morning of the seventh day, Holley awoke to a strange sound, coupled with a different motion of the ship. The familiar throbbing of the engine was missing, and the regular pitching of the tired ship had been replaced by an undisciplined rolling motion. Reaching the deck with all possible speed, he discovered to his horror that the paddle shaft had broken just outside the hull and that one wheel was banging against the planking with every roll of the vessel as she wallowed in the trough of the high seas. Captain Ludlow tried desperately to hoist enough sail to get steerage way but the foreyard snapped in two. The ship now had sprung a leak and the bilge pumps were started to keep the water down. The gale rose to hurricane force and the ship barely was staying afloat. The terrified passengers huddled in small groups and endeavored to keep together as the ship lurched and shivered from the impact of the waves and wind. When it seemed certain that she could not survive another day, the storm suddenly subsided and the bruised and broken vessel limped back to Queenstown. Holley's next attempt to reach home was without incident.[13]

The railroad book bearing the long title *The Permanent Way and Coal-Burning Locomotive Boilers of European Railways; with a Comparison of the Working Economy of European and American Lines, and the Principles upon which Improvement Must Proceed* appeared late in 1858 and immediately attracted attention. It was a handsome volume of 192 pages of text, 260 illustrations, and 51 engraved plates.[14]

Colburn's responsibility had been primarily the section comparing the operation of English and American railways. He showed that the annual operating expense of American railways was $120 million against $80 million for the same mileage in England; that the maintenance of the road bed in America came to $33 million, whereas the cost in England was only $12 million; and that fuel cost was $18 million against $7.5 million in England.[15]

The British press was delighted with the book and gave it, as well as the discussions that followed its publication, much attention. Struck to the quick, American editors printed a variety of excuses for the differences.

Some said that the English lines had been built expensively and that what was saved in current expenses had been paid out in interest on the first cost. Others thought that labor and materials were cheaper on the other side of the ocean and that this could account for the variation.[16] Henry Varnum Poor, editor of the *American Railroad Journal,* was much disturbed. Enraged, he termed the book an insult to American railroads and launched a counterattack. Both Holley and Colburn already were known to Poor. Although Poor could not still have resented young Holley's endeavor to collect his fee for his articles on knife making, Colburn had left Poor's journal under circumstances that were far from friendly. More than likely that was what formed the basis for the vitriolic attack. When the article, deriding the book and its editors, appeared in the issue of Poor's paper for 5 February 1859, the ordinarily calm Holley threatened to sue Poor for libel unless he made a public apology. The irate Poor had written:

> The simple fact that Messrs. Holley and Colburn have done what they could to disparage our roads is the great reason why their report has been so warmly commended in England. They are held up as *experienced* and *conscientious* engineers; while, in fact, neither of them is, nor ever has been, a railroad engineer, either by experience or training. . . .
>
> We know nothing against his [Holley's] character as a man. It is well known, however, that Mr. Colburn is an empiric; ignorant, conceited and superficial. For years he drifted round from shop to shop and from place to place. Wherever employed, the parties found themselves anxious to get rid of him. At last he conceived the idea of getting up a railroad paper upon the Spread Eagle plan. In this he was joined by Mr. Holley, which was his first appearance on the railroad stage. They floundered around at a great rate for a while. . . . They abandoned their paper between two days, apparently. as it contained not the slightest hint of its approaching decease. It fell from sheer inanition. Our railroad companies would have nothing to do with these *experienced* and *conscientious* engineers. The first thing we heard from them afterwards, was their wonderful book, to which the article copied refers, and in which they have done what they could to revenge themselves upon our railroad companies for the cold support received from them. Those who know Mr. Colburn will readily understand his motive. He has, to a certain extent, accomplished his object, for we can bear testimony that his book has excited a powerful influence in discrediting our railroads in England.[17]

Colburn also was furious and urged Holley to start suit immediately. No matter what his faults may have been, Colburn had served the railroad industry well. At eighteen he had written the book *The Locomotive Engine*, he had worked on locomotives in Boston, and he had set up the department to build locomotives at the well known Tredegar Iron Works in Richmond. Later he had been superintendent and consulting engineer at the

New Jersey Locomotive Works in Paterson. All of this Poor certainly must have known.

Shortly thereafter, Colburn returned to England. A joint action was not possible legally, and in England Colburn merely could boil with anger. Holley, however, did take action and wrote in a letter to Poor:

> You will recollect that I agreed not to take any legal steps about your article of February 5 till you had heard from me again. You propose to publish a fair argument about the merits of the book and to correct the general impression that I am a tyro and an imposter, provided upon inquiry, you find my statements corroborated — but you do not agree to make such absolute retraction and denial of your statements as would convince your readers that your attack on me was totally groundless, and a pure invention on your part. No mere counter statement would establish such an impression in the minds of every person who had heard of and read the article. They would still believe there was some ground for your attack, whereas, you acknowledge your total ignorance as to my antecedents and position, of which you speak so confidently.
>
> Mr. Poor, it would be gross injustice to myself and to my profession, to allow you, at the cheap rate you propose, to indulge in statements which would totally and irretrievably damn me, professionally, — if they were true. After further reflection and consultation, therefore, I am obliged to say, altho, I regret its necessity, that I shall commence legal proceedings. All I ask, however, is to be put in a position, which will enable me to afford to loose [sic] the confidence and patronage of those who have been influenced by your attack on me. I have no desire to push to extremes, such legal proceedings, and their attendant publication of your position, as would at least be lawful. I have no revenge to satisfy, and do not ask compensation for sacrifice of feeling. If possible, I would secure a simple justice, without injuring you in the slightest degree. Should you think my conduct severe, you must remember that I am the injured party, and that you are not the proper person to prescribe what I shall be satisfied with. I should be entitled to the most revengeful attempt at punishment, so it were legal — while even this enclosed pamphlet shows at a glance that you have not the shadow of defence. I must refer you, with reference to further steps in the matter, to my attorney, Mr. S. D. Cozzens, No. 7, Wall Street.[18]

The enclosure was a twelve page booklet that described the volume in question. The booklet contained the opinions of both English and American engineers and railway managers as well as reviews from the British press. The letters from Americans showed without doubt that the executives of the railroads did not share the opinions of Mr. Poor.[19]

The outcome of the affair was that Poor finally printed a short apology in an inconspicuous position in his journal. Although it was far from satisfactory in Holley's opinion, he accepted it and took no further action.

The next few years were difficult ones for Holley. With little financial backing and with only the experience gained during the demise of their two papers, Holley and Colburn had issued a book that would have strained the resources of an established publisher. Holley often has been referred to as a wealthy man, despite what is known about his continuing bouts with creditors during these years. Even the sizeable estate left him by his mother had been more than half-spent by his twenty-fifth birthday.[20] Nor could the elder Holley, while well off, have been considered a rich man. When his father, John Milton Holley, died in 1836, his estate amounted to $114,000, of which young Holley's father inherited $15,000. Through his careful management of the iron interests and astute handling of his other affairs, this amount probably had been increased modestly.[21]

Since his college years, difficulty in managing his finances had plagued young Holley continually. His father often had been forced to rescue his son from the threats of insistent creditors, although the sums paid usually did not take the form of gifts. Notes for varying lengths of time piled up steadily. In 1856 Holley's father had been elected governor of Connecticut. Aware of the dignity and integrity demanded by his office, he had become more severe in his attitude toward his son's financial affairs. At his father's insistence, Holley made periodic efforts to keep records of his expenditures and income. However, these spurts of bookkeeping never were maintained consistently, and before long he again would be in debt.[22]

Young Holley now was twenty-seven, and his chances of a career seemed as uncertain and remote as they had been the day he had left Corliss and Nightingale with the intention of spending the rest of his life working on locomotives. He had conceived working for the New York Locomotive Works as his goal but, when the opportunity of purchasing a paper was offered him, he had not resisted. With Colburn, he had issued an impressive book on railways that had been received enthusiastically. Characteristically, he had spent $8,000 for engravings, fine printing, and superior binding. It was a handsome publication and a proud achievement, but the bills were, for the most part, unpaid. Of 1,200 copies printed, 1,016 already had been sold in 1858, but not all had been paid for. Holley hoped to meet his obligations when he sold the rest and had collected for those already delivered.[23]

His relations with his father again were strained. The older man no longer concealed his impatience at his son's frequent pleas for money. His letters usually contained little news, but instead a request that a note be taken to his father's Iron Bank for discounting. Or, if that were not acceptable, his son would suggest that a new note be substituted for a past due obligation.[24]

As the situation worsened, Holley became depressed about his inability to acquire or manage funds. In despair, he wrote, "I get so tired and miserable chasing around after money that I have no heart or strength to write."[25]

Even an accountant at that time would have had difficulty analyzing Holley's affairs. Debts were shifted from note to note and frantic letters dispatched to purchasers begging for immediate payment. Once books were borrowed from early purchasers and resold. Discovery of this piece of legerdemain violated his father's sense of honesty. He was outraged.

In a lengthy letter to his father, Holley explained that he had the sheets for 1,220 books but the engravings for only 1,000 copies. Thus, he had 1,000 complete books on hand at the beginning and part of 220 more, and it was necessary to have prints made for the additional 220 copies. He endeavored to convince his father that his friends who owned copies didn't mind giving up the books temporarily since Holley would replace them from the additional printing—which he did.[26]

Colburn now was in London editing the *Engineer* and, although he did nothing to distribute *The Permanent Way,* he continued to demand what he thought should be his share of the proceeds. A side of his nature that Holley had not experienced previously now began to appear. Colburn chose either to ignore Holley's letters or to make unfair accusations in the replies that he did make. Holley, in New York, was in a whirl simultaneously trying to secure money to satisfy Colburn's demands, to pay something on his own debts, and to reassure his worried father.[27]

Alex and Mary Holley had been living in a boarding house on Fifth Avenue near Thirtieth Street. Although the twenty-five dollars weekly room and board was not excessive, often Holley had difficulty finding the money. Young newly married couples in New York frequently lived in such a fashion, postponing household responsibilities.[28] Holley's father considered this mode of living too expensive, and his son had promised that as soon as his wife's health and his own affairs would permit, he would look for a small house. He told his father that the artist, Frederick Church, and his friend, Sam Cozzens, might like to live with them and share the expenses.[29] In addition, Holley was forced to admit that he owed large tailor bills. His fondness for expensive clothes had been a continuous source of annoyance to his father, who saw it as another manifestation of his son's ability to spend money much faster than he was able to earn it.

In his efforts to add to his income, Holley sought out Henry J. Raymond, editor of the *New-York Times.* From a four-page paper, first issued in 1851, the *Times* had increased steadily in size and circulation. Among the newspapers of the day, it was unique for its coverage of the arts and sciences.

Late in 1858, Holley submitted his first article to Raymond and, with its acceptance, began a relationship with the paper that continued until 1875. Inasmuch as Holley signed a limited number of the articles and all his editorials were unsigned, it is difficult to determine exactly how much material he contributed to the paper. A close friend estimated that during the five years that spanned the largest portion of his activity, he probably wrote well over 200 articles. They covered all aspects of engineering, and for this Holley was paid eight dollars per column.[30]

In addition to his work on the *Times*, Holley kept abreast of announcements of new inventions and ideas. He was especially attracted to a steam carriage designed by J. K. Fisher. Holley worked assiduously to obtain financial backing for the construction of a test model. The carriage had been awarded the highest premium in the Twenty-eighth Fair of the American Institute, an organization devoted to the encouragement and promotion of industry in the United States. The enthusiasm with which Holley could approach a new project and his ability to bring others to the same frame of mind were displayed well by this episode. With barely enough money to pay his board bill, he was thinking in terms of thousands of dollars for the building and exploitation of the carriage. His description of the bright future of the steam carriage and the merit of Fisher's example were so convincing that his father agreed to contribute a substantial portion of the money needed to build the trial model.[31] (Fisher's invention proved unsuccessful and the money was lost.)

Fisher also had designed a new type of fire engine that proved to be almost as impractical as the carriage. Holley was more realistic concerning this invention and wrote about it at length in the *Times*. He analyzed Fisher's engine with that clarity of perception that he showed about things mechanical, but that never seemed to function with respect to things monetary.

"The size of a fire engine," wrote Holley, "is limited by the size of the men who can handle the hose. We can build an engine that will throw a river over a mountain—who will hold the butt? Boilers which can raise steam from cold water in three minutes can do almost anything else in the next three minutes if they are not closely watched."[32]

Despite its chilling effect on his finances, *The Permanent Way* helped to expand Holley's professional reputation. By exerting every effort, finally he had disposed of the entire edition, although he was still in debt for some of the printing. In the short time since its publication, the railroads had introduced enough changes and improvements to convince Holley that a second volume was desirable. Although Colburn was still in England and easily could have collected the additional information now needed, he refused to have anything to do with the project.

At the same time, he would not release his share of the copyright unless Holley paid him an additional fee. That he was also claiming sole credit for the book is evidenced by an angry letter that Holley wrote him recounting in detail the work on the book by each of them. Colburn had written the financial section and a good portion of the rest of the book but, Holley pointed out, it was he who had interviewed engineers, ferreted out the facts, supplied at least half the drawings, and secured all the money from outside sources and most of the subscriptions.

"All this is said to you and to you only," concluded Holley. "In view of it, I hope you will not find it expedient to say anything more about me, or about your principal authorship of the book. Let me alone, and we are forever square. Were I to die today, and hence lose all power to defend myself, I cannot see that you would gain anything by any attempt to deprive me of my rightful share of the book. I think you have got enough out of me to be satisfied. I do not wish to be your enemy if you do not insist upon it. Simply let everything rest and I shall never molest or trouble you, but am always ready to help you if I can. . . ."[33]

From this time on Holley saw little of Colburn. In writing about the episode to George W. May, an associate who knew Colburn, Holley said, "He is a queer fellow. I wish he would be more faithful to those who have been faithful to him."[34]

Colburn's unwillingness to cooperate made it necessary for Holley to undertake another European trip, one that he could ill afford. To economize Holley moved to cheaper lodgings and his wife went to visit an aunt in Baltimore. Importuning letters poured from his downtown office as he endeavored to collect outstanding accounts owed to the defunct firm of Holley and Colburn. In addition, he solicited advance subscriptions for the new volume of *The Permanent Way*. His efforts to borrow money were not especially successful, although his good friend Sam Cozzens lent him $500. This frenzied activity eventually did bring in enough money for Holley's trip and permitted his wife to accompany him as well.[35]

Having learned that Henry Raymond was to sail 28 May on the *Orago*, a new steamship of the U.S. Mail Line, Holley booked passage for the same trip. He wrote his father that he meant to cultivate the *Times* editor, indicating that, although he had written considerable material for the newspaper, he had not become well acquainted with its editor.[36] Raymond's health had been poor and he was taking the trip with an old friend, Judge James Forsythe of Troy, in the hope that the sea air would prove beneficial. Raymond thoroughly enjoyed the thirteen-day crossing. He spent the mornings in bed and the afternoons on deck. At night, there were endless games of whist and eating and drinking.[37]

Holley must have had many opportunities to visit with Raymond inasmuch as later he carried a letter dated Paris, 21 August 1859. The letter stated that Mr. Holley was connected with the *New-York Times* and was visiting England for the specific purpose of writing upon engineering and scientific subjects, and that Mr. Raymond, as editor, would appreciate any aid given to Mr. Holley.[38]

A most unusual sight awaited Holley in London: the colossal ship *Great Eastern*, then the world's greatest tourist attraction. A product of the skill and genius of two of the leading engineers of the day, Isambard Kingdom Brunel and John Scott Russell, the ship had been under construction since 1854.[39] In contrast to his size, Brunel, whose nickname was Little Giant, never in his life entertained a small thought. In the 1830s he had built the Great Western Railway that ran from London to Bristol and boldly had constructed a two-mile tunnel for it near Bath. He was the designer of the Lord of the Isles, a famous example of outstanding locomotive design. As early as 1835, Brunel had drawn plans for the Great Western, the first true Atlantic steamship, which was built in 1838. Later he was to design the Great Britain, which was both the first large iron steamship and the first large ship to be driven by screw propellers.[40]

Brunel had built or helped to build 25 railroads in England and elsewhere. Eight piers and drydocks, 5 suspension bridges, and at least 125 railway bridges were engineered and supervised by this energetic man. Brunel encompassed and absorbed any project with which he was involved. He not only engineered the rail routes and supervised the construction of the railroads he built, but also chose the colors of the cars that would run on his tracks.[41]

It is safe to say that if Brunel considered building a large ship, the project could be expected to reach extraordinary proportions. As one present-day writer has observed,

> The Great Eastern had nothing that other vessels didn't have, she simply had a lot more of it. She held the record of being the largest ship for a half century, being exceeded in length only by the Oceanic, built in 1899, and in sheer size by the Lusitania, built in 1906. ... With such an enormous increase in size everything had to be scaled up to suit: machinery, masts, rigging and equipment, launching arrangements, all had to be designed from data derived from smaller vessels but which had not hitherto been proved for the sizes now contemplated. In short, all design data had to be extrapolated from existing engineering tables."[42]

John Scott Russell, in whose shipyard on the Isle of Dogs in the Thames the ship was nearing completion, was responsible for the hull as well as the paddle wheels. For some years Russell had studied the effect of wave formation on the resistance of ships. He had come to the conclusion that reversing the curve of the bow would reduce resistance. He had built several ships using this principle, of which the Great Eastern was to be the outstanding example.

Holley probably had seen the ship on his first trip to England. He had been in London at the time of the first effort to launch her in November 1857. Christened Leviathan, the vessel started down the ways but, after moving three feet, stuck. It was not until January 1858, that she finally reached the water. The launching difficulty occurred partly because the ship was built broadside to the river. Although this method of building and launching ships was not unusual in the Great Lakes area in the United States, it had been used in England only for small vessels.

Now she was being prepared for her first acceptance trial trip to Holyhead. Holley, who had established a deep and warm friendship with Scott Russell, was invited to go along. While he waited in London for the fateful day, Holley wrote at length about the vessel. A stream of articles describing the ship's structure and machinery appeared in the *New-York Times.* His reports of both the technical and the commercial aspects of this new giant of the seaways were so thorough that the *Times*'s coverage of the event was more comprehensive than that of many English newspapers.

Although christened the *Leviathan,* she never was called anything other than the *Great Eastern.* She was 692 feet long, 83 feet wide, 120 feet across her paddle wheels, and was designed to carry 6 masts and 5 funnels. She had bunker space for 12,000 tons of coal and could go to the East Indies and Australia without refueling. Built without ribs, she had two hulls, one inside the other. Supplying iron plates, 30,000 of them, already had required placing the British iron industry on a wartime footing. Holley estimated that 1,000 tons of iron and 3 million rivets had gone into the construction of the hull.[43]

On 1 October 1859, the *New-York Times* announced the appointment of Holley as its special correspondent on board the *Great Eastern.*

> The Special Correspondent of the *Times* on board the *Great Eastern* gives the public today some very interesting and important points concerning the structure of this vessel and the points established by her recent trip from the Thames to Portland. He writes, as his readers will readily see, more for the instruction than the amusement of the public; and while his letters may consequently lack something of the glow and factitious interest which belongs to most of the exaggerated descriptions that have been given of the vessel and her performances, they will furnish, we venture to say, a much better basis of forming a reliable judgment concerning the character and prospects of this new and important experiment in Ocean Steam Navigation than has hitherto been afforded the public on either side of the Atlantic.[44]

The trial trip, which began on 7 September 1859, turned out to be a trial in every sense of the word. As Holley wrote, "For eight years the ship has been the favorite topic and for three years she had been the daily sight of these people, at times almost an eye sore, — but now her own mistress, she was leaving forever her old haunts for a field of boundless seas and a career of fame."[45]

The ship left her moorings and was towed down the Thames to anchor for the night near Purfleet. Early the next morning, accompanied by all the fanfare that such an occasion demanded, she left the river and slid past the treacherous Goodwin Sands into the English Channel. By breakfast time she had achieved twelve knots an hour, proving beyond doubt that she was the fastest ship afloat. She sailed all day in a stormy, choppy sea amazing the guests aboard with a steadiness never before experienced in a channel voyage. Her crystal chandeliers hardly swayed with the motion of the ship.

Such auspiciously smooth sailing could not last for a ship born to trouble. On the evening of the ninth, the guests had just finished dinner when an explosion took place. Holley, with a half dozen friends, had gone forward and was lying in the extreme angle of the bow under the bulwarks for protection from the strong head wind. "We were all looking aft listening to Mr. Naysmith, a noted engineer, and Lord Alfred Paget, who were discussing the safety and prospects of the ship," reported Holley, "when suddenly with mingled roar and crash of a battery of artillery and a line of muskets, up shot the great forward funnel of the ship in two pieces, 30 feet in the air amid a shower of splinters and pipes and a volume of steam smoke." At first it was thought that a boiler had exploded, a common occurrence in those days, but instead it was found that an escape cock had been closed, which caused steam to build up in the funnel casing.

The wounded ship limped into Portland Bill near Weymouth, a sad sight with her forward funnel blown out and her elegant saloon a shambles. Five of her firemen were dead and others of the crew had been burned fatally. The Little Giant, already gravely ill from strain and worry, died four days after he learned of the tragedy.[46]

Holley believed that now it was only barely possible that the ship would be able to make her trial trip to New York. He warned his father that if she did sail he might be out of touch with him for as long as six weeks.[47]

From Weymouth, the great ship was sent around to Holyhead for repairs and Holley was invited by Scott Russell to make the trip.[48] At Holyhead the vessel once again became a tourist attraction. Thousands of curious sightseers came on board. Soon the owners realized that the ship would be in no condition to make a voyage to the United States that fall, and she was moved to Southhampton for the winter. Holley returned to the United States on 26 October on the *Vanderbilt*.

Her maiden voyage finally was rescheduled for 9 June 1860, and Holley returned to England to cover it for the *New-York Times*. It was not until the seventeenth that the *Great Eastern* succeeded in leaving Southampton.[49] Holley's description of boarding the vessel the day before sailing, set the scene for what was to follow.

It was confidently believed that the vessel would sail at high tide this afternoon. Circumstances, however, decidedly unavoidable — a forecastle full of at least hilarious firemen, the approaching darkness, the narrow Solent choked with shipping and the rugged spurs of the Needles capped with mist and storm before us — most properly detained her. So night is again creeping over the great bulk and clouds hang about her, silent and gloomy as on the brow of some sea-beaten headland. But within her iron sides, in gilded apartments, expanded and multiplied by mirrors into endless suites of saloons, music and wine are making the night merry.

The day has not been without incident. Our party left Southampton at nine o'clock. The great ship five miles below and looking like any other steamer a mile away, gradually loomed up before us filling the horizon as the Lilliputian tug hovered under her lea. . . .[50]

The embarkation, as Holley reported it in his day-to-day account of the voyage, was far different from that of her previous voyage.

The final embarkation — the real trial trip — the first ocean voyage — the test journey of a ship which has been the parent of more talk, speculation, wonder, and world-wide interest, than any craft that has floated since Noah's Ark — the very birth-day of the *Great Eeastern*'s practical career, could hardly have been accomplished with less ceremony and public demonstration. One poor little faithful tug, which had come alongside to take the last messages and letters, with half a dozen shivering gentlemen on her paddle boxes, followed us down to the Isle of Wight, reminding us, the few "foolhardy" who were venturing on an "unfortunate and ill-fated ship" — clinging to the howling rigging under that Wintry sky — of the picture of "the last mourner," familiar to our youth — the drunkard's dog following his body — all alone — to the Potter's Field. One English cheer from the pilot's boat, as we cast it adrift, was the only sound of comfort. Under such auspices did the *Great Eastern* start for New York.[51]

The passengers were lost in the cavernous public rooms and on the long decks. Holley discovered a favorite spot that soon enthralled his companions. The guardwalks on the top of the paddle wheels, which extended fifteen feet outside the vessel, became a rendezvous for many passengers who spent hours there watching the giant hulk slip through the water with scarcely a ripple. Holley wrote rapturously of the view from the platform at the top of the main mast. The passengers stayed up half the night watching the phosphorescent wake receding in the dark. There were musicales at night; the few passengers soon became comrades-in-arms betting "whole vineyards" on the speed of the great ship. Frequently bored, they devised games and races for amusement.[52]

The ship's arrival in New York City on 28 June was a gala occasion. She was decked out fore and aft with flags. Steamers loaded with sightseers began to appear and as they came closer there was all of the cheering and flag waving that has become traditional for such occasions.[53]

Holley was involved with the great ship once more on the occasion of an excursion from New York to Cape May on August 2 and 3 to show off the vessel to the Americans. Two thousand tickets were sold for the two-day voyage at ten dollars each although, because of lack of funds, there were sleeping accommodations for only 300. The trip was a nightmare of which mismanagement, bad food, and unruly passengers were only part.[54]

The entire affair was described by Holley in his dispatch to the *Times*. By the time the ship reached Cape May, many of the passengers were surfeited and left the vessel there to return to New York by train, where they had ample time to air their grievances before the ship's arrival.[55]

NOTES

1. "Obituary of Zerah Colburn," *Van Nostrand's Eclectic Engineering Magazine* 2 (June 1870): 654–55. Inconstant, erratic, Colburn was a brilliant engineer as well as writer. His fitful career was destined to be short. He was a nephew of the famous child mathematician, Zerah Colburn. Known as the "Calculating Boy," he was exhibited in England and on the Continent during the early 1800s. Because of the similarity of their names, the uncle and nephew often have been taken for one another. The nephew's astonishing memory, which enabled him to memorize with ease all the dimensions and parts of a standard locomotive, may have been inherited from his amazing uncle. Holley recalled that, as publisher of the *Railroad Advocate*, Colburn kept no books for months; he simply remembered correctly when payments for subscriptions and advertisements were due.

2. *Memorial of Alexander Lyman Holley* (New York: American Institute of Mining Engineers, 1884), p. 116.

3. Ibid., p. 117.

4. ALH to father, 12 May 1857, Connecticut Historical Society (hereafter cited as CHS).

5. *Memorial of Alexander Lyman Holley*, p. 117.

6. Diary of Alexander Hamilton Holley, 13 October 1857, Holley Family

7. Circular advertising railroad book. They were the Pennsylvania Central, New York and Erie, Hudson River, Illinois Central, Galena and Chicago, Philadelphia, Wilmington and Baltimore, and the Georgia Central. It is not clear in what form the railroads supported the endeavor. Evidently some gave money while others guaranteed the purchase of a specified number of the finished report. To this support, Holley and Colburn found it necessary to add funds of their own (Scrapbook, Author's Collection [hereafter cited as AC]).

8. ALH to Maria Holley, 13 November 1857, CHS.

9. The Institution of Civil Engineers, whose membership was composed of the elite of the profession, enjoyed prestige and even political power that were recognized widely. It was not unusual for the papers of the day to print "Great George Street thinks" or "The influence of Great George Street was evident" in referring to the activities of the institution.

10. James Dredge, "Holley Memorial Address," *Transactions of the American Institute of Mining Engineers* 20 (1891): xxviii.

11. The two merrymakers took great sport in sketching each other in ridiculous situations. Holley's small drawings at first are conventional, but rapidly they take on a relaxed point of view. Most are cartoons that illustrate engineering terms. A glass of liquor is *Colburn's Gauge*. A man on a rotating wheel is *Colburn on Rotary Engines;* a man on a railroad track with a bottle tilted to his mouth is captioned *Colburn on Permanent Way;* a man stretched out under a table loaded with bottles is titled *Colburn's Experiments on Ports;* and, finally, a bottle of cognac bears the legend *Index to Colburn's Works*. Colburn's series begins, as did Holley's, with the same side-wheeler, landscape, and architectural renderings; but Colburn's drawings rapidly become less dignified. A man falling onto the

sidwalk is titled *Holley on the Neutral Axis;* a man draining a bottle becomes *Holley on Drainage;* and so on (Scrapbook, AC).

12. Dredge, "Holley Memorial Address," p. xxvii.

13. Scrapbook, AC.

14. Zerah Colburn and Alexander Lyman Holley, *The Permanent Way and Coal-Burning Locomotive Boilers of European Railways; with a Comparison of the Working Economy of European and American Lines, and the Principles upon which Improvement Must Proceed* (New York: Holley and Colburn, 1858).

15. *Memorial of Alexander Lyman Holley,* p. 119.

16. Ibid., p. 120.

17. "Comparative Productiveness of English and American Railraods" (editorial), *American Railroad Journal* 15 5 February 1859): 89-90.

18. ALH to Henry Varnum Poor, 23 February 1859, Letterbook, AC.

19. Sales of the book at ten dollars a copy had been good. The New York Central, Michigan Central, and some thirty other managements each had bought between ten and twenty copies. An executive of the Galena and Chicago Union Railway Company wrote that from his examination of the book he believed that the three hundred dollars would be more than repaid in savings. There were similar letters from other leading railroad managers (Scrapbook, AC). Years later a railway executive said, "I keep the book in my office still; and frequently, when inventors call on me with their new ideas about rails and joints, and sleepers, and boilers, and so on, I open Colburn and Holley, and show them their inventions, already described and discussed" (*Memorial of Alexander Lyman Holley,* p. 120). W. P. Shinn, who was to work with Holley in later years, first became aware of him through the book. He remarked that it was of the greatest value and almost a revelation to him. No other such study of American railroads had been made. Many now-familiar appliances described in the report then were unheard of in the United States and many established American railroads first learned how to use them from Holley and Colburn's report (Ibid., p. 26).

20. Father to ALH, 23 July 1857, CHS.

21. Holley Papers, CHS.

22. ALH to father, 19 November 1858, Letterbook, AC.

23. ALH to Maria, 18 May 1858, CHS.

24. ALH to father, 26 November 1858, Letterbook, AC.

25. ALH to father, 22 November 1858, Letterbook, AC.

26. ALH to father, 4 December 1858, Letterbook, AC.

27. ALH to father, 10 December 1858, Letterbook, AC.

28. Frances Trollope, *Domestic Manners of the Americans* (New York: Vintage Books, 1949), p. 283.

29. ALH to father, 26 November 1858, Letterbook, AC.

30. *Memorial of Alexander Lyman Holley,* p. 121. Rossiter Raymond itemizes the subjects of the articles that Holley wrote for *Times* as follows:

I have found 276 articles from his pen, published in that paper, of which about 200 appeared between 1858 and 1863, and the remainder at rarer intervals to 1875, the last being the leading editorial of April 27th, 1875, on the recently appointed United States Testing Board. The range of these articles is indicated by the following classification: Setting aside 52 miscellaneous articles (descriptive, political, etc.), and 30 which may be called 'scattering,' though devoted to engineering topics, we have 194, divided as follows: Railways (including street railways), 49; steam navigation, 42; war ships and armor, 30; the Stevens battery, 22; arms and ordnance, 19; boiler explosions, 11; and steam engines, 7. The most important and remarkable of these articles were, perhaps, those on the *Great Eastern,* written under the signature of "Tubal Cain."

31. Ibid., p. 122.
32. Ibid., p. 123.
33. ALH to Zeral Colburn, 18 April 1859, Letterbook, AC.
34. ALH to George May, 18 April 1859, Letterbook, AC.
35. ALH to numerous persons, April 1859, Letterbook, AC.
36. ALH to father, 13 May 1859, Letterbook, AC.
37. Ernest Francis Brown, *Raymond of the* Times (New York: W. W. Norton, 1951), p. 169.
38. Henry J. Raymond to To Whom It May Concern, 21 August 1859, Scrapbook, AC.
39. James Dugan, *The Great Iron Ship* (New York: Harper & Bros., 1953), pp. 5, 19.
40. Aubrey F. Burstall, *A History of Mechanical Engineering* (Cambridge: M. I. T. Press, 1965), p. 271.
41. Dugan, *The Great Iron Ship*, p. 22.
42. John Guthrie, *Bizarre Ships of the Nineteenth Century* (New York: A. S. Barnes and Co., 1970), p. 132.
43. *New-York Times,* 30 June 1859.
44. Ibid., 1 October 1859.
45. Ibid., 25 September 1859.
46. Ibid.
47. ALH to father, 26 October 1859, CHS.
48. Certain authorities seem not to have welcomed Holley's presence on the ship inasmuch as Scott Russell sent a note to Holley advising that he was to go as private secretary to Scott Russell.
49. People had so little faith in the announced sailing of the jinxed vessel that, when she finally got under way, the *Great Eastern* carried only 35 paying passengers, 8 company officials, and a crew of 418. According to one author, the cargo consisted solely of 500 gross of London Club Sauce (Dugan, *The Great Iron Ship,* p. 55).
50. *New-York Times,* 29 June 1860.
51. Ibid.
52. Ibid.
53. Ibid.
54. Holley regretted that the wine steward had permitted 300 tons of ice, which had been taken aboard to cool the champagne, to melt. It was neither used for the wine nor to provide ice water for the thirsty passengers. The English manager of the excursion later admitted that he had no idea that the Americans entertained such a passion for ice water.
55. Holley depicted with zest, the lack of planning, the shocking behavior of the crew, and the outrageous prices charged for hot water in which to wash. A letter to the editor of the *Times* added that even the towels brought a price. Logically, a clean one was more expensive than one that had been used (*New-York Times,* 2 and 3 August 1860).

V · The Civil War
and the Battery

I n 1860 Holley's fortunes entered a phase in which they took a definite turn for the better. Although he started the year in debt (principally for clothes and printing), the amount was considerably less than it had been in prior years. His articles in the *Times*, as well as the publication of *The Permanent Way*, had established him as a skillful writer on engineering subjects. In January he was offered the editorship of the mechanical department of the *American Railway Review*, a position that he accepted with alacrity. It would be time-consuming work and he was aware that he would be kept very busy, inasmuch as he already was devoting most of each day to preparation of the new edition of the railroad book. In his continuing struggle to meet incessant deadlines he spent many nights in his office, in a constant race against time.

Years later, his friend Rossiter Raymond described Holley's method of work:

> An examination of this paper [*American Railroad Review*] and the *Times* throws an interesting light on the manner in which he contrived to do so much literary and professional work. He made one hand wash the other. The topics treated for the general public in the *Times* were served up in more technical form in the *Review*. The anonymous editor in the *Times* frequently quoted and commented upon the avowed editor of the *Review*, and vice versa. But all this was merely the incidental though necessary occupation of the period. He wrote to earn money, and he wrote with the rapidity and versatility of a Bohemian; but all the time his eye was upon his profession and his articles were but the chips thrown off in the labor of preparing his *American and European Practice*.[1]

Between October 1859, and his trip to England in June 1860, for the maiden voyage of the *Great Eastern*, Holley spent uncounted hours preparing the new edition. He must have regretted having undertaken such a vast work, for he now realized that even if the entire edition were sold, there could be no profits; indeed, he would be fortunate to meet expenses.

Coal-burning freight engine, drawn by A. L. Holley, about 1860. Frontispiece from *American and European Railway Practice.* Produced by the Rogers Locomotive and Machine Works, Paterson, New Jersey.

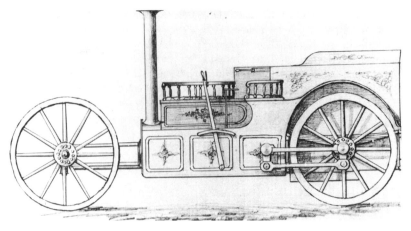

Design for a steam carriage by Alexander Lyman Holley. Drawn in 1860.

Remembering the success of traveling from one railroad office to another in selling *The Permanent Way*, Holley repeated this routine. In a few weeks of travel he secured orders for 250 copies. However, orders for 350 were needed to meet expenses. Fortunately, a few months later, the publishing firm of Van Nostrand and Company agreed to distribute the volume, thus relieving Holley of an onerous task.[2]

Although in the spring of 1860 political unrest was sweeping the country, Holley seems to have paid little attention to it. He wrote his father on the first of June from aboard the *Asise* on his way to Europe that he knew very little about political matters but it appeared to him that Seward had been sacrificed for Greeley for purely spoils motives and that Lincoln would not carry as many votes as Seward would have done. "I rather believe that I am getting as much down on one party as another, though fortunately, I have no political influence and don't want anything to do with the whole business. Of course I shall vote the Republican ticket."[3]

The son seems not to have inherited his father's keen interest in politics. Previously a Whig, the older man had become a Republican when the new party was formed. In spite of young Holley's lack of political interest, upon returning from his book selling expedition, he reported to his father that he thought "there would be trouble down south if Lincoln was elected." He went on to say that, in his opinion, Congress was dealing too extensively with the slavery question and neglecting other matters that were equally injurious to the future of the United States. His solution to the slavery issue was simple and direct. "The States should hold a national convention and let the slaves go," he wrote his father.[4]

The calm good judgment with which Holley could approach the slavery question was shattered when he turned his attention to the safety factors of engineering. An editorial that he wrote for the *Times* of 3 May 1860, illustrates the extent to which his feelings could be aroused.

> As the law-makers tread the decks of the steamers, they will need no expert to unfold the probable causes of danger. Let them but look and tremble; a single compartment with a wooden shell ... weak and rotten — quivering with each stroke of the engine ... fires roaring and lamps swinging within the very embrace of seasoned combustibles—boilerplate of questionable tenacity, straining under pressure—firepumps and hose, bilge-pumps and life-boats, lashed and stowed away for the voyage like so much cargo ... sailing vessels, without light or signal, bearing down into our very path. ... The chances are, that there will be a collision—that a hole in any part of the hull will admit the sea ... that the furnace-fires will be quenched before the steam-pumps can be set to work; that the ship will founder before all the boats can be cleared and launched; that all the boats will not hold half the people; that

there is no material for rafts; that the vessel will take fire; that the fire will get the advantage before the pumps can be got to work and the hose adjusted; that the deficient means of overcoming danger ... will give accident and carelessness the winning start of defence and caution.

Urgent and reasonable as are the demands of the case, it is found practically impossible to meet them fully and fairly with legislative enactments. Sectionalism, State rights, spoils, corruption, selfishness, and the whole catalogue of personal and party motives and influences, menace alike the interests of the white man and the negro—the African traveller and the Red River slave.[5]

As John Guthrie wrote, "A ship in those days died fighting, usually taking most of her crew and passengers with her. The record of sailing casualties on the Atlantic passage in the 1860s was appalling."[6]

Another sizeable commission came to Holley in March 1861, from the publishers of a revised edition of *Webster's Unabridged Dictionary*. He was to write and review the definitions of 1,200 engineering terms and to provide the necessary illustrations. For this work, which he eagerly accepted, Holley received a mere $200 and the promise of appropriate credit in the dictionary. To obtain the definitions Holley was obliged to visit machine shops and to consult with outstanding mechanics. Coleman Sellers, owner of a well-known machine shop in Philadelphia, tells of being asked for a definition of a set screw that could be used in connection with a law suit. Sellers assured the inquirer that the definition could be found in *Webster's Dictionary* and added, "The reason I knew it, was because Mr. Holley had put it there. Mr. Holley called on me once to have one of those little chats that we used to enjoy so much ... for the purpose of getting my help in preparing the technical terms on mechanical science for *Webster's Dictionary*. So we went through the shop to get these words, and ... the definition of a 'set screw' was one of the terms which Mr. Holley and I worked out together."[7]

By taking any position that came his way, Holley gradually was solving his financial problems. At about the same time he was engaged by Edwin Stevens of the Camden and Amboy Railroad to alter one of its wood burning locomotives to a coal burner. The job was to pay him eight dollars a day and he expected that it probably would take a month to make the change.[8]

In March 1861, the post of United States Supervising Inspector of Passenger Steamers was opened to candidates and Holley pursued it with every means at his command. The security of the government appointment had many attractions for the harassed Holley. He submitted a formidable array of letters and signatures. Leading figures in engineering and railroading such as George Corliss, Peter Cooper, Abram Hewitt, Cyrus W. Field, Erastus Corning, Edgar Thomson, and forty-two others vouched for Holley's capabilities. He wrote to his father early in April that he had been in Washington about the inspectorship but did not expect to get it. "I have as fine a list of engineering names as could be offered," he said, "But I have no hard working New York City political backers, except Mr. Raymond [the *Times* editor]. Another candidate is more of a politician than an engineer, and he will probably get the place. However, I may get it. I shall certainly not go and bore the President and departments. If they can't take my papers as evidence they ought not to take my statement." After the other candidate did win the appointment, Holley asked for the return of the letters. On the final page of the scrapbook in which they are mounted he wrote in large letters, "What came of it? Mr. Holley was not appointed." [9]

The Civil War had brought the North to a high pitch of excitement and Holley was deeply stirred by the outbursts of patriotic fervor. Toward the end of April 1861, in a letter to his father, he conveyed his reaction to the ardor for the Union cause that was sweeping New York.

I suppose you will see in the papers, what a fearful excitement we are in, but I must assure you from my personal observations, that New York is on fire for war, from the ultimate top stories of Trinity to the cellars of the Five Points. The whole town is in a blaze of enthusiasm—to protect and honor the stars and stripes and to uphold the government. What a day this has been! Hundreds of thousands of people in the streets,—three New York regiments marching down and embarking on the *Baltic*—a Massachusetts and a Rhode Island regiment, arriving, marching, embarking,—and now at a late hour of the night, three thousand troops from Boston marching down to the wharves amid the cheers of ten thousands of men and women, and at this moment artillery rumbling by my window—the brazen dogs of war glistening in the moonlight. Five or six steamers are now under steam at their docks, and are fast filling with troops. God help the union men of the South—the traitors will get their deserts. Van Nostrand (the New York publisher) was in Philadelphia Friday—he says the whole town is a camp. Boston the same. How nobly she has responded. I hope the administration will have the pluck to make the war terrible and hence short and decisive. We are in the greatest anxiety about Washington. I think the traitors will come there in full force with fearful accession in Virginia before we can defend the capital. But we shall hear in due time. I am anxious to do something as an engineer and believe I can, granted a chance, with reference to the steam navy. . . . Mary is in Baltimore and I have felt alarmed about her. But her father and mother think she is all right there—as safe as she could be even if the city is bombarded, for her uncle lives two miles out of town. I can't get there, but she has troops of friends there and could better get here. Her friends with whom she is staying openly declare themselves not only unionists but 'Republicans'. . . . I shan't volunteer without letting you know and taking advice. I had rather get less glory and do more good in a quiet engineering way, if I can. [10]

On 23 April 1861, young Holley wrote to Secretary Chase and to the
War and Navy Departments. After mentioning his application for the
post of United States Supervising Inspector of Steamers, he continued:

> In the present state of affairs, I am anxious to bring such engineering
> facilities as I may command into the service of the Government. I have an
> intimate acquaintance with Mr. Scott Russell and many other English engineers
> and shipbuilders, having been sent to Europe several times on engineering
> business.
>
> If I can do the Government any good with reference to the construction or
> purchase of ships or arms, the construction of batteries, gun-boats, or iron
> vessels of light draught, the application of armor to ships, etc., I hereby offer
> my services, *without compensation*, desiring only that my necessary expenses
> be paid, as I have not the money to pay them.
>
> I know that I have peculiar facilities for gaining the latest information
> relative to the construction of iron ships and boats and their engines, without
> waste of time.
>
> I refer you to the names appended to my application aforesaid. I am ready
> at an hour's notice. [11]

The offer to be ready at an hour's notice was not acknowledged. A
few months later John Ericsson received the same treatment when he
too volunteered his services. Holley accepted the incident philosophically
and continued working away at his multitudinous assignments. Any
bitterness must have been forgotten quickly when he began to work with
the locomotive for the Camden and Amboy Railroad. He relished the
opportunity once more of ending a day spattered with oil and grease.
The exhilaration of associating with other engineers and sharing mechanical problems reminded him of the old days in the Corliss shop. [12]
Although in the future Holley would have little to do with locomotives,
he never lost his affection for the engines. [13]

Holley was exceedingly pleased to have been engaged by a person of
such stature as Edwin Stevens and proud to have been entrusted with
the responsibility of converting a locomotive from a wood burner to a
coal burner on a railroad as important as the Camden and Amboy. The
association with Edwin Stevens also was to lead to Holley's next important
commission, a fact-finding trip to study European armament.

The Stevens brothers first had become interested in projectiles during
the war of 1812. The success of their experiments at that time led them
to develop a shot- and shell-proof vessel of Robert Stevens's design called
the *Battery*. First designed for defending New York harbor, this vessel
had a long and complicated history. The brothers struggled for nearly
a quarter of a century to complete it. Although the idea for the *Battery*
had been conceived in 1837, it was not until 13 August 1841 that the

brothers petitioned the secretary of navy for permission to build an armored vessel. After a period of testing, a bill was drawn up by Congress on 14 April 1842 authorizing the secretary of navy to make a contract under an appropriation of $250,000 to the Stevens brothers. The brothers contributed large amounts of money to the completion of the ironclad as well. When these amounts were exhausted, Robert Stevens petitioned for additional funds but was refused. Until his death in 1856 he continued to battle with successive incumbents in the Navy Department in an unsuccessful effort to complete the vessel. Throughout these years the *Battery* was the subject of endless interest, although few observers gained admittance to the covered shed in Hoboken, New Jersey, where she lay.[14]

Vast changes had taken place in warship construction between 1841 and 1861, both in the size of vessels and in the weight of their armament. The *Battery* in 1861 little resembled the ship first projected in 1841. She was now 420 feet long with a beam of 53 feet. Furthermore, up-to-date projectiles would make it necessary for her to carry armor plate heavier and thicker than that called for in the original plan.[15]

Robert Stevens had devoted the last years of his life to the futile effort to finish the ironclad. After his death, Edwin Stevens regarded its completion as a sacred trust. Although the *Battery* had preceded French and British ironclads by more than ten years, Stevens believed that by making certain changes in the plans it was possible to finish her. In his eagerness, Stevens several times petitioned the government but to no avail until 1861, when the Civil War aroused new interest in the vessel. At that time engineers speculated about what the course of events might have been if "Stevens Folly" had been completed instead of having been permitted to lie for so many years in her Hoboken dock.[16]

The famous battle of the armored ships *Monitor* and *Merrimac* took place in Hampton Roads on 9 March 1862.[17] The *Monitor* had been built in the record time of four months. On the following 25 February she was commissioned, and a week later fought her famous battle.[18]

Chagrin at the speed with which the *Monitor* had been built and at Ericsson's having been retained to construct two vessels of similar design caused Edwin Stevens to turn again to plans for continuing construction of the *Battery*. He was familiar with Holley's reports on the *Great Eastern* and respected his engineering ability. He immediately offered Holley a commission to investigate the use of armor plate by European navies, observe shipbuilding, and inquire into armament production abroad. From Holley's report Stevens hoped to gain information that he could utilize in completing the *Battery* in such a manner that the obstinate Naval Commission finally would be convinced of her worth.[19]

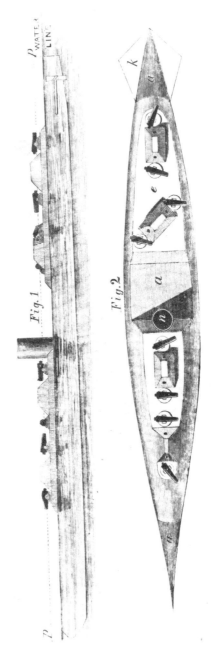

Fig. 1

Fig. 2

The *Stevens Battery*. From *Scientific American*, n.s.5 (31 August 1861): 129.

NOTES

1. *Memorial of Alexander Lyman Holley* (New York: American Institute of Mining Engineers, 1884), p. 122.
2. ALH to father, 1 October 1860, Connecticut Historical Society (hereafter cited as CHS).
3. ALH to father, 1 June 1860, CHS.
4. *Memorial of Alexander Lyman Holley*, p. 127.
5. *New-York Times*, 3 May 1860.
6. John Guthrie, *Bizarre Ships of the Nineteenth Century* (New York: A. S. Barnes and Co., 1970), p. 155.
7. *Memorial of Alexander Lyman Holley*, pp. 88, 123.
8. Ibid., p. 124.
9. Scrapbook, Author's Collection.
10. ALH to father, 21 April 1861, CHS.
11. *Memorial of Alexander Lyman Holley*, p. 127.
12. *Engineering and Mining Journal* 27 (24 May 1879): 368. Holley's sense of fun led him to wager with some of the men that he could run his locomotive a mile without fire, water, or steam. The locomotive was taken cold from the shop and towed some distance by another engine to a level stretch about a mile long to the point where the trial was to begin. Holley rode the cold locomotive in solitary grandeur; his friends were in the towing locomotive. They did not suspect that Holley had so arranged the driving apparatus that, during the trip, the motion of the drivers and pistons stored the boilers with compressed air. By the time the engine reached the starting point power had accumulated with the result that his bewildered companions saw Holley triumphantly riding the mile without power.
13. In a speech that Holley made in 1876 he said:

   The thoughtful locomotive-driver is clothed upon, not with the mere machinery of a larger organism, but with all the attributes of a power superior to his own volitions. Every faculty is stimulated and every sense exalted. An unusual sound amid the roaring exhaust and the clattering wheels tells him instantly the place and degree of danger, as would a pain in his own flesh. The consciousness of a certain jarring of the footplate, a chattering of a valve stem, a halt in the exhaust, a peculiar smell of burning, a sudden pounding of the piston, an ominous wheeze of the blast, a hissing of a water-gauge, warning him respectively of a broken springhanger, a cutting valve, a slipped eccentric, a hot journal, the priming of the boiler, high water, low water or failing steam. These sensations, as it were, of his outer body, become so intermingled with the sensations of his inner body, that this wheeled and fire-feeding man feels rather than perceives the varying stresses upon his mighty organism (Alexander Lyman Holley, "The Inadequate Union of Engineering Science and Art," *Transactions of the American Institute of Mining Engineers* 4 [1876]:194).
14. Archibald Douglas Turnbull, *John Stevens: An American Record* (New York: American Society of Mechanical Engineers, 1928), p. 415.
15. "The Stevens Battery," *Scientific American*, n.s. 5 (31 August 1861): 129–32 passim.
16. Ibid.
17. Turnbull, *John Stevens*, p. 425.
18. The effect of the battle on the European governments has been greatly exaggerated, according to James Phinney Baxter 3d, (*Introduction to the Ironclad Warship* [Cambridge, Mass.: Harvard University Press, 1933]). Lord Sydenham expressed the English

point of view when he commented that the *Monitor* may have given fresh impulse to the use of armor plate but it certainly did not provide the initiative. In his opinion, it was the use of the revolving armored turret on the *Monitor* that made the deepest impression. But the battle's outcome made a decided impact on the American people, since they had been devoting their attention to the locomotive and the opening of the West and paying little heed to the subject of armored warfare.

19. The success of the *Monitor* venture was due mainly to the engineering genius of John Ericsson, but the financial risks borne by ironmaster John Winslow and banker John Griswold, both of Troy, New York, cannot be ignored (*Memorial of Alexander Lyman Holley*, p. 129).

VI · Henry Bessemer

During August of the, year 1856, while Holley was editing the *Railroad Advocate* and making every effort to repay his creditors, an event occurred in Cheltenham, England, that was to influence his future profoundly. The British Association for the Advancement of Science was meeting in that small fashionable resort town beneath the Cotswold Hills. Its members were to hear papers on numerous technical subjects and, at the same time, enjoy the mineral springs and the Promenade, for both of which Cheltenham was famous.

When the members of the Mechanical Section of the association had assembled on the morning of August 11 in the meeting room of the Queen's Hotel, the president introduced the author of the session's first paper. Henry Bessemer rose from his chair, walked to the front of the room, and read his paper, "The Manufacture of Malleable Iron and Steel without Fuel."

As the speaker began, the men shifted in their seats and listened with obvious skepticism. By the time he had finished, however, the group was so affected that no effort was made to discuss the paper. James Nasmyth, the inventor of the steam hammer, did rise to express approval of the principles of the new process and a few other members spoke of trying it in their plants. Whatever their attitudes, it is unlikely that anyone present recognized the sounds of both the death rattle of the Iron Age and the birth cries of the Steel Age.

Henry Bessemer's paper set in motion a chain of events that would leave its author a rich but embittered, suspicious, and often unhappy man. Within the year, the worth of the patents described in the few pages would be challenged amidst controversy, heartbreak, and unceasing conjecture. Bessemer was soon to realize that he had not yet brought his new process to a point at which it would function successfully for him or for anyone else. He was beset also by the claims of an American inventor, William Kelly, for a process that was similar to the one for which Bessemer had gained his English patents. Hovering over both Bessemer and Kelly would be the tragic figure of Robert Mushet, who would haunt Bessemer's conscience and dog his steps to the end of his days.

Since 1856 thousands of words about the controversy have been written by the three protagonists and by other partisans. The debate has been marred by contumely and by references to theft, espionage, and double-dealing; a detached and impartial analysis of the role of the three men never has been achieved. It may be possible now to approach the controversy objectively and to concede that each of the three — Bessemer, Kelly, and Mushet made his own particular contribution to the development of the steel industry. Standing with them as the catalyst who was to combine, perfect, and coax their accomplishments into the phenomenon that the Bessemer process became in America certainly would be the figure of Alexander Lyman Holley.

The mistaken belief that Henry Bessemer's origins were incongruous with his accomplishments long has been accepted, particularly in the United States. The facts of his life as he related them in his autobiography both refute that allegation and show specifically how he acquired his knowledge of steelmaking. [1]

Anthony Bessemer, Henry's father, was born in London, but was taken to Holland at the age of eleven and articled as apprentice to a machinist. At twenty-one he moved to Paris. Only five years later, he was elected to the august Academy of Science in a recognition for his improvement of the microscope. While working in the Paris mint, he invented a machine known as a portrait lathe on which medallion dies of a specified size could be engraved in steel from an enlarged model. When Robespierre came to power, Bessemer was transferred from the mint to the management of a public bakery. Accused of making the loaves underweight, Bessemer was thrown into prison. He escaped and fled to England. Although he landed with no means of support, he managed to obtain a post with the English mint.

Using his knowledge of gold working, Bessemer subsequently prospered by manufacturing fine gold chains. He observed that jewelers removed discolorations on the gold articles they manufactured by placing the objects in a mixture of alum, salt, and saltpeter. Bessemer discovered that the solution removed not only the stains but also a thin layer of gold. He secretly devised a method of recovering this gold from the liquid, which he purchased from the unsuspecting jewelers.

In about 1797, alarmed at Napoleon's threat to invade England, Bessemer converted his property into cash and retired to a small estate in Charlton, Herfordshire, forty miles from London. But retirement for such a talented person was impossible. In France Bessemer had cut many type fonts for the celebrated firm of type founders Fermin Didot. He now turned again to this activity, spending the next several years cutting a type series for the distinguished designer, Henry Caslon. In his experience, Bessemer had observed that if small amounts of tin, copper, and bismuth were added to the antimony and lead of the generally used type metal, an alloy could be produced that lasted twice as long as the ordinary sort. The business that he established based upon this new metal soon was flourishing.

The relationship between Caslon and Anthony Bessemer also prospered so that, when a son was born to Bessemer in 1813, Caslon acted as godfather and the child was given his name. Henry Bessemer received only the limited education that was available in the local schools. Like Alexander Holley, he displayed an early aptitude for machinery. His inventiveness probably was inherited from his father, from whom he learned many skills.

When Henry was seventeen, the family returned to London. He found himself in the position of "knowing no one and myself unknown—a mere cipher in a vast sea of human enterprise."[2] In his loneliness, Bessemer wandered the streets of London day after day, fascinated by everything he saw. He was not particularly unhappy, but he was deeply aware that he was a tiny part of a huge entity and that if he was not to be swallowed up, he must make his mark in some fashion.

He tried to develop a multiplicity of ideas and often was successful. Among these was the reproduction in metal of objects of intricate structure. After spending months trying to cast a rose, he hit upon the idea of changing the molding materials and succeeded in reproducing a rose so perfect that the fine pile and the tiny thorns on the leaf and stem were visible. Now he wanted to improve the dull gray appearance of the flower by coating the delicate castings with copper or brass. After many trials, he found a means to copper-coat the flowers.

His research led Bessemer to develop a new way to make revenue stamps fraudproof. At that time the British government was losing large amounts of income because it was possible to reuse the stamps from discarded documents. Bessemer worked out a method of producing the stamps in which the principal feature was the use of movable dates. With his development, stamps no longer could be salvaged from old documents, and the increase in revenue to the Stamp Office was considerable.[3]

Bessemer's interest in invention continued. He developed a sawing machine to cut graphite for pencils. While working on this invention he discovered that the practice of hand sawing, the only method then known, produced large amounts of dust that could be purchased very cheaply and molded by hydraulic pressure, an early application of powder metallurgy.

Next he devised a machine for casting type. When completed, it was sold to a well-known company of type founders in Edinburgh. According to W. T. Jeans, "The machine did produce the most accurate type up to that time but the valve through which the metal was injected into the mold used to fail after casting some six or eight thousand types and owing to this defect the machine was eventually abandoned."[4]

At about the same time, Bessemer purchased a machine that would do engine turning. With this beautiful mechanism, it was possible to engrave delicate, lacy lines and designs. Plates of this kind frequently were used in printing labels for patent medicine bottles, foreign bonds, and in other fine printing. But the curved lines in the prints were not as clear as Bessemer wanted. Recognizing that the deeply incised plate lines caused the metal to bur and that the quality of the metal was responsible, Bessemer developed an alloy that avoided the difficulty.

Although it would appear that Bessemer was doing exceedingly well for a young man still in his twenties, he was not content. "One branch of trade seemed to lead imperceptibly to another; but I was always waiting and looking forward for the establishment of the one large and steady branch of business that I hoped would some day allow me to drop the many schemes which my versatile mind so easily created, seized upon, and engrafted on the business I was carrying on; but this one great branch of trade, so earnestly desired, had not yet manifested itself."[5]

In addition to developing his own numerous inventions, Bessemer also devoted time to working on the ideas of others who consulted him. One such client was James Young, a merchant from Lille, who paid Bessemer a guinea per day to work on the development of a machine that would set type by pressing keys similar to those of a piano. Inventors for some time had worked on such devices and at least one typesetting machine had been patented as early as 1822. The young man wrestled with this difficult problem for fifteen months before he produced a machine that could be considered successful.[6]

One invention that was to provide Bessemer with a steady income for many years produced an imitation of the magnificent, old Genoa fabric, with its embossed velvet figures on a satin background. The fabric was both exceedingly fashionable and expensive. The imitation was to utilize cheaper modern material, known as Utrecht velvet, which was deep piled, harsh, and difficult to work. Part of the complexity of the invention stemmed from the need to devise a machine that could produce the embossing on a commercial scale. As with many of his earlier inventions, it was necessary for Bessemer to develop a special alloy for the rolls, since the cast iron of his time could not withstand both excessive pressure and high temperature. The new product had been found practical as a covering for railroad and omnibus seats as well as for furniture, and enormous quantities of the Utrecht material were treated by the new process.

The development of Bessemer's next invention called forth all of the inventor's perception, insight, and ingenuity and, at the same time, drew

upon a synthesis of his past experiences. A successful machine for producing bronze powders at a cost far below that of the market established
Bessemer on a scale much larger than he had known previously. Profits
were enormous. By employing only relatives he succeeded in keeping
the method of manufacture secret for forty years. Knowing that if patented, his process could be duplicated easily, Bessemer never did so.
Henry Bessemer now had a steady business and a large income and felt
freer to turn to other inventions. Some concerned the manufacture of
paints, oils, and varnishes. Others dealt with improvements in railway
carriages, another with mine ventilation. For a machine to extract the
juice from sugar cane, Bessemer won the gold medal offered by Prince
Albert. During the time when he was working on the sugar cane machine,
Bessemer explained the advantages to a creative engineer of having
worked in diverse areas of industry and of having been free to range
widely among them.

> How often it has occurred to me and how often have I expressed the opinion
> that, in this particular competition — as in many other cases — I had an immense
> advantage over many others dealing with a problem under consideration,
> inasmuch as I had no fixed ideas derived from long established practice to
> control and bias my mind, and did not suffer from the too general belief that
> whatever is, is right. Hence I could without check or restraint, look the question
> steadily in the face, weigh without prejudice, or preconceived notions, all
> the pros and cons, and strike fearlessly in an absolutely new direction, if I
> thought desirable. [7]

Bessemer also made known his opinions about the social worth of
patents for inventions that also had been developed by their creators.

> There is one point in connection with patented inventions upon which I
> have always felt strongly. I have maintained that the public derives a great
> advantage by useful inventions being patented, because the invention so secured
> is valuable property, and the owner is necessarily desirous of turning that
> property to the greatest advantage; he either manufactures the patented article
> or he grants licenses to others to do so. In either case the public reaps the
> advantage of being able to purchase a better or a cheaper article than was
> before known to them, due to the inventor's perserverance in forcing his
> property upon the market. But if a novel article or manufacture is simply
> proposed by a writer, and published in the technical press or in newspapers, as
> a rule (almost without exception) no manufacturer will go to the trouble and
> expense of trying to work out the proposed invention. He says to himself "I
> shall not risk the expense necessary to develop this new idea, for it may entirely
> fail, or even if I succeed, its development will cost me more than it will cost
> other manufacturers, who will immediately avail themselves of it if I succeed;
> no let someone else try it and so the invention is lost to the world in conse-
> quence of being given away." [8]

NOTES

1. Inasmuch as no full-length biography of Henry Bessemer exists, I have relied on the autobiography [Henry Bessemer, *Sir Henry Bessemer: An Autobiography* (London: Offices of *Engineering*, 1905)], the manuscript for which was unfinished at the time of Bessemer's death in 1898. There are several matters of interest respecting the authorship and verity of that work.

First, no Bessemer archive of any sort ever has been discovered, either by me in my researches in the United States and England for this book or later by Alan Birch for his *The Economic History of the British Iron and Steel Industry, 1784–1879* (London: Frank Case and Co., 1967). Bessemer's work on his book ended with the year 1872, twenty-six years before his death. A final chapter to the book was added by a son, who commented, "The unfortunate destruction of my father's copious notes relating to those years of his life after he had retired from active business but not usefulness, has made my task a difficult one, because I have to rely on memory, aided by some memoranda and letters; and because I was not at that time in constant touch with my father, as he resided in London and I in a rather distant part of the country." Indeed, one verifiable section of the added chapter, having to do with the United States steel industry, surely suffered from the unavailability of Bessemer's notes.

The book was published by *Engineering* seven years after Bessemer's death. Bessemer himself acknowledged that he wrote it at the urging of James Dredge and William H. Maw, coeditors of that journal. It is possible that Dredge may have had a hand in writing it as well. It was Dredge who delivered the Bessemer Memorial Address in 1898, in which he referred both to Bessemer's unfinished manuscript and to other sources of information for the address, although the latter have never been found or even identified. Dredge's close association with the book may have been responsible for the disappearance after its publication of the notes on which it was based. These notes may have been the "other sources of information" to which he refers.

Another person who may have had access to the initial sections of Bessemer's manuscript was J. S. Jeans whose *Steel: Its History, Manufacture, Properties and Uses* (London: E. and F. N. Spon, 1880) gives a detailed description of Bessemer's work in developing his process as well as extensive material about Bessemer's background and early life. Since, in the preface to his book, Jeans acknowledges his obligation to Bessemer for the aid he provided, it is possible that Bessemer lent Jeans part of his manuscript dealing with his youth since the two accounts are similar. Other information may have come from personal interviews.

Another book by W. T. Jeans, *The Creators of the Age of Steel* (New York, Charles Scribner's Sons, 1884) also contains similar material. The disappearance of the Bessemer notes and manuscripts has insured that there is but a single version of Bessemer's early life and this may well be what he intended. All works that have appeared in the field since J. S. Jeans's contain identical descriptions of Bessemer's early life and work. But there was little Bessemer could do to influence accounts of his later life, inasmuch as after he had read his paper at Cheltenham he became a public figure whose activities were a matter of record.

2. *Bessemer, Autobiography,* p. 5.

3. Ibid., p. 6. Some forty years later Bessemer was offered a decoration by the French government, but his own government refused to give him permission to wear it. In a fury, Bessemer wrote a seething letter to the London *Times*. A soon as he saw the letter in print, he realized it contained serious accusations and he thought it wise to write to Prime Minister Beaconsfield to explain the circumstances. Beaconsfield

acknowledged the letter and a knighthood for Bessemer followed shortly thereafter. Contrary to popular belief, he was honored primarily for the stamp innovation of his youth rather than for the steel process that bears his name.

4. W. T. Jeans, *Age of Steel*, p. 21.
5. *Autobiography*, pp. 4, 41.
6. Ibid., p. 43. According to Bessemer, a typesetter could set as many as 6,000 letters per hour, in contrast to the 1,700 or 1,800 letters ordinarily set by hand. He believed that typesetting would be a desirable occupation for women. But strong objections developed not only to the labor-saving device itself but also to the threat of competition from female labor. The machine eventually died a natural death.
7. Ibid., p. 93.
8. Ibid., p. 117.

VII · Development of the Bessemer Process

n 1851, before the beginning of the Crimean War Bessemer
had obtained approximately twenty-five patents. They covered a
wide range but he had not yet entered the area of armor or
projectiles. The annihilation of the Turkish fleet at Sinope
demonstrated the havoc that shellfire could inflict upon wooden ships
and provoked incessant discussion about the merits of armored vessels
and their resistance to shot and shell. Elongated projectiles also were
discussed frequently, but, although American inventor John Stevens
had been interested in them forty years earlier, there still was no rifled
ordnance in which to use this type of missile.

Intrigued by the problem, Bessemer theorized that rotation could be
accomplished with an elongated missile fired from a smooth-bore gun.
He believed that this could be implemented by making "small passages
lengthwise through the projectile, and open at the end nearest the breech
of the gun. Through these passages a part of the exploded powder found
its way, and being emitted from the opposite sides of the projectile,
the reactive force of the exploded gunpowder produced the rotary motion
required."[1] Bessemer believed that this type of shell could be adapted
for use in any smooth-bore gun, and he applied for and received such a
patent in November 1854. When he presented his theory to the British
War Department, he brusquely was turned away. Convinced that his
invention had merit, Bessemer built a miniature testing ground in his
garden and was able to demonstrate that when he fired the projectile
from a smooth-lined mortar it did revolve and did proceed end-on in
its flight.

Some months afterward, Bessemer was in Paris with his friend Lord
James Hay. He joined a house party at the home of Hay's daughter where,
at a large dinner party, Prince Napoleon was a guest. After dinner, as
the men smoked their cigars in the library, the conversation turned to
artillery. Seizing the opportunity, Bessemer mentioned his invention for
firing elongated projectiles from smooth-bore guns. The prince expressed
curiosity and Bessemer, who *happened* to have a working miniature of
the invention in his pocket, demonstrated the principle of its use to the

distinguished guest. The prince recognized the missile's possibilities and promised Bessemer that he would tell his cousin the emperor about it.

Prince Napoleon kept his word and an invitation to appear before Napoleon III followed. At the audience Bessemer explained the principle of his invention while the French ruler examined closely the small replica. Impressed with the projectile's potential, the emperor offered the Englishman the use of the armory at Vincennes on the outskirts of Paris for conducting suitable tests and more extensive trials. But Bessemer believed that he could work to better advantage on his home ground at Baxter House, and obtained permission to do so. He was to return to France when the missile reached a stage suitable for trial. In a few weeks, Bessemer returned to Paris with the projectiles, which weighed twenty-four and thirty pounds and were shaped to fit the 4.75-inch twelve-pounder smooth-bore French guns. He went directly to Vincennes to meet Commander Minié, who was to superintend the scheduled trials on the Polygon of the fortress. Targets were set up and the projectiles, which had been coated with black japan, were fired. From the manner in which the japan was scraped off in spiral lines as the projectiles penetrated the planks and ricocheted in the deep snow, it was established without question that the shells had rotated from one-half to two and one-quarter times as they traveled the length of the gun. Still resentful of his treatment by the British authorities, Bessemer commented, "Evidence was thus afforded that the dogmatic way in which the invention had been ignored by our military authorities was in no way justified. Whatever the real merits or demerits of my invention may have been, it was at least shown that, at a time when we had no established rifled system, this early attempt at a solution of the difficulty had sufficient merit to render it worthy of a trial."[2]

After the trials, Bessemer and the officers, cold and tired, returned to the warm officers' quarters in the Vincennes fortress. There, over cups of mulled claret around a blazing wood fire, they discussed the trials and the various problems related to the use of the elongated projectile. Commander Minié, an inventor of the expanding bullet that gave his name to any rifle in which it was used, was a man of considerable intellectual attainment. He commented that, although the trial shots had revolved with sufficient rapidity and had penetrated the targets point forward, he mistrusted the guns that the French were using and did not consider it safe practice to fire a thirty-pound shot from a cast-iron gun intended to fire a twelve-pound shell. He believed the real question to be: "Could any guns be made to withstand the shock of firing such heavy projectiles?"[3]

Who could describe the effect of Minié's words on Bessemer better than the inventor himself? In the light of the controversy that followed, it is important to know how Bessemer reacted initially. Although his description was written many years after the event, it is obvious that Bessemer remembered the occasion clearly.

> This simple observation was the spark which had kindled one of the greatest industrial revolutions that the present century has to record, for it instantly forced on my attention the real difficulty of the situation, viz: How were we to make a gun that could be strong enough to throw with safety these heavy elongated projectiles? I well remember how, on my lonely journey back to Paris that cold December night, I inwardly resolved, if possible, to complete the work so satisfactorily begun, by producing a superior description of cast-iron that would stand the heavy strains which the increased weight of the projectiles rendered necessary. At that moment I had no idea whatever in which way I could attack this new and important problem, but the mere fact that there was something to discover, something of great importance to achieve, was sufficient to spur me on. It was indeed to me like the first cry of the hounds in the hunting field, or the last uncertain miles of the chase to the eager sportsman. It was a clear run that I had before me — a fortune and a name to win — and only so much time and labour lost if I failed in the attempt. When only a few days later, I personally reported to the Emperor the results of the trials at Vincennes, I told his Majesty that I had made up my mind to study the whole question of metals suitable for the construction of guns, a proposal which he encouraged by many kind expressions, and a desire that I should communicate to him the result of my labours.
>
> My knowledge of iron metallurgy was at the time very limited, and consisted only of such facts as an engineer must necessarily observe in the foundry or smith's shop; but this was in one sense an advantage to me, for I had nothing to unlearn. My mind was open and free to receive any new impressions, without having to struggle against the bias which a lifelong practice of routine operations cannot fail more or less to create.
>
> A little reflection, assisted by a good deal of practical knowledge of the properties of copper and its several alloys, made me reject all these from the first, and look to the metal iron, or some of its combinations, as the only material suitable for heavy ordnance. At that time nearly all our guns were simply unwrought masses of cast iron, and it was consequently to the improvement of cast iron that I first directed my attention.
>
> The experiments at Vincennes took place on or about the twenty-second of December 1854 and before the close of the year I found myself once more at Baxter House, busy with plans for the production of an improved metal for the manufacture of guns, which improvement in the quality of the iron I proposed to effect by fusion of steel in a bath of molten pig iron in a reverberatory furnace. I soon determined on the form of furnace, and applied for a patent for my "Improvements of Iron and Steel," which was dated as early as January tenth, 1855 — that is, within three weeks after the experiments in the Polygon at Vincennes.[4]

Bessemer now had made up his mind to explore the field of ferrous metallurgy. As he was the first to admit, this was a subject about which he was poorly informed. Little as he knew about the manufacture of iron and steel, from his earliest experiments and inventions he had been familiar with the wide range of possible uses for the various metals. In his boyhood, he had watched his father develop a superior metal for type in the furnace at Charlton and had used this same metal in casting small parts for his models. He had worked with various alloys in developing a metal for his medallion castings. He had been aware of the need for a special alloy for the hard and tough rolls that he had used in the velvet embossing process as well as in producing the bronze powders. Given his experience in working with furnaces and high temperatures, it was natural that he should lose himself in this new problem.

In 1854 the science of metallurgy was in its infancy. If Bessemer had little knowledge of the field that he was about to enter, he had plenty of company in the large number of men already working in the iron industry who never were quite sure of what they were doing or why the results of their labors succeeded or failed.

Great Britain in 1854 was producing a little over three million gross tons of pig iron and only forty to fifty thousand tons of steel.[5] Wrought iron was the most commonly used form of iron, although the brittle cast iron was used for many purposes as well. The metal that went into bridges, ships, railway rails, wheels and axles was iron, principally wrought iron.

Wrought iron is the oldest known form of iron. No record exists of its crude beginnings. A new method of producing iron had been developed about seventy-five years before. In 1784, an Englishman from Lancashire, Henry Cort, purified iron using the reverberatory furnace, invented by John Rovenson around 1613. The process was used first in 1766 by the Cranege brothers, Thomas and George. They patented a process using raw coal in a reverberatory furnace but, for reasons now unknown, the brothers never used their patented process commercially. Thus, Henry Cort is credited for the success of what later came to be known as the dry puddling process.[6]

The reverberatory furnace was so called because the heat of the furnace was reflected from its roof. The furnace had three sections, the grate or fireplace located at one end, the neck or flue leading to the stack at the other, and the hearth or puddling basin in the center. These parts were covered by an arched roof that sloped at an angle of ten degrees from the fireplace to the flue.

Cort described his process in a 1784 patent application.

> For the preparing, manufacturing, and working of iron from the ore, as well as from sow and pig metal, and also from every other sort of cast iron ... I make use of a reverberatory or air furnace or furnaces of dimensions suited to the quantity of work required to be done, the bottoms of which are laid

hollow or dished out, so as to contain the metal when in a fluid state. My furnace for the first part of the process being got up to a proper degree of heat by raw pit coals, or other fuel, the fluid metal is conveyed into the air furnace by means of ladles or otherwise. . . .

After the metal has been some time in a dissolved state, an ebullition, effervescence, or such like intestine motion takes place, during the continuance of which a bluish flame or vapour is emitted; and during the remainder of the process the operation is continued (as occasion may require) of raking, separating, stirring, and spreading the whole about in the furnace till it loses its fusibility, and is flourished or brought into nature. . . . [7]

In 1947 the metallurgist James Aston described Cort's process as follows:

He hollowed out the hearth of the furnace so as to make a "puddle" of molten iron which was stirred to speed up the refining operation. The hearth was lined with sand which fused with some of the iron oxide resulting from the partial oxidation of the iron to form a siliceous slag. Coal instead of charcoal was used as fuel. . . . In his furnace the hot gases from the burning coal passed over and fused the charge of iron removing most of the metalloid impurities by oxidation.

The process was wasteful, however, both in time and in metal loss. The yield was often only about 70 percent of the metal charged. [8]

It remained for Joseph Hall in 1830 to modify the process and increase its efficiency. John Percy, who knew Hall well, described the improvements.

Mr. Hall sought to establish his reputation as an inventor on three grounds: —first, the substitution of cinder for sand upon the plates for the puddling-furnace bottoms, to which he was accidentally led by attempts to save and work up the residue, rich in iron, of forges . . . secondly, an improved construction of iron bottoms, consisting of a frame of plates, united one to another, all round the inside of the furnace, of a height of about 15 inches or 16 inches, with a projection of fire-brick hanging over the top of the plates . . . and thirdly, the application of roasted tap-cinder or "bull-dog." [9]

The metal loss was cut to about 10 percent, the time of the heat shortened, and the use of a wider range of grades of pig iron made possible. The Cort process became known as dry puddling because very little slag was formed, the slag-forming impurities having been eliminated in the refinery. The Hall method, because of the large amount of slag that formed and boiled vigorously during the refining operation, at first became known as the pig boiling or wet puddling process and later simply as the puddling process.

After Hall's time the process changed very little. Cold pig iron was charged into a coal-fired reverberatory furnace. When it melted, iron oxide was added in the form of cinder or dry roll scale detached from bars in rolling, in order to make a basic slag. Then the batch was stirred thoroughly by a puddler using a long-handled bar called a rabble. The oxidizing reaction, principally between the iron oxide in the slag and the lining, almost completely eliminated the impurities present in the pig iron. The silicon, manganese, phosphorus, and sulphur all were eliminated before the carbon, inasmuch as it was the carbon that protected the iron from oxidation. This stage of the process was called clearing. The beginning of the carbon oxidation was marked by the appearance of small flames called puddler's candles on the surface of the slag.

As the refining process proceeded, the impurities content decreased as these elements left the metal and joined the slag, and the melting point of the metal rose rapidly. Even though the covering layer of slag remained molten during the entire heat, the highest furnace temperature that could be maintained was too low to keep the metal in a molten state.

As the metal assumed a pasty consistency, it "came to nature," a phrase used by the puddlers who then kept the molten mass uniform in temperature and composition by rabbling or stirring it. One writer described the appearance of the metal at this stage as "small globules, like butter in churned milk, each globule representing a portion of the iron that had become decarburized."[10] Overman in 1854 thought of the sight as "iron forming small and to all appearances round particles of the size of peas, which swim in the cinder."[11]

When all the heat had come to nature the puddler divided the pasty metal and slag into balls, each weighing approximately 100 pounds. Dripping fiery slag, the balls were removed from the furnace one by one and delivered to a squeezer that removed surplus slag. Enough remained, however, so that when the ball was passed between grooved rolls and formed into a bar, elongated strings of slag extended in the direction of the rolling. In well-made wrought iron there may have been as many as 250,000 of these glass-like slag fibers per square inch.[12] At first their presence was considered detrimental, but later it was recognized that they were responsible for the desirable properties of wrought iron.

Hand puddling operations disappeared from the scene many years ago. In 1930 a commercial process for the manufacture of wrought iron came into use. It duplicated the essential elements of hand puddling but made it possible to increase production, control the operation, and, most important, secure a more uniform product than was possible using old-fashioned puddling methods.[13]

At the midpoint of the nineteenth century, steel was being produced by the cementation, crucible, or natural steel process. It is not known when the cementation process first was used, but it did flourish during the eighteenth and nineteenth centuries in England and even was practiced until fairly recent times to a limited degree.

In the cementation process, a low carbon ferrous product such as wrought iron was heated to a red heat in contact with a carbonaceous material such as charcoal, and the metal absorbed the carbon. The extent of the absorption would vary with the length of time the metal was in contact with the carbon and with the temperature at which the process was carried on.

The 1855 converting furnace consisted of two simple firebrick troughs twelve feet long, three feet deep, and three feet wide. The fire room was placed between them and the whole was covered by an arched vault so that the heat could be distributed equally around the trough. Wrought iron, in the form of bars, was placed in the troughs in alternate layers with charcoal broken into pieces about the size of beans. When filled, the troughs were covered with sand or loam, which partly vitrified and caked to form a seal through which no air could enter. Such a furnace could hold about twenty tons of iron. The fire then was lighted and sixty to seventy hours later the iron became fully heated. Then the conversion began.

Charles Sanderson, writing in 1855, explained the conversion: "The pores of the iron being opened by the heat, the carbon is gradually absorbed by the mass of the bar, but the carbonization or conversion is effected, as it were, in layers. ... let me suppose a bar to be composed of laminae — the combination of the carbon is first effected on the surface, and gradually extends from one lamina to another, until the whole is carbonized. To effect this complete carbonization the iron requires to be kept at a considerable uniform heat for a length of time. Thin bars of iron are much sooner converted than thick ones."[14]

In later years the degree of carburization was determined by a fracture test on a bar withdrawn through a small opening provided for that purpose. In earlier times the workman had to judge when the time had come to cool the furnace until the bars could be handled.

The surfaces of the withdrawn bars were likely to be covered with blisters from the expansive force of the carbon monoxide formed when the carbon reacted with the oxides of the slag. Thus the steel produced by this process was known as blister or cemented steel. Because the carbon was absorbed at the bar's surface and decreased toward its center, the steel was not uniform. To remedy this, at least in part, the blister bars were reheated, hammered and rolled. Such steel was used for coach springs and agricultural implements like spades and shovels.

The making of steel in crucibles, although an ancient practice, had fallen into disuse. In 1742 it was rediscovered by Benjamin Huntsman, an English clockmaker who was searching for a way to produce superior steel for clock spring material. He was an ingenious workman whose patient experiments produced a steel known as cast steel, which was suitable for his trade. In 1855, Huntsman's son still was carrying on the father's business.

The manufacture of cast steel is in itself a very simple process. Bar steel is broken into small pieces, which are put into a crucible, and are melted in a furnace about 18 inches square and 3 feet deep. The crucible is placed on a stand 3 inches thick, which is placed on the grate-bars of the furnace. Coke is used as fuel, and an intense heat is obtained by having a chimney about 40 feet high. Although a very intense white heat is obtained, yet it requires 3½ hours to perfectly melt 30 lbs. of bar steel. When the steel is completely fluid, the crucible is drawn from the furnace, and the steel is poured into a cast-iron mould. The result is, an ingot of steel, which is subsequently heated and hammered, or rolled, according to the want of the manufactureres. Although I stated that the melting of cast steel is a simple process, yet, on the other hand, the manufacture of cast steel suitable for the various wants of those who consume it requires an extensive knowledge; a person who is capable of successfully conducting a manufactory, must make himself master of the treatment, to which the steel in manufactures will be submitted by every person who consumes it. Cast steel is not only made of many degrees of hardness, but it is also made of different qualities; a steel maker has, therefore, to combine a very intimate knowledge of the exact intrinsic quality of the iron he uses, or that produced by a mixture of two or three kinds together. [15]

Natural or raw steel was produced from pig iron with a content of about 4 percent each of carbon and manganese. The design of the furnace for producing raw steel was almost everywhere similar to that of a common charcoal refinery. The method of producing it varied according to the composition of the pig iron and the habits of the workmen in the countries that produced it, principally Germany, France and Austria. Sanderson describes the process as it was carried on in the mid-nineteenth century.

When the fire is hot, the first operation is to melt a portion of pig iron, say 50 to 70 pounds, according as the pig contains more or less carbon; the charcoal is then pushed back from the upper part of the fire, and the blast, which is then reduced, is allowed to play upon the surface of the metal, adding from time to time some hammer slack, or rich cinder, the result of the previous loop. All these operations tend to decarbonize the metal to a certain extent; the mass begins to thicken, and at length becomes solid. The workman then draws together the charcoal and melts down another portion of metal upon the cake; this operation renders the face of the cake again fluid, but the operation of decarburization being repeated in the second charge, it also thickens, incorporates itself with the previous cake, and the whole becomes hard; metal is again added until the loop is completed. During these successive operations, the loop is never raised before the blast, as it is in making iron, but it is drawn from the fire and hammered into a large bloom, which is cut into several pieces, the ends being kept separated from the middle or more solid parts, which are the best.

This operation, apparently so simple in itself, requires both skill and care; the workman has to judge, as the operation proceeds, of the amount of carbon which he has retained from the pig iron; if too much, the result is a very raw, crude, untreatable steel; if too little, he obtains only a steelified iron. . . .

The raw steel, being so imperfect, is not considered so much an article of commerce with the manufacturer, but it is sold to the steel refiners, who submit it to a process of welding.[16]

Such was the state of iron and steel production when Bessemer began his study of it. All the present-day methods for testing the quality of steel then were unknown. Metallography, the study of the microscopic structure of steel, also was unknown. Although Henry C. Sorby first applied the microscope to the study of metals in 1863, thirty years elapsed before others recognized the importance of his discovery that steel has a grain structure.[17] The importance of metallography would not be understood until the appearance in 1891 of the first textbook on the subject, the first work to combine scientific metallography with what then was known about metallurgy.[18]

In more recent times steel would be ordered in many ways: by product specification, by chemical composition and quality designations such as SAE and AISI numbers, by mechanical properties such as tensile strength, by hardenability limits as typified by the standard H alloy steels, by requesting the steel producer to furnish steel to make an identified part, or by specifications such as those written by the armed forces.[19] But in 1855 the steel purchaser's expectations were based on faith that the quality of the purchase would conform to that of a previous satisfactory order. The workmen who were responsible for quality developed an uncanny ability to use both their eyes and a special second sense that the old timer brought to his trade. But soon that would not be enough.

Bessemer had knowledge of the puddling process used in making wrought iron and he understood the methods by which steel was being produced. At the beginning of his studies he had no intention of making steel. He wanted to produce a better quality iron than so far had been possible, a metal that would have the characteristics of wrought iron or steel, but that would be run into a mold or ingot in a fluid state instead of being forged.

By 1831, the engineer William Fairbairn, working in Manchester and later in London, had been experimenting extensively in an effort to produce an improved cast iron. At that time he was building iron ships, bridges, and heavy machinery. Aware of the limitations of cast iron for these uses, he had tried to improve the quality of the metal by melting malleable iron along with pig iron and scrap in a cupola. The sulphur content of the finished product was so high, however, that its quality was affected. When Fairbairn endeavored to make a better product, he did not know how to eliminate this unwelcome element.[20]

The two furnaces in most frequent use were the previously mentioned reverberatory and the cupola. Foundries used the cupola to melt their pig iron for castings. Its cylindrical shape made it similar to a blast furnace. Open at the top, it was constructed of boiler plate lined with brick. The charging took place through doors about halfway up the shell and consisted of alternate layers of coke and iron, with a little limestone added near the end of the heat or just before the molten metal was run out of the furnace into the sand molds. Air to burn the coke was blown in through nozzles called tuyeres, which were located near the bottom of the cupola. The heat of the resulting combustion was sufficient to melt the iron.

Since Fairbairn used the cupola in his foundry, logically he used it in his unsuccessful efforts to improve the quality of cast iron. Profiting from Fairbairn's trials, Bessemer decided to try the reverberatory type furnace rather than the cupola in his experiments. He altered its dimensions so that the grate was wider than the hearth and arranged the hearth so that, at the back of the bridge, a plentiful supply of air met with the gases from the furnace, causing a continuous and intense heat to sweep over the surface. From there the heat flowed to a down shaft that led to the chimney. The hearth was filled with a bath of molten pig iron to which had been added broken pieces of blister steel made from Swedish and other charcoal iron. This charge was melted in the bath and had no contact with the fuel.

During the weeks following his trip to Paris, Bessemer conducted innumerable experiments. He rebuilt the furnace many times in his efforts to secure the most efficient proportions and heat. The resulting furnace design and method of operation formed the first of a long line of patents, covering iron and steel production, that were issued to Bessemer between 1855 and 1880. Dated 10 January, 1855, the first patent bore the simple title, Improvements in the Manufacture of Iron and Steel.

Some of the metal produced during these experiments appeared to have a fine grain and great strength when it was cast into ingots. Bessemer cast a small model gun from the molten metal and noticed, when he turned it in the lathe for the finishing process, that the shavings were slightly curled. He thought that they resembled the turnings from a steel ingot. Later, when the metal was polished, his first observation was confirmed, since the metal appeared to be as white and as close-grained as any steel he had seen.

Satisfied with his efforts, Bessemer took the small gun to Paris and presented it to Napoleon. According to Bessemer, the emperor was much pleased with the gun and commented that some day it would have historic interest. He immediately approved the erection of a furnace at the government foundry at Rouelle, near Angoulême, so that Bessemer might continue his work. Plans were made for the construction of the furnace and, upon his return to England, Bessemer sent the government foundry several thousand special fire brick for the furnace lining. [21]

Meanwhile, he resumed his experiments with the reverberatory furnace at Baxter House. Very soon an incident occurred that halted the projected work in France and, indeed, altered his present investigations and future plans. Bessemer happened to notice several pieces of pig iron that lay at one side of the bath, unmelted by the intense heat of the furnace. In order to increase the combustion, he turned on more air. Half an hour later he opened the furnace door to check on the progress of the heat and observed that the two pieces of pig iron still remained unfused. With the intention of pushing them into the molten metal, he poked them with an iron bar. To his surprise, he saw that they had become thin shells of decarburized iron. Ths phenomenon led him to understand that atmospheric air alone was capable of decarburizing pig iron completely and of converting it into malleable iron.

The incident started Bessemer off on a new train of thought. Bessemer wondered whether a conversion into malleable iron would result if he could bring air into contact with a sufficiently extensive surface of molten crude iron. To experiment, Bessemer ordered crucibles made of fire clay with dome-shaped perforated covers. He also obtained some fire clay blow pipes, which he joined to a three-foot length of one-inch gas pipe. The opposite end of the gas pipe was attached by rubber tubing to a fixed blast pipe. With this kind of connection, the blow pipe easily could be introduced into and withdrawn from the crucible. He half-filled the crucible with ten pounds of pig iron and, after blowing air into the vessel for thirty minutes, Bessemer thought he had converted the gray pig iron into soft malleable iron. Of course, fuel had been used to maintain the heat of the air being blown into the crucible, but now he wondered if sufficient internal heat could be maintained in the charge of metal to decarburize it completely.

For this experiment, he built a cupola with an interior height of four feet similar to the kind used by iron founders. He reasoned that this size would permit only the sparks and heated gases to escape through the central hole in the top. The converter, [22] as Bessemer now called it, had six horizontal tuyeres arranged around the lower part connected by six adjustable branch pipes, which in turn received their air supply from an annular rectangular chamber that extended around the base of the converter. [23]

Bessemer first ran about seven hundred pounds of molten pig iron into the converter through a spout in one side of the vessel and then turned on an air blast of ten or fifteen pounds. For about ten minutes everything was quiet. A few sparks accompanied by hot gases were discharged lazily through the opening in the top of the converter, just as he had expected. Shortly thereafter, the converter became a veritable Mount Etna in eruption. The sparks increased and an enormous brilliant white flame shot from the top. Explosions rent the air and molten slag and metal splashed in all directions. For a few minutes Bessemer feared the shop itself would be consumed by the holocaust. The explosion continued at such a rate that it was impossible to approach the vessel in order to turn off the pressure. In about ten minutes, the fireworks stopped, the flame disappeared, and Bessemer tapped the converter. He poured the molten metal first into a ladle and then into an ingot mold. He believed that he had a casting of wholly decarburized iron.

For his second trial of the vessel, Bessemer devised what he thought would be a method for controlling the top explosion. He procured a cast iron coal-hole plate, about a foot in diameter, which he suspended from a chain about eighteen inches above the top opening. This he thought would prove sufficient to control the slag and molten splashes of metal from erupting to such distances. When the pressure was turned on the performance was repeated, except that the suspended plate melted and disappeared, leaving only the chain that had held it dangling as mute testimony to the failure of the idea.

Bessemer then tried numerous other ideas such as reducing the number of tuyeres, decreasing their diameter, or even lessening the blast pressure. Instead of improving the process, the changes made it fail completely, since they reduced the temperature, which in turn prolonged the process. In one trial, the process was blown an hour and the result was a solid mass of brittle white iron. This convinced Bessemer that the success of the operation depended on the rapidity of the process and that the fireworks were an integral part of it.

Experiments were started immediately to change the shape of the converter in order to eliminate the metal splashings and volcanic uproar. Bessemer decided to design the next converter with the opening at one side instead of in the top so that the action could be directed against a wall that in turn would direct the escaping metal into a pit. The new converter had a sealed-over upper chamber. The metal now could expend itself against the roof of this chamber and fall back into the converter. The flame and part of the slag could find its way through two small openings in the top portion. He also arranged a means whereby the molten metal, after it was ready for tapping, could flow from the tapping hole in the bottom of the converter directly into the mold. The converter now had become a combined converting and casting apparatus.

Those who knew the inventor as a self-contained rather taciturn person would have been surprised could they have read his description of this last experiment. Written many years later, it describes the progress of his work

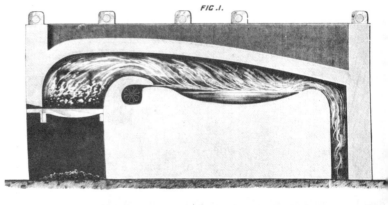

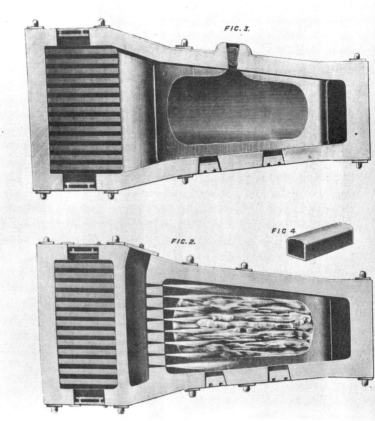

Early Bessemer furnace and converter. Fig. 1, early form of reverberatory furnace used by Bessemer, vertical section; fig. 2, reverberatory furnace, horizontal section; fig. 3, reverberatory furnace, horizontal section; fig. 4, shell of decarburized iron; fig. 5, air furnace containing crucible, vertical

90

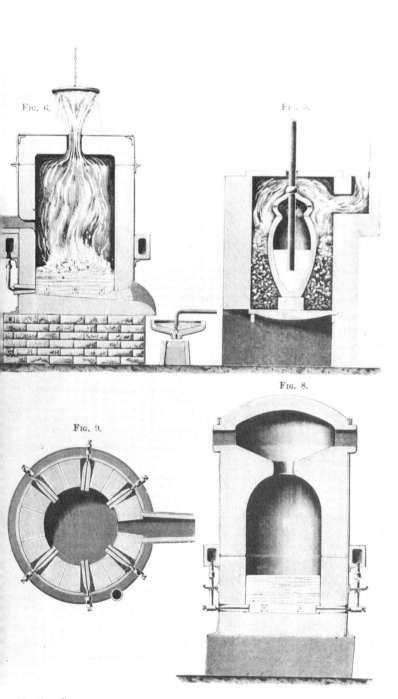

FIG. 6.

FIG. 5.

FIG. 8.

FIG. 9.

(*Continued*)
section; fig. 6, cylindrical converter, vertical section; fig. 8, converter with
upper chamber, vertical section; fig. 9, tuyere pipes fitted to openings left
in the lining. From *Engineering*, 11 December 1896.

with care and precision. When the inventor comes to the final step, his description no longer is so circumspect.

It is impossible for me to convey any adequate idea of what my feelings were when I saw this incandescent mass rise slowly from the mold. The first large prism of cast malleable iron that the eye of man had ever rested on. This was no laboratory experiment. In one compact mass we had as much metal as could be produced by two puddlers and their assistants, working arduously for hours with an expenditure of much fuel. We had obtained a pure homogeneous ten inch ingot as the result of thirty minutes' blowing, wholly unaccompanied by skilled labour or the employment of fuel; while the outcome of the puddler's labour would have been ten or a dozen impure, shapeless puddle-balls saturated with scoria and other impurities, and withal so feebly coherent, as to be utterly incapable of being rendered, by any known means, as cohesive as the metal that had risen from the mold. No wonder then, that I gazed with delight on the first born of the many thousands of the square ingots that now come into existence every day.

I had now incontrovertible evidence of the all-important fact that molten pig iron could without the employment of any combustible matter, except that which it contained, be raised in the space of half an hour to a temperature previously unknown in the manufacturing arts, while it was simultaneously deprived of its carbon and silicon, wholly without skilled manipulation. What all this meant, what a perfect revolution it threatened in every iron-making district in the world, was fully grasped by the mind as I gazed motionless on that flowing ingot, the mere contemplation of which almost overwhelmed me for the time, not-with-standing that I had for weeks looked forward to the moment with a full knowledge that it meant an immense success, or a crushing failure of all my hopes and aspirations. [24]

Since Bessemer was color blind, [25] he probably never saw, except as a shade of gray, the reddish brown flame that indicated the silicon was leaving the molten metal. But could see, the final, triumphant, brilliant white flame with its golden yellow showers of sparks that heralded the completion of a heat, and he was satisfied.

In testing the first ingot, Bessemer dealt it three severe blows with a carpenter's ax. If it had been cast iron, pieces would have chipped off, but this did not occur. Instead, the ax penetrated the soft metal, indicating to Bessemer that the metal was malleable.

Not completely trusting himself, the inventor decided to demonstrate the process to an outsider who could view the operation with more detachment than he, himself, could master. He invited George Rennie, a noted engineer, to come to Baxter House to witness it. Rennie's enthusiasm was immediate, and he insisted, "This is important, the invention must not be kept secret another day." Bessemer replied that since it was not yet a commercial success perhaps he had better perfect it before allowing it to be seen. "Oh, all the little details requisite will come naturally to

the ironmaster," he assured Bessemer. "Your great principle is an un-
questioned success; no fuel, no manipulation, no puddle-balls, no piling
and welding; huge masses of any shape made in a few minutes."[26] Rennie
then suggested the reading of a paper at Cheltenham. As President of
the Mechanical Section of the British Association for the Advancement
of Science, he promised to arrange for placing Bessemer's paper first
on the program. Carried away by Rennie's insistence, Bessemer reluctantly
agreed.

Having gone down to the meeting from London, Bessemer took rooms
at the Queen's Hotel. The morning of the meeting at breakfast, he met
William Clay, manager of the Mersey Forge at Liverpool, whom he
knew quite well. Bessemer has described what followed.

> A gentleman who turned out to be a Mr. Budd, a well-known Welsh iron-
> monger, came up to the breakfast table, and, seating himself opposite my
> friend, said to him, "Clay, I want you to come with me into one of the Sections
> this morning, for we shall have some good fun." The reply was, "I am sorry
> that I am specially engaged this morning or I would have done so with plea-
> sure." "Oh, you must come Clay," said Mr. Budd. "Do you know, that there is
> actually a fellow come down from London to read a paper on the manufacture
> of malleable iron without fuel? Ha, Ha, Ha." "Oh," said Clay, "That's just
> where this gentleman and I are going." "Come along, then," said Mr. Budd
> and we all left the table and proceeded towards the room occupied by the
> Mechanical Section. . . . The room was well filled. The paper had had only a
> brief advance notice but the sensational title had attracted considerable
> attention and many people were there for the same reason as Mr. Budd—to
> ridicule the author of such an absurd proposition. . . . I ascended the raised
> platform and was cordially received by the President. Soon after . . . Mr.
> George Rennie, stood up, and in a few appropriate words explained that, at
> the eleventh hour, he had become acquainted with the fact that a most
> important discovery had been made in the manufacture of iron and steel, and
> he had considered it desirable that a paper describing the invention should
> be read at the meeting. . . . He had ventured on a step which he hoped would
> be excused by all those gentlemen who had favored them by preparing papers
> for that occasion. He considered that the paper about to be read was too
> important to be put at the tail end of the list, and, as the only alternative,
> he had ventured to put it at the head.[27]

The group accepted Rennie's introduction and sat back to hear what
the tall Englishman with the attractive and intelligent face had to say.
Bessemer read his rather long paper and drew sketches of the converter
on the blackboard as he proceeded. Samples of the metal produced from
the converter were displayed on a table in front of the platform.

When Bessemer had finished, the first person on his feet was James
Nasmyth. The famous engineer hefted a small piece of the metal in his
hand and said, "Gentlemen, this is a true British nugget."[28]

After Nasmyth, a somewhat chastened and embarrassed Budd arose to offer his compliments to the inventor and to volunteer the facilities of his ironworks. A London *Times* reporter approached Bessemer and confessed that his notes for the first portion of the talk were incomplete since he had been swayed by the obvious skepticism of the ironmasters when Bessemer had begun to read. Realizing his error, he asked to take the paper with him. The story appeared on 14 August, 1856, in the *Times* (London) and two weeks later in the *Illustrated London News*.

Subsequent events, however, caused the British Association for the Advancement of Science not to print the paper. Some technical journals expressed varying opinions regarding its worth. *Practical Mechanics' Journal,* discussing it at length, commented that there was something very like a mare's nest about it.[29] The *Engineer* expressed the opinion that an M. V. Avril had a similar process that was much better.[30] A letter to the editor of the same paper stated that Captain Franz Uchatius, engineer-in-chief of the gun foundry in the Imperial Arsenal in Vienna was producing steel of a superior quality there.[31] The *Engineer* also devoted considerable space to a comparison of Bessemer's patents with those of Joseph Gilbert Martien of Newark, New Jersey, who had taken out his first patent on 23 August, 1855.[32]

At the time Martien was living in London.

> The essence of his invention was explained as: a process which had for its object the purification of iron when in the molten state from the blast or refining furnace, either by air or steam, or vapour of water applied from below, so that it might rise up amongst and completely penetrate and search every part of the metal previous to congelation, and prior to its being run into a reverberatory furnace for puddling. By this means the manufacture of wrought iron by puddling and the manufacture of steel from cast iron in the ordinary manner were believed to be greatly improved. In carrying out his invention Martien employed channels or gutters, so arranged that the numerous streams of air, of steam, or of vapour of water were passed through and amongst the melted metal, as it flowed from the blast furnace. ... In this process, the novelty claimed was that of purifying iron from a blast furnace while still in a molten state, without the intervention of fuel, thus preparing it for the puddling process in a state of greater perfection than by the old process.[33]

John Percy, a contemporary metallurgist and author, was one of those who held that Martien was not a threat to Bessemer, but that, except for a mishap, one George Parry very well might have been. As Percy described the situation regarding the Martien and Parry patents:

> It is perfectly clear from the specification, that the patentee did not propose to effect by his process the conversion of pig iron, whether unrefined or refined, either into steel or malleable iron; and it is equally clear that he simply intended

it to be used as accessory to the ordinary process or processes in common use for effecting the conversion of pig-iron into malleable iron. The expression "purifying iron" as used in the patent is ambiguous, and I do not know whether it was meant to signify merely decarburization, or the elimination of such matters as silicon, sulphur, and phosphorus. The essence of the patent seems to consist in exposing a stream of molten pig-iron *in its course* to the action of jets of atmospheric air, of steam, or of a mixture of both, introduced at the *bottom* or *below* the surface of the liquid metal. This, so far as relates to atmospheric air, had not, I believe, been previously done, at least *designedly* or otherwise than accidentally. The patentee emits not the slightest hint to show that he was aware of the fact, that by blowing atmospheric air through molten pig-iron sufficient heat would be developed to keep it in a state of liquidity even for a very short time. Air and steam are spoken of precisely as though they were similar agents and would produce similar effects; whereas their effects would be radically dissimilar. Air would, by oxidizing both the carbon and the iron of the molten metal, not only maintain, but greatly raise the temperature; but steam would cause an immediate reduction of temperature and tend speedily to solidify the molten metal. As we shall presently see, Martien never either did or contemplated doing what Bessemer has done; and it is idle to pretend the contrary. However, in October or November 1855, that is two or three months prior to the publication of Bessemer's first patent, in which he first announced that he could perfectly decarburize molten pig-iron by blowing air through it without the further application of external heat, the following remarkable experiment was proposed and conducted by Mr. George Parry of the Ebbw Vale Iron-Works: and if an accident had not unfortunately occured, Bessemer might have been shorn of his glory.

In the bed of a reverberatory furnace several wrought-iron pipes, about 1 in. in diameter, were laid parallel to each other, and about 3 in. apart, in the direction of the long axis of the furnace. The pipes were all put in connection with the blast apparatus. Their upper surfaces were perforated with holes, about 3 in. apart, of which there were about 80 or 100 altogether; and wires having been first stuck in these holes, the pipes were covered solidly over with fire-clay. When the clay bottom thus formed, had become dry, all cracks in it were carefully filled up with fire-clay; and when this also had become dry, the wires were pulled out. The furnace was very gradually heated, and then about 1½ ton of pig-iron from No. 1 blast-furnace, at the Victoria Works was run in, the blast having been previously led into the pipes. Vigorous action occurred, when, by some mishap, the molten metal escaped from the furnace into the road. The then managing director of the works was unwilling that the experiment should be repeated, and the furnace was dismantled, happily for Bessemer.[34]

William Durfee, writing in 1891 on the history of steel production said:

Success is always perilously near to failure. All great inventions and discoveries have usually more than one claimant, and this revolutionary process [Bessemer] is no exception to the rule — a rule which is so universal that it almost justifies the belief that when in the fullness of time the world is prepared for a decisive advance in the sciences or the arts, an overruling power indicates simultaneously to minds separated oftentimes by continents and oceans some way to satisfy the growing needs of the world, and all to whom such revelations are given, who contribute to their promulgation and success, are entitled to an honorable recognition and reward commensurate with the value of their services to mankind. [35]

NOTES

1. W. T. Jeans, *The Creators of the Age of Steel* (New York: Charles Scribner's Sons, 1884) p. 26.
2. Henry Bessemer, *Sir Henry Bessemer: An Autobiography* (London: Offices of Engineering, 1905), pp. 131, 132.
3. Ibid., p. 135.
4. Ibid., pp. 136, 137.
5. *Proceedings of the American Iron and Steel Association* (Philadelphia: AISA, 1877), p. 53. Charles Sanderson, "On the Manufacture of Steel, as Carried on in This and Other Countries," *Journal of the Society of Arts* 3 (1855):451. Sanderson was a highly respected steel manufacturer in Sheffield.
6. W. K. C. Gale, *Iron and Steel* (London: Longmans, Green and Co., 1969), p. 32. J. M. Camp and C. B. Francis, *The Making, Shaping, and Treating of Steel,* 4th ed. (Pittsburgh: Carnegie Steel Co., 1919), p. 213.
7. John Percy, *Metallurgy: Iron and Steel* (London: John Murray, 1864), pp. 627-28.
8. James Aston and Edward B. Story, *Wrought Iron* (Pittsburgh: A. M. Byers Co., 1947), p. 11.
9. Percy, *Metallurgy,* p. 670.
10. Camp and Francis, *Making, Shaping, and Treating of Steel,* p. 220.
11. Frederick Overman, *The Manufacture of Iron,* 3d ed. (Philadelphia: Henry C. Baird, 1854), p. 268.
12. Mineral Industries Extension Division, *Ferrous Metallurgy* (State College, Pennsylvania: School of Mineral Industries, Pennsylvania State College, 1938)1:435.
13. Aston and Story, *Wrought Iron,* p. 31.
14. Sanderson, "On the Manufacture of Steel," p. 454.
15. Ibid., p. 455.
16. Ibid., p. 453.
17. Cyril Stanley Smith, *A History of Metallography* (Chicago: University of Chicago Press, 1960), pp. 172-73.
18. Ibid., pp. 237-40 passim.
19. Charles M. Parker, "How Steel Producers View Steel Compositions and Specifications," *Steel* 124 (14 March, 1949):91-98.
20. Sulphur always has been a notoriously harmful element in iron and steel, and one that demands careful consideration. A high percentage of sulphur in steel makes the resultant product "hot short," which means that the steel will crack and tear in rolling or forging. In iron manufacture, sulphur is present in the coke entering the furnace.
21. Bessemer, *Autobiography,* p. 140. One serious problem that confronted Bessemer was

the size of the furnace he needed for his experiments. If a furnace were to be built with a capacity sufficient to cast a five- or ten-ton gun, a much higher temperature than was possible in the small experimental furnace would be necessary. He hypothesized that if the ashpit were enclosed and air forced into it, he could find a means of increasing the temperature. In accordance with his practice of patenting every development as he went along, this phase of production was patented on 17 October, 1855, some months after the casting of the small gun.

22. W. M. Lord, "The Development of the Bessemer Process in Lancashire, 1856-1900," *Transactions, The Newcomen Society* 25 (1946-47):165. Mr. Lord suggests that "the fixed converter with an auxillary above the main reaction vessel bears a strong resemblance to Bessemer's furnace for the manufacture of optical glass and it is probable that this plant being on hand at Baxter House was adapted for the early steelmaking trials."

23. An interesting observation was made by Henry Howe. "As far as my observation goes, metallurgical writers almost invariably use the word 'converter,' while in the steel works the word 'vessel' is almost always used. Vessel has, of course, a generic sense, but it has acquired a distinct specific meaning—the Bessemer converter. It seems to me high time that this objectionable word should be recognized. Indeed, as the briefer name, and as the one in actual use, it seems on the whole preferable to converter" (Henry Marion Howe. *The Metallurgy of Steel* [New York: Scientific Publishing Co., 1890], p. 339).

24. Bessemer, *Autobiography*, pp. 143-46, 153.

25. *Engineer* 2 (17 October, 1856):559.

26. Bessemer, *Autobiography*, p. 154.

27. Ibid., p. 156.

28. W. M. Lord in his paper on the development of the Bessemer process in Lancashire, wrote that Nasmyth must have listened to the Bessemer paper with mixed feelings for he himself had anticipated Bessemer's own invention to some extent. He had patented a process for puddling iron by means of a jet of steam, and this process had been tested satisfactorily in Bolton. Lord, "Development of the Bessemer Process," p. 164.

Nasmyth later wrote to Bessmer about the Bolton trials.

My friend Thomas Lever Rushton, proprietor of the Bolton Ironworks was so much impressed with the soundness of the principle, as well as with the great simplicity of carrying the invention into practical effect that he urged me to secure the patent, and he soon after gave me the opportunity of trying the process at his works. The results were most encouraging. There was a great saving of labor and time compared with the old puddling process and the malleable iron produced was found to be of the highest grade as regarded strength, toughness and purity. My process was soon adopted by several iron manufacturers with equally favorable results. Such however was the energy of the steam that unless the workmen were most careful to regulate its force and the duration of its action, the waste of iron by undue oxidation was such as in great measure to neutralize its commercial gain as regarded the superior value of the malleable iron thus produced.

Before I had the time or opportunity to remove this commercial difficulty Mr. Bessemer had secured his patent. The results were so magnificently successful as to wholly eclipse my process and to cast it completely into the shade; at the same time I may say that I was in a measure the pioneer of his invention; that I initiated a new system and led up to one of the most important improvements in the manufacture of iron and steel that has ever been given to the world (*James Nasmyth Engineer: An Autobiography*, ed. Samuel Smiles [London: John Murray, 1883], pp. 366-70).

W. M. Lord adds that Bessemer recognized the truth of these statements and offered Nasmyth a one-third share of his patent. But Nasmyth was about to retire from business at the early age of forty-eight, and declined the offer.

29. J. S. Jeans, *Steel: Its History, Manufacture, Properties, and Uses* (London: E. and F. N. Spon, 1880), p. 56.
30. *Engineer* 2 (9 September, 1856):511.
31. Ibid., p. 515.
32. Ibid., p. 511.
33. J. S. Jeans, *Steel,* pp. 106, 107.
34. Percy, *Metallurgy,* pp. 811-12.
35. William F. Durfee, "The Development of American Industries since Columbus: VIII. The Manufacture of Steel," *Popular Science Monthly* 39 (October 1891):744.

VIII · Bessemer Process:
Failure and Success

Some years before, shortly after establishing the bronze powder factory, Bessemer had realized that he was spending at the drawing board time that he could ill afford. A friend, Mr. Bunning, the London city architect, had recommended to him a young man named Robert Longsdon. Longsdon's interview with Bessemer proved to be so satisfactory that he was engaged. Bessemer rented an office in Queen Street Place where his new assistant could handle architectural and engineering problems and thus free Bessemer to devote more time to inventions.

The arrangement had worked out well, and, when some time later Longsdon married Bessemer's sister, the relationship was cemented further. It will be recalled that when it had been necessary to keep secret the details of the bronze powder process, Bessemer had employed his wife's three brothers to operate the business. There had been a fourth brother, William, who was then too young to be employed. He went to live with the Bessemers, however, and received the benefits of all the teaching that Bessemer could give him. The relationship was exceedingly close, and, as William Allen grew older, Bessemer came to rely to a great extent on his youngest brother-in-law's judgment. Only Longsdon, who had aided in the development of some of the inventions and had assisted from the beginning in the work on the converter process was equally trusted.

Upon Bessemer's return to London from Cheltenham, the effect of the paper that he had read became evident immediately. He quickly realized that he had been unwise in presenting the paper so soon. Giving his idea to the public in this fashion was a two-edged sword. It had brought him money to fight his cause, but it also had brought a pack of hungry wolves snapping at his heels. Sensing that a process such as this might have the fabled qualities of the alchemist's furnace, numerous pseudoinventors set to work to try to imitate the process that Bessemer had described. Others hoped to use the process without paying. Bessemer described one such incident.

> I visited a gentleman in his London office and found him with a packet of papers a foot high getting all the cases he could against me for repealing

by *scire facias* the whole of my patents. He was employed by a company of ironmasters to do so; and he told me candidly afterwards, "When I had gone through the whole of your patents and about seventy patents which they said more or less anticipated you, I found that they had not a leg to stand upon, and I advised them to come to you for a license." There had been a good deal of scurrilous writing against me by one of the parties connected with the firm, and I said, when they applied for a license, "I know you only come now for a license because you cannot upset the whole of my patents, but I shall not refuse you on that ground, but I refuse until I have a letter of apology from one of your people—such a letter as will show that the statements made against me were without foundation."[1]

The apology was written and the license obtained.

Bessemer's custom of patenting every phase of his process and the machinery for its operation proved to be of the utmost importance to him. After the Cheltenham meeting, other inventors scurried to the Patent Office to search the files for some possible loophole that Bessemer might have overlooked, but the thorough inventor had plugged them all. Between January 1855, the date of his first patent, and January 1857, he took out 12 patents in England alone. During his lifetime he acquired 115 patents.[2]

Bessemer now was visited by many of the leading English ironmasters, all of whom were interested and eager to see the process in operation. It was not unusual for groups as large as eighty to gather to watch the converter go through its paces. Oddly enough, no steelmaker showed an interest in what he was doing.

Bessemer was in a quandary as to how best to handle the use of his process. He did not want to sell it outright because it was possible that future profits would be large. He did not want to license it indiscriminately for fear that it would not be controlled properly. Nor did he feel equal to entering business himself and competing with the entire iron industry. After due thought and consideration of the various facets of the problem, he and Longsdon decided on a solution. They would divide the country into five sections, and a special interest would be given to one ironmaster in each district. Not wishing to lose any of the patents, he decided that a royalty of ten shillings per ton for malleable or wrought iron would be fair. However, to the first applicant in each district he would give the right to make a specified number of tons per year at a royalty of one farthing per ton during the entire period the patents were in force. This right would be purchased by paying at once a ten shilling royalty on the annual amount agreed upon. Bessemer believed that with this arrangement he would have sufficient cash on hand to fight any patent battles that might arise, as well as to encourage the licensee to protect his own interests.

Shortly after the August meeting, Bessemer sold the first of the five rights for £10,000 to the Dowlais Iron Works, then the largest iron manufacturing company in the world. Soon afterward, the Butterly Iron Company, John Brown of Sheffield, Dixon of Govan, and a South Wales tinplate company each paid licensing fees, thus completing the group of five. Before the month was out, a total of £27,000 was paid in royalties by the group using the Bessemer process. During that period Bessemer had an unpleasant encounter with Thomas Brown of the famous Ebbw Vale Ironworks in Wales. In highhanded fashion, Brown offered £50,000 for the patent rights outright for all of Britain with no further discussion possible. Bitterly disappointed by Bessemer's refusal to part with the process, Brown departed in a dudgeon muttering, "I'll make you see the matter differently yet."[3]

The next few years were to be the most difficult that Bessemer was to know. He had tasted glory and acclaim as important men listened to his words with respect and awe. But when the small upright converters were built at several of the licensed plants, they all failed. When Bessemer first started his experiments, he had, with unconscious irony, used the term *purgatory* in describing the vessel used in his process. His life for the two years to follow must often have reminded him of that term.

For the experiments at the Govan Iron Works, a small cupola to melt the metal, an engine of equally small capacity, and a miniature converter that looked very much like a detached cylinder of an upright stationary engine were constructed. The ingots poured from the converter weighed approximately 100 pounds each. They first were placed in a ball heating furnace, after which they were hammered. The impact of even very light blows of the hammer broke the metal as though it had been a brick.

Exhausted, sad, and haggard, Bessemer watched gloomily as heat after heat crumbled and disintegrated on the forge. The trials continued for a month, often under unpleasant circumstances. On several occasions, the antagonistic foreman instructed the shingler at the steam hammer to strike full blows, which splintered the ingot into myriad fragments, endangering not only the shingler but also the foreman who, in his eagerness to make the threatening process fail, seemed heedless of the danger.

The results were a disaster. With complete candor, Bessemer admitted, "The transition from what appeared to be a crowning success to one of utter failure well nigh paralyzed all my energies. Day by day reports of failures arrived; the cry was taken up by the press; every paper had its letters from correspondents and its leaders, denouncing the whole scheme as the dream of a wild enthusiast, such as no sensible man could for a moment have entertained."[4]

In the United States, Abram Hewitt of Cooper & Hewitt, agents for the Trenton Iron Company owned by Peter Cooper, became interested in Bessemer's process from accounts in the journals that reported the Cheltenham meeting. Hewitt immediately wrote to Peter Cooper's son Edward, a close friend of Cyrus Field who was then in London in connection with the laying of the Atlantic cable, and asked him to make inquiries concerning the process. Following an interview with the inventor, Field wrote Hewitt, "If you feel sure that Mr. Bessemer's invention for making iron is valuable you had better come out here to close a contract with him. I have said and done all that I can, and your firm will have the first offer of his patent for the United States."[5]

Hewitt was unable to go to England then, but a short time later Bessemer's agent arrived in the United States to file an application for patents in Washington. After discussing details with the agent, Hewitt instructed the superintendent of the blast furnace at Philipsburg, New Jersey, to construct a small converter for a trial of the Bessemer process. Two trials took place, one in December 1856 and the other late in January 1857. Both were unsuccessful.

Peter Cooper, who at that time was in England and thus was in a position to hear the discussions and to read the newspaper articles concerning the failure of Bessemer's invention, reported to Hewitt. His own experiences, coupled with the letters from abroad, persuaded Hewitt to discontinue his experiments, and he abandoned any idea of securing the American rights.[6]

As Bessemer's world crumbled around him, he arrived at a difficult decision. He had spent two years of endless labor and had devoted large amounts of money to the development of the process. Despite its failure, he still was convinced of its worth. Of the £27,000 he had received for the rights, he settled £10,000 on his wife as protection for his family. The remainder of the fund would be used to discover why the process had failed. Beset by the press and his antagonists, Bessemer must have felt as though Longsdon and young Allen were his only remaining allies.

Bessemer's first act in his fight to reinstate his name and reputation was to engage Dr. T. H. Henry, an outstanding English professor of chemistry, to analyze all the materials that had been used in the experiments. Henry's investigations were supplemented by the work of Edward Riley and Dr. John Percy. Much information also was gathered from publications and from previous research done by Robert Hunt of the Record Office of the School of Mines.[7]

By these concerted efforts, the culprit eventually was discovered—it was phosphorus that, then as now, caused brittleness in the final product. The old timers in the business minced no words in describing such metal. They spoke of it as "rotten." Bessemer already had been aware that sulphur caused problems in his work, but not that phosphorus was part of his trouble. These two elements have one common characteristic; they are not reducible in the converter. In fact, their proportion increases slightly in the finished heat because a certain amount of iron always is lost by oxidation or ejection from the mouth of the converter. But the two elements more than hold their own; consequently the furnace man can do little about them once they have entered the converter.

British pig iron was highly phosphoric in compostion. When Bessemer had ordered small amounts of pig iron from a London dealer for his experiments in Baxter House he had not specified a particular kind. He had not even thought of doing so. By sheer chance, he had been supplied with Blaenavon pig iron, a variety that was extraordinarily free of phosphorus.

Having learned the cause of his failure, Bessemer set to work. For more than a year he built, altered, and tore down converters. He would spend weeks building a vessel, only to discover, in the first hour of its operation, that again he had constructed a failure. Experiments were conducted almost daily as he increased the size of the charge to as much as two tons. Against the urgings of his friends, he continued the struggle. Living in solitude, seeking no capital, and making no effort to explain, Bessemer worked in an atmosphere of resentful loneliness. The strange attitude of the ironmasters was a source of wonder to Bessemer. There was only silence from the group whose encouragement he had expected.

In his search for a phosphorus-free pig iron, Bessemer learned that the purest metal was Swedish. Almost as a last resort, he had a quantity of pig iron delivered from Sweden. He anxiously tested its qualities and, as if by a miracle, all the routine of the process returned to its original pattern. Bessemer now was confident that his name would be cleared. He announced to a skeptical world that, after several years of experiment, the process now was successful and available for licensing. The iron industry, mindful of the past catastrophe, paid no attention. Not a single iron or steelmaker was interested in attempting the process a second time.[8]

In order to produce tangible evidence of the converter's new success, Bessemer used the best Swedish pig iron to produce a few ingots of tool steel quality. He took them to Sheffield, where he had the steel rolled into bars identical in size with those commonly used. Next he took the bars to his friends the Galloways in Manchester. During a two-month trial of Bessemer's steel, the workmen at the Galloway plant had no knowledge of its special nature and were unaware that they were using steel in any way different from that to which they were accustomed. There was one important difference, however; the steel to which they were accustomed had cost sixty pounds per ton as against six to eight pounds for Bessemer's product.

The Sheffield steel manufacturers remained unimpressed. Some of them expressed an interest in obtaining a monopoly on the rights, but they had no interest in licensing arrangements. Bessemer now had reached the point of no return. Physically as well as financially, he had expended all his resources, yet he could not turn back. He decided to do what he had sworn he never would do. With the support of the faithful Galloways, who contributed £5,000 to the enterprise; his brother-in-law William Allen, who offered £500; and with the £6,000 that he and Longsdon got together, Bessemer made plans to invade the territory of the steel-makers. In Sheffield, famous as a cutlery center since the Middle Ages,[9] he built a plant. Realizing that the licenses that he had sold might give undue advantage to the original five licensees now that the process had been perfected, Bessemer repurchased the rights that had lain unused since failure of the process.

Longsdon designed a neat white brick "range" of buildings, and in less than a year the first Bessemer steel plant was established. Bessemer acknowledged that there were risks in such a maneuver. The workmen of Sheffield were notoriously tough, and under the circumstances would have been quite capable of "rattening."[10] Bessemer was of the opinion that the only reason he escaped such treatment was the conviction on the part of both the workmen and the owners that the Bessemer process offered no competition.

While the works were being built in Sheffield, Bessemer scoured the country looking for a source of phosphorus-free iron. After a long search, he found that the iron ores used by the Workington Iron Company were extremely pure, but that their pig iron was heavy with phosphorus. He visited the officers of the company and assured them that, if the phosphorus could be eliminated, there would be a large market for their pig iron. The owners were at a loss to explain why, if the ores contained no phosphorus, the pig iron carried such a large percentage. Bessemer obtained their permission to watch their production process to see if he could determine where the element was making its entrance. After inspecting the raw materials and studying the process, Bessemer was no wiser than the management. As he returned to the office with the foreman, they passed a large heap of slag and cinder. Bessemer inquired about the pile and was told that it was flux for the furnace. "What is it?" pursued Bessemer. "Why it's puddle furnace cinder, we use it to make the furnace work smooth and the iron fluid," replied the foreman. "We send some of our ore into Staffordshire for use in their puddling furnace and they return the cinder for us to use." Bessemer realized at once that this innocent looking pile contained phosphorus picked up

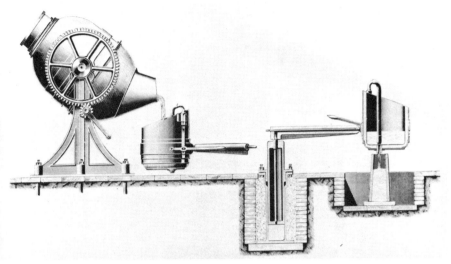

First movable converter at Bessemer's steelworks in Sheffield. It was moved by hand gearing. From *Transactions of the American Society of Mechanical Engineers* 18(1897): 474.

in its journey through the Staffordshire puddling furnace and that the unsuspecting Workington plant was contaminating its finest ores every time it used the cinder as a flux. Workington substituted black shale for the cinder, and Bessemer at last had found a supply of good phosphorus-free pig iron.

The first converter was placed in operation in 1858. All through these months of experimentation, Bessemer constantly had been changing the working of the vessel. As soon as the idea of blowing air through the bottom of the converter occurred to him, Bessemer had realized the importance of preventing the molten metal from running out of the bottom holes while the vessel was being charged. This he achieved by swinging the converter on trunions so that the vessel lay on its side while it received the hot metal. Then, as it was swung into its operating position, he turned on the air. The blast was enough to start the chemical action and the air pressure prevented the metal from pouring through the bottom.[11]

For more than a year, the plant turned out tool steel of quality good enough for use by such plants as those of Sir Joseph Whitworth, Sir William Fairbairn and Company, Platt Brothers, as well as the arsenal at Woolwich. However, there were still no steel licensees of the patent, and the orders were for small amounts of steel—orders for twenty-eight or fifty-six pounds at a time were not uncommon.

On 24 May 1859, the inventor again read a paper, this time before the Institution of Civil Engineers. He presented his new discoveries under the title "Manufacture of Malleable Iron and Steel" and stated that the primary elements of the process had not changed. The facts had been neither altered nor modified during the last three years of work. The same apparatus that he had described then would do the job now. Bessemer did permit himself the pleasure of a small amount of sarcasm.

> It is singular to observe how prone the practical man is to deny to the inventor of a new process that very practical knowledge which he himself so much values. If the inventor cannot show in the first week of his apprenticeship the skill which it is well known can only be acquired by years of practice, it suffices to condemn the new system which, in its mere infancy, is expected to be as perfect in all its details as that which the manufacturers have grown gray in the daily practice of. The same conviction of the truth on which the new process in based, and which led the author to bring it before the British Association, has since determined him (in spite of the opinions loudly expressed against the process) to pursue one undeviating course until the present time, and to remain silent for years under the skepticism of those who predicted its failures, rather than again to bring forward the invention until he had himself practically and commercially worked the process, and produced by it iron and steel of a quality which could not be surpassed by any specimens of those metals made by the tedious and expensive processes now in general use.[12]

Bessemer then reviewed all the criticisms that had been leveled against the process and refuted them one by one. He spoke of the significant chemical analyses that finally had pinpointed the culprit. He told of the search for pig iron in places as far away as India and Nova Scotia, but that ended finally in Sweden. He presented facts concerning the tensile strength of his steel, which had been tested at the Royal Arsenal as well as at the Mersey Works of William Clay. He enlarged upon his belief in the importance of using this steel for ordnance. Through the years of trial, since that cold day at Vincennes, this had been the objective that had driven him onward.

Much to his amazement and pleasure, Bessemer's paper was received with enthusiasm and extended applause, a rare occurrence at these meetings. Possibly the engineers were not so much applauding the content of the paper as paying tribute to a man who, faced with disgrace and oblivion three years before, had asked no quarter and had fought his way back alone. However, the discussion that followed was a continuation of the old skepticism about the ultimate worth of his process. The general reaction of the group was due both to distrust and to continuing inability to understand the process. In speaking against the process, Bessemer's longstanding enemy, Thomas Brown, freely questioned

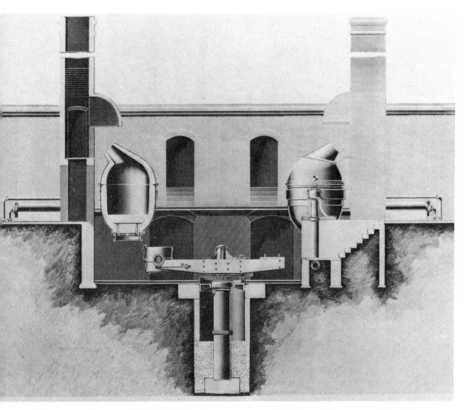

Bessemer plant at Sheffield, showing converters, ladle and crane, and casting pit. From *Sir Henry Bessemer: An Autobiography* (London: *Offices of Engineering*, 1905), pl. 17.

Bessemer's figures, such as those for materials cost. He also asserted that there was a waste of nearly 40 percent of the metal during the operation. He added, as a final barb, that his agent had informed him that, on more than one occasion, the converter had produced nothing more than cinder.

From the beginning of its operations, Bessemer's plant had undersold the Sheffield market, sometimes by as much as twenty pounds per ton. At this time John Brown, owner of the Atlas Works and Bessemer's neighbor, was about to expand his steelmaking activities. At first, he had intended to use the special crucible process developed by Krupp in Germany, but he decided to investigate Bessemer's method first. As he watched the eruption of cinder from the mouth of the converter, followed by the typical brilliant white flame, Brown's interest quickened. When finally the converter slowly turned on its axis and disgorged its

Henry Bessemer in 1866 at the age of fifty-three. From *Cassier's Magazine* 10 (September 1896).

searing torrent of incandescent metal, Bessemer's neighbor was convinced. Bessemer had found his first steel licensee.

When Bessemer was able to add a large converter to his plant in 1859, it became possible to consider expanding the use of his steel to a much wider variety of products.[13] That, in turn, justified a higher royalty. He now set the royalty at two pounds per ton for all products except rails, for which the rate was one pound.

During the first two years, the Bessemer Steel Works operated at a loss, £729 in 1858 and £1,093 in 1859. In 1860 the works made a profit of £923 and never again suffered an annual loss. At the end of the partnership fourteen years later, each partner in the original group had received eighty-one times the amount of his subscribed capital from the profits of the company, quite apart from the income from the licensing of patents.[14]

Shortly after his plant was built, at Bessemer's suggestion Colonel Eardley Wilmot, superintendent of the Royal Gun Factories at Woolwich Arsenal, went to Sheffield to study the process. After observing many trials, he invited Bessemer to Woolwich with a view to adapting the plant there for the process. Bessemer submitted plans and an estimate of the cost of installing suitable apparatus. After a week or more had passed and Bessemer had heard no word from Wilmot, he went to Woolwich to find the reason for the delay. To his surprise, he was told that Wilmot had been replaced in his post by Sir William Armstrong, whose company, Elswick Ordnance, was producing a gun that was built up of wrought iron rings shrunk upon an inner steel barrel. Loaded at the breech, it was rifled and threw an elongated projectile. Bessemer was astonished and said, "Why go to my rival? Why go to the gentleman who has a scheme of his own to carry out instead of mine?" The decision had been made and Bessemer left Woolwich in disgust.[15]

Years later, while the rest of the world was using steel for its ordnance, Britain was still hobbling along with this type of gun. The *Scientific American* commented in 1863, "Instead of calling all scientific heads together to watch events with the Ordnance Office, Sir William Armstrong was shoved up the ladder alone. His own committee approved his own guns, his own factory at Elswick turned them out, on his own evidence, without sufficient proof and trial."[16]

NOTES

1. Henry Bessemer, *Sir Henry Bessemer: An Autobiography* (London: Offices of *Engineering*, 1905), p. 164.
2. James Dredge, "Henry Bessemer, 1813-1898," *Transactions of the American Society of Mechanical Engineers* 19 (1898):960-64.
3. Bessemer's old friends, the Galloways, had taken a license to manufacture malleable iron by his process in Manchester, plus a ten-mile surrounding area, before he read his paper at Cheltenham. According to Charlotte Erickson in her *British Industrialists: Steel and Hosiery, 1850-1950* (Cambridge: At the University Press, 1959), p. 142, Brown did stop off to visit Robert Mushet on his way back to Wales from Cheltenham and probably possessed some valuable information about how to make the Bessemer process work when he made his offer.
4. Bessemer, *Autobiography*, p. 170.
5. Allen Nevins, *Abram S. Hewitt, with Some Account of Peter Cooper* (New York: Harper & Brothers, 1935), p. 128.
6. Ibid., p. 129.
7. Dr. Percy was offended when Bessemer failed to acknowledge the help that the chemists gave him. Percy writes, "No one is more willing than the author of this work to concede to Mr. Bessemer what his ingenuity and perseverance justly entitle him to expect. But the manner in which Mr. Bessemer refers to the chemical part of the investigation might lead to the inference that this had been done exclusively either by himself or his partner. He has, however, received valuable aid from more than one expert analyst, which it would have been generous on his part to have

acknowledged before such an audience as that of the Civil Engineers" (John Percy, *Metallurgy: Iron and Steel* [London: John Murray, 1864], p. 820).

8. Bessemer seems not to have been aware of the superiority of Swedish pig iron even though Sheffield steel producers long had used it to make their steel. They were importing as much as 24,000 tons a year and converted virtually all of it into steel (J. S. Jeans, *Steel: Its History, Manufacture, Properties, and Uses* [London: E. and F. N. Spon, 1880], p. 20).

9. While the Galloways had been unable two years before to operate the process satis-factorily, they now were convinced of its practicality. It was mutually agreed that they now should give up their original license for Manchester and join Bessemer as partners.

Sir Robert Hadfield, the eminent British metallurgist noted for his development of manganese steel, knew Sheffield well. In a paper delivered before AIME he presented some interesting statistics concerning the number of steel manufacturers there near the close of the eighteenth century.

That Sheffield can pre-eminently claim the title "Steelopolis," is not less from its modern development than from its longstanding and traditional associations with the early developments of the metallurgical industry of iron and steel, is shown in an interesting manner by the same directory. We find that there were some half-dozen manufacturers of adzes and hammers, about 50 makers of edge-tools; not less than 40 engaged in file-making; over 300 in pen-, pocket-, and table-knife manufacture; at least 50 in razor-making; close upon a 100 in scissors; and some 60 or 70 in the manufacture of scythes, sickles and shears. Many of these were, no doubt, small workers rather than owners of large concerns; but it will be seen that here was the center for a considerable employment of steel (R. A Hadfield, "Benjamin Huntsman, of Sheffield: The Inventor of Crucible Steel," *Transactions of the American Institute of Mining Engineers* 10 [1894], pp. 178-79).

10. The expression *rattening* came into use during the 1830s when machinery first made its appearance in industry. Usually it involved stealing tools or destroying equipment in order to force payment to a union or compliance with its rules, but in Sheffield it frequently took the form of a bottle of gunpowder in the furnace flues. A man no sooner made a new tool or machine than jealous and acquisitive rivals sought to steal the fruits of it before and after it was patented, and fearful workmen were ready to smash and burn it. The inexperienced commonly were unable to word their specifications so that the patent was of any value to them. Kay, Boulton and Watt, Heathcoat, and many others suffered from these difficulties. Many an inventor showed greater wisdom by silence concerning his plans and secrecy in using his machine, and if he did take out a patent he made the specifications deliberately vague. Today that work safely can be left to one of the many experienced patent agents (C. R. Faye, *Round about Industrial Britain, 1830-1860* [Toronto: University of Toronto Press, 1952], pp. 26, 65).

11. W. T. Jeans, *The Creators of the Age of Steel* (New York: Charles Scribner's Sons, 1884), pp. 68-69.

12. Bessemer, *Autobiography*, p. 180.

13. J. S. Jeans, *Steel: Its History, Manufacture, Properties, and Uses* (London: E. and F. N. Spon, 1880), pp. 63, 68-76 passim.

14. Ibid., p. 67.

15. W. T. Jeans, *Age of Steel*, pp. 90-91.

16. *Scientific American*, n.s. 9 (7 November 1863): 403. Although the first use of Bessemer steel plates in a ship was in the *Jason* built on the Thames by Sameda and Company in 1859, greater interest may be associated with the use of Bessemer steel in the *Banshee*

built in Liverpool. The first steel vessel to cross the North Atlantic, she sailed from
Liverpool on 3 March 1863 on the first of her eight trips to the United States as a
blockade runner. Captured by the federals, she was converted into a gunboat (W. M.
Lord, "The Development of the Bessemer Process in Lancashire, 1856–1900," *Transactions, The Newcomen Society* 25 [1946–47]: 171, 179).

IX · Kelly-Bessemer Controversy

News of Bessemer's paper reached the United States within a month after the Cheltenham meeting. In the *Railroad Advocate* of 20 September 1856, Holley quoted from a London *Times* article about the process and also reprinted a letter about it written by W. Bridges Adams, a leading English engineer. In the issue of the following week, Holley mentioned the claim of a Joseph Martien regarding Bessemer's patent.

Scientific American, also taking note of the event, reprinted an excerpt from Bessemer's paper in its issue of September 13 and quoted from the London *Times* article the following week. On September 27, *Scientific American* came out in support of Martien's claim and on October 11 mentioned that *Mechanics Magazine* of London also had joined the Martien side.

These articles, particularly those in *Scientific American,* attracted widespread attention and brought forth an unexpected letter. It was signed William Kelly, a name little known to the iron producers in the United States and not at all to those in England. Published in *Scientific American* in the issue of October 18, Kelly's letter launched a dispute that has not been resolved yet.[1]

> In November 1851, I commenced a series of experiments with a view of converting fluid pig metal into malleable iron, with the aid of a strong blast of air, and without the use of fuel, which process I term "air boiling." My object was to drive off the carbon in the iron and to make powerful blasts of air do the work of the fire and the manipulation of the puddler's bar in the puddling process. My first efforts were quite satisfactory, as with a blast taken from my furnace and introduced into a suitable cupola with liquid metal taken directly from the furnace I produced a fair article of malleable iron. I found when using gray iron cold blast answered my purpose, but when the metal was white I found hot air had a better effect. I therefore had a small furnace erected to heat the air in blast pipes.
>
> My experiments were conducted publicly at this establishment; hundreds of persons called to see the trials I made, and the subject was discussed amongst the iron masters, etc. of this section, all of whom are perfectly familiar with the whole principle and object I had in view, as discovered by me nearly five years ago.
>
> I was surprised to notice in *Scientific American* of the 13th of September

William Kelly. From Herbert N. Casson, *The Romance of Steel* (New York: A. S. Barnes & Co., 1907), opp. p. 4.

an account of a similar process of converting pig iron into malleable iron, claimed as the discovery of Mr. Bessemer of London, and made within the past two years, the process not differing in the slightest from that I had in practical operation nearly five years since.

I have reason to believe my discovery was known in England three or four years ago, as a number of English puddlers visited this place to see my new process. Several of them have since returned to England and may have spoken of my invention there.

A charcoal furnace such as I have—using cold blast—produces various grades of metal, that I found had to be treated in the air boiling process with some variation; this caused difficulties which I have succeeded in removing and expect shortly to have the invention perfected, and bring it before the public.

William Kelly
Suwannee Iron Works, Eddyville, Kentucky
30th September 1856[2]

At that time Bessemer already had applied for an American patent, which was granted bearing number 16,082, on 11 November 1856. Claiming priority of invention, Kelly applied for his patent the following year. In support of his claim, Kelly submitted a file of affidavits to the

U.S. commissioner of patents that provided evidence about the timing of Kelly's first work.

Strother I. Smith, constructor, builder of blast furnaces, and a machinist well known in the Eddyville area, attested that, in 1847, Kelly had described the new process to him in great detail. He wrote that he was able to fix the time so exactly because it was the same year that an unusual flooding of the Cumberland River occurred.

Jeremiah Tiley, a forge man at Kelly's Union Forge (located about three miles from Suwannee furnace) described how Kelly had made some large drawings and placed them in the forge so that the workers might study them and understand what he wanted to do. Tiley said that he saw both the drawings and the furnace that Kelly built in 1847.

A physician named Alfred H. Champion testified that in the fall of 1851 he was present, with two or three ironmasters, when Kelly said that he had just finished a new furnace for making iron without fuel and invited the group to visit his ironworks to witness the operation for themselves. The ironmasters were skeptical and condemned the process sight unseen, saying that it immediately would chill the furnace. Dr. Champion testified, "The company present all differed in opinion from Mr. Kelly and appealed to me as a chemist in confirmation of their doubts. I at once decided that Mr. Kelly was correct in his theory and then went on to explain the received opinion of chemists a century ago on this subject, and the present received opinion which was in direct confirmation of the novel theory of Mr. Kelly. I also mentioned the analogy of said Kelly's process in decarbonizing iron to the process of decarbonizing blood in the human lungs."

William Soden, another witness, remembered that in November 1851, he had cast pipes for Kelly to use in his "air boiling furnace." He said that he often was present when the air boiling furnace was in operation and frequently assisted in ladling the molten metal from the blast furnace into the air boiling furnace. It was Soden who mentioned that he knew of seven air boiling furnaces that Kelly had built, all alike in principle but varying in the number of tuyeres. But Soden was the exception. Most of the men who submitted affidavits mentioned that they either had been told of Kelly's new process or had helped build one of the furnaces in which it had been used, but admitted that they actually had not seen the process in operation.

One witness, Andrew A. Johnson, testified that he had been in the employ of Kelly and Company as a forge man at the Union Forge when Kelly had described the new process. Johnson said that Kelly had promised him a share in the invention if he would carry out the necessary mechanical details. Although he had agreed to do so, he stated in his affidavit that he had not carried out the work, but gave no reason for not having accepted Kelly's offer. He did say, however, that every practical workman, with the exception of himself and Jeremiah Tiley, condemned the process. He also was under the impression that two or three English puddlers at the Tennessee Rolling Mill, located some twelve miles away, had viewed the process with some favor.

John P. Evans, the forge manager at the Union Forge in 1851, testified that he never had agreed with Kelly's theory. He said he often had seen the furnace that had been built at the forge in 1847, and in 1851 Kelly had told him of the furnace being built at Suwannee. Evans went on to say that a tuyere had been sent to Suwannee to be installed in the new furnace. "The same tuyere came back to the forge stuck in a chill of iron which he had blown into it, it was partly malleable and partly refined pig iron. At another time I conversed with persons who told me that they had seen and worked some of the wrought iron made by Mr. Kelly by his new process and that it was good wrought iron."[3]

Following hearings in Washington, the acting commissioner of patents

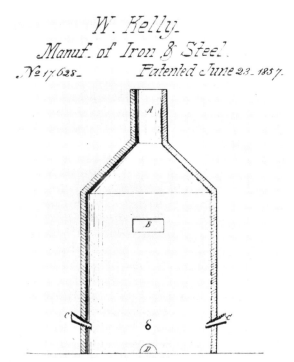

Only known design for Kelly's air boiling furnace, from U.S. patent number 17628. *A* is the flue to carry off the carbonic gas formed in decarburizing the iron; *B* is the port through which the charge of fluid iron is received; *C* and *C'* are the tuyeres; *D* is the tap hole for letting out the refined metal.

acknowledged Kelly's priority. On 13 April 1857, he ordered a patent issued to Kelly subject to appeal by Bessemer within sixty days. No word of testimony had been offered either by Bessemer or by his attorney, R. H. Eddy, to impeach Kelly's testimony, and no appeal was taken by the English inventor.

Bessemer was permitted to retain the patent covering the machinery he had devised for operating the converter. Kelly, who had done little development of the mechanics of his invention, had used a stationary converter. Bessemer had carried his ideas much farther. He had developed a tilting converter whose enormous weight was as sensitive to the touch as a fine small mechanism.

The loss of the American patent occurred at a time when Bessemer's process was under fire in England and delayed any immediate use of his patents in the United States. This must have been a bitter blow to Bessemer, both emotionally and financially. For many years, for reasons that are unknown, Bessemer offered no public explanation of the strange similarity between the two inventions. [4]

After the decision of the U.S. Commission of Patents became known, partisans of both Kelly and Bessemer made their views known. One such expression came from Joseph Martien, now returned from London, in a letter dated 29 May 1857, to Munn and Company, patent agents and publishers of *Scientific American*. [5]

> Gentlemen: Relative to the claim of William Kelly, of Lyon County, Kentucky, and myself to certain improvements in the manufacture of iron without the use of fuel, I take this occasion of informing you and the public generally that agreeable to appointment for the examination of various witnesses in the above case for priority of invention between myself and Mr. Kelly I have found and have been made perfectly satisfied from the ample testimony laid before me in the case that Mr. Kelly is honestly the first and original inventor of the said process of manufacturing iron without fuel. I find, moreover, that he has quietly been and is making improvements and advancing with his invention in a very praiseworthy manner, and of which the public will be put in possession in a short time. I have therefore deemed it a duty that I owe myself and the original inventor to come out and publicly state the above facts and give publicity to the same, trusting he may receive the due honor therefor and be richly rewarded for his genius. [6]

In one historian's opinion, "The validity of this claim [Kelly's] cannot be impeached. But it must also be said that Mr. Bessemer . . . successfully employed the principle in the production of steel and that Mr. Kelly did not. The Kelly process produced refined iron of good quality. Furthermore, the machinery with which Mr. Kelly operated his process was not calculated to produce rapidly or at all the large masses of even refined iron; whereas Mr. Bessemer's machinery was successful after it was perfected in producing steel in large quantities and with great rapidity."[7]

Dr. Hermann Wedding, a distinguished contemporary chemist and engineer, said in relation to the controversy, "It is no real invention to have the idea of turning a known natural law to account without being in a position to indicate the right means for carrying it out."[8]

Even *Scientific American* commented editorially, "A certain fixed time should be allowed for an inventor to apply for a patent for his invention and if he does not do so within that period, if his invention has been publicly used by himself, it should become public property, otherwise he should not be allowed to subvert a patent granted to another, who has taken proper measures to put the public in possession of the invention."[9]

Kelly belonged to no scientific society, nor was it his custom to seek publicity. The only extended statement Kelly ever made was in an article that he wrote for James Swank, secretary of the American Iron and Steel Association. In the first edition of his history of the iron industry, which appeared in 1884, Swank wrote that Kelly had prepared the account of the development of his process especially for that book.[10] But the same statement had appeared earlier in a volume of statistics on the iron and steel industry that Swank compiled in his capacity as special agent for the superintendent of the census of 1880.[11]

The staunchest believer that Kelly alone had invented the process that now bears Bessemer's name was John Newton Boucher, a local historian who lived near Pittsburgh. Boucher never knew Kelly but, based on numerous stories that Kelly's widow told him, he wrote a book that strongly supported the claims of the American inventor.[12]

The salient facts of Kelly's life have been established fairly well. Only later, when the subject of his invention enters the picture, does the atmosphere begin to fog. Kelly's father was born in Ireland in 1782 and became an ardent Irish patriot. To escape punishment for his part in the rebellion against British rule, he emigrated to the United States in 1801, where he settled in Pittsburgh.

By 1811, when his son William was born, Pittsburgh was becoming an iron center. The first iron foundry had made its appearance around 1805, and in the next few years it was followed by three nail factories and two more foundries.[13]

Boucher claimed that Kelly, even in his youth, had a natural bent for scientific research, particularly in the field of metallurgy. In spite of this proposed interest in ironmaking, Kelly did not seek work in that rapidly growing industry. Instead he joined his brother-in-law and brother John in opening a wholesale dry goods establishment.[14]

As the business prospered, Kelly's principal duty was to travel for the firm, collecting accounts and taking orders for new shipments. Once a year, he made a round of the customers who, by this time, were scattered over an area as far west and south as Indiana and Tennessee. On one of these trips, early in the summer of 1846, Kelly journeyed to Nashville, Tennessee, where the commencement exercises of a young women's seminary were the best amusement that friends could offer him on a warm June night. Here he met Mildred Gracy, the youngest member of the graduating class and, in Kelly's eyes, the most beautiful. She, in turn, was much attracted to the tall, energetic man with blue eyes and close-clipped beard. At that time she was sixteen and he was thirty-five. [15]

After her graduation, the young girl returned home to Eddyville, Kentucky, where her father was a well-to-do tobacco merchant. Kelly followed her to the small town on the Cumberland River, and extended his stay as he searched for some way of establishing himself there in order to marry Mildred Gracy. [16]

Kelly was attracted to a nearby iron forge and furnace that, to his unpracticed eye, appeared to be an excellent investment. The supply of red hematite ore that lay on the surface of the ground seemed boundless, and the timberland stretched farther than the eye could see. He prevailed upon his brother to come out to survey the prospect. When John Kelly arrived, he was equally pleased with what he saw. Transferring the mercantile business in Pittsburgh to their brother-in-law, the Kelly brothers bought the furnace and its 14,000 acres of ore and timberland. [17]

The new company produced pig metal and charcoal blooms from ten forge fires and two large finery fires. A finery fire was a primitive process, but an effective one, in which about 1,500 pounds of pig iron was sandwiched between layers of charcoal which then were ignited and the blast turned on. As the charcoal burned, an additional amount was added from time to time until the iron eventually was refined. The hematite ore of the neighborhood was of good quality, and there seemed to be an ample supply easily available on the surface. At the time of the purchase, Kelly had been pleased to learn that there also was a valuable large reserve of ore lying beneath the property. [18]

Soon the brothers were able to build another furnace, the Suwannee, close to the timberland and about seven miles from Eddyville. About a year later, when the surface ore began to play out, the Kellys arranged to mine their reserve supply. Although to his untrained eye, the ores above and below ground appeared identical, when the mined ore was used in the furnace, its behavior was vastly different. Heavily impregnated with a black flint known as shadrach, it would not burn any more than would Shadrach of biblical fame when he, Meshach, and Abednego were cast into the fiery furnace. The poor quality of the deposit forced

Kelly to search for other ore in order to continue operations. Since the timberland also was being depleted, the need to transport both ore and fuel threatened eventual bankruptcy. [19]

Boucher theorized that this situation may have necessitated Kelly's innovation in his process, using a blast of cold air as fuel. However, there is nothing in Kelly's background that would indicate earlier experimentation with metals. Whereas Bessemer, in his autobiography, describes just how he conceived and carried out his theory of making iron without fuel,[20] Kelly does not do so in the comparatively brief statement he wrote for James Swank. Just as Bessemer's autobiography was written many years after he invented his process, so was Kelly's description of his invention. It is quite possible that his memory of the details was colored by subsequent events.

> To the process of manufacture, I gave my first and serious attention: and after close observation and study, I conceived the idea that, after the metal was melted, the use of fuel would be unnecessary—that the heat generated by the union of the oxygen of the air with the carbon of the metal would be sufficient to accomplish the refining and decarbonizing of the iron. I devised several plans for testing this idea of forcing into the fluid metal powerful blasts of air; after making drawings of the same, showed them to my forgemen, not one of whom could agree with me, all believing that I would chill the metal, and that my experiment would end in failure. . . .
> [Kelly described the new furnace as a] small blast furnace, about 12 feet high having a hearth and bosh like a common blast furnace. In this I expected to produce decarbonized metal from the iron ore; but, if I failed in this, I could resort to pig metal and thereby have good fluid metal to blow into. The novelty of this furnace was that it had two tuyeres, one above the other. The upper tuyere was to melt the stock; the lower one was fixed in the hearth near the bottom, and intended to conduct the air-blast into the metal. That portion of the hearth in which the lower tuyere was placed was so arranged as to part from the upper portion, and consisted of a heavy cast-iron draw, lined inside with fire-brick, so that, when the iron was blown to nature by the lower tuyere, the draw could be run from under the hearth, and the iron taken out, carried to the hammer, and forged.[21]

Kelly averred that he began his experiments with this type of furnace in October 1847, but, because Suwannee was in the process of construction, had little time to pursue them. It was not until four years later, in 1851, that he renewed his efforts. Hoping to use the molten metal from the blast furnace at Suwannee, he built a furnace adjacent to it. The outer portion of the new structure was square and of brick construction. The inner part was a circular chamber with a concave ladle-like bottom in which he fixed a circular tile made of fire clay that was perforated for the tuyeres. Under this tile he placed the air chamber, which was connected with pipes to the blowing engine.[22]

Kelly considered the first trial to be satisfactory. He termed the iron "well refined and decarbonized." He was certain that, if he persisted, he would succeed in producing malleable iron. For the next eighteen months, he conducted a variety of trial runs in his effort to improve the final product. At one time, he did succeed in obtaining a heat of metal that forged very well, but that was the exception and not the rule.[23]

Kelly believed that if he were unable to produce malleable iron at least he would be able to turn out a refined metal. Providing sufficient air pressure was his biggest problem. In an effort to do so he built a converting vessel out of boiler plate and placed it near the tapping hole of the blast furnace. The new vessel was round, about five feet high, with an inside diameter of eighteen inches. Kelly applied the blast through four three-quarter-inch holes in the side instead of through the bottom. The vessel held approximately 1,500 pounds of metal, which could be refined in five or ten minutes. Kelly claimed that he had produced iron in this fashion for several years and that it became well known in the Cumberland River district as a product of his air boiling process. He said that the blooms were in high repute and were used almost entirely for making boiler plates.

Kelly did not apply for a patent on the process and anyone in the area was free to watch it being carried out in his furnace. Kelly explained, "The reason why I did not apply for a patent for it sooner than I did was that I flattered myself I would soon make it the successful process I at first endeavored to achieve, namely, a process for making malleable iron and steel."[24]

The origin of Kelly's process has remained a subject of speculation. How did he conceive the idea of blowing cold air into the molten charge? Explanations from numerous sources include stories of a fortuitous circumstance, of Kelly's desire to solve certain labor problems, and of similar processes then in use in the Far East, where the idea had been known for 300 years.

Boucher is the source for the statement, made years later, that the process originated by chance when employees failed to replenish the charcoal beneath one of Kelly's furnaces. He discovered that the cold air that was reaching the molten metal was not cooling the charge, but, to his amazement, was heating it. Boucher also relates that work at the first furnace was carried on primarily by slave labor. It was virtually impossible to hire white men, either to do the menial labor or to work along with the slaves. Thus Kelly had to buy slaves or to contract with local slaveholders when he needed additional labor. This was a costly source of labor due to the ease with which slaves were able to escape across the Ohio River only forty miles distant.[25]

Kelly is said subsequently to have tapped a quite different source of labor through the efforts of an acquaintance in Philadelphia, a tea merchant with business connections in China. Kelly was able to import a group of Chinese whom he found to be both excellent laborers and efficient iron workers.[26]

A story told in the western Kentucky area helps to explain their particular competence. They were said to have been selected by an American consul in China not only for their suitability as laborers but also for their experience in the working of iron. This they presumably had acquired from having worked in one of the many forges that dotted the Chinese countryside, supplying the needs of rural areas. There was speculation at the time about the possibility that Kelly may have obtained the idea for his process from the Chinese laborers among whom it may have been an accepted practice.[27]

In order to expand the iron business, Kelly had borrowed large sums of money from his father-in-law. Gracy was a practical business man and had little sympathy for Kelly's experiments. Having become financially involved in the iron business, Gracy considered it his privilege to establish a few ground rules. He ordered that Kelly stop any further experimentation with the process, which his son-in-law now called the "pneumatic process." Seemingly agreeing, Kelly, nevertheless, continued his experiments whenever he could find time.[28]

Another attempt to discourage Kelly's use of the new process came from a different quarter, the firm of Shreve, Steele & Company in Cincinnati, which was one of Kelly's best customers for the large iron kettles used in sugar making. An official of the company wrote Kelly that he understood the kettles that the company was buying were being made by some "newfangled" process and he wanted to state categorically that, in the future, if the company did not receive kettles made in the usual way, there would be no more orders.[29]

Shortly after he received this ultimatum from one of his best customers, Kelly was visited by two young Englishmen seeking employment at the furnace. Around their subsequent association with Kelly and the circumstances of their departure has grown up a story that has become part of the Kelly legend and, though it may be apocryphal, cannot be ignored in an account of Kelly's relationship to the Bessemer process.[30]

Boucher relates it in this way. Two Englishmen came to Kelly's furnace and expressed great interest in what he was doing. Since Kelly had been receiving small encouragement, the unappreciated inventor not only welcomed them, but also hired them. Neither man ever was referred to by name, but they were described by Kelly as strongly built and very intelligent, with an unusual knowledge of machinery. Practically everyone in the small community was aware of Kelly's process since he never had carried it on with any degree of secrecy, and as employees the Englishmen could not help but know of it.[31]

Crowds witnessed demonstrations of the process. In later years the son of the original owner of the furnace described the first demonstration he witnessed as a small boy. The molten metal in the furnace, the flow of cold air into the seething mass, and the pouring of the metal into molds made an indelible impression upon the boy's memory. When the metal had cooled, he saw with his own eyes, how the blacksmith took a portion, placed it on an anvil, and hammered out a horseshoe. This horseshoe then was fastened on the foot of a horse belonging to one of the onlookers. The crowd was amazed when they saw that what had been pig iron a few minutes before now was malleable iron. A few skeptics remained unconvinced. "Some crank'll be burnin' ice next," one commented.[32]

Boucher relates that, one night, not long after this demonstration, the Englishmen disappeared. According to Kelly's son, they left so hastily that neither collected the wages due him. Curiously enough, Kelly set out in pursuit, although they owed him nothing and were not in possession of secret information. Kelly's son recalled that, with the aid of bloodhounds, his father traced the fleeing men to the wharf at Eddyville where he learned that they had taken a boat up the Ohio River to Pittsburgh. Kelly is said to have learned later from an unexplained source that the two traveled on the Pennsylvania Railroad to New York and boarded a steamer for England.[33]

The identity of the two strangers never has been ascertained, nonetheless the legend persists that one of them was Henry Bessemer. Mrs. Kelly herself asserted that her husband, when shown a photograph of Bessemer, excitedly identified the picture as that of one of the Englishmen who had departed so suddenly several years earlier.[34]

Bessemer almost certainly never visited the United States. Although professional societies invited him there on numerous occasions as an honored guest, he never ventured the Atlantic crossing. All his life, he was subject to such violent seasickness that even a channel crossing was enough to incapacitate him for weeks.[35]

Bessemer received a cordial invitation to visit America at the time of the unveiling of the Holley Memorial in 1890. In declining he wrote James Dredge, "Nature has interposed an insuperable barrier between us by giving me a constitution that does not permit me to make the shortest sea voyage without absolute danger to life but for this circumstance . . . I would have gone to the U.S."[36]

Although Bessemer cannot have visited the United States, his brother-in-law, William Allen, did so in 1854 when he supervised the installation of some sugar cane crushing and refining machinery made by

Bessemer. At the time Bessemer was carrying on his experiments with the converter. But where Allen went, how long he stayed, or whether any significance for Bessemer's work can be attached to the visit cannot be ascertained.[37]

Discussing invention, French scientist Henri Poincaré said that there are four stages of discovery: Preparation, Incubation, Illumination, and Verification. Throughout each of them runs the enigma of human memory with its power of storing up latent images.[38] It is conceivable that at some time during his American trip Allen either saw or heard of Kelly's work, since it was sugar machinery that Allen was installing. Later, when Allen observed Bessemer's struggle with a recalcitrant process, he may have recalled that someone in America was blowing air through hot metal and told his brother-in-law, who then experimented similarly. Yet Bessemer was not alone in these endeavors.

One of the earliest accounts of the combustion of iron in a blast of cold air appeared in 1773 in a book by the razor maker, Henry Horne. From that time forward many inventors tried the use of air, occasionally of steam. Zerah Colburn described Horne's work and also the work of others attracted by the use of oxygen.

It was the creed, however, of most of the authors of the many schemes for purifying iron, which were published [about 1850] that air alone, and plenty of it, was the thing needed for their purpose. And so it was, and these clever gentlemen stumbled only in the mode of its application, where the great specific was more commonly administered in overdoses and, from this cause, with fatal effect. ... It is useless, except as provoking to better effort in future, to recall how close, twenty years ago, were ironworkers and metallurgists to the greatest discoveries, and yet how they rested and were thankful with the knowledge they possessed. For there is now a chance that the whole art of ironmaking may be changed, and this by the working out of a principle which had lain so long in the path of the iron trade that it is the greatest wonder it was not stumbled upon years ago—a principle of that degree of simplicity which strikes most minds as ludicrous, and the very discovery of which convicts mankind of previous wilful blindness. As plainly as nature, interpreted by chemistry, could speak to man, she has, for the last century said, "Let me work for you. I will fine your iron, only do not stop the way of my oxygen. Let it in—let it in." But the sons of Vulcan have shaken their heads, and replied, "No, no; let it stop outside, and we will set a luckless drudge at work with his 'rabble,' and bring the iron out, bit by bit, to meet your chemical stuff." This literally bringing the "iron to nature," as the puddlers say, instead of permitting nature to come to the iron. And all because the ironworkers had not learned to regulate the dose of oxygen.[39]

The decision of the commissioner of patents in 1857 had vindicated Kelly in the eyes of his family, no matter how skeptical others may have remained. However, in the same year economic crisis swept the country.

In New York, Holley was to lose his railroad journal in the financial upheaval that followed; in Eddyville, Kelly likewise was forced into bankruptcy. According to Boucher, Kelly had borrowed large sums from his father-in-law and from his own father. The father had given Kelly the inheritance that eventually would have been his. Not wanting his patent to fall into the hands of creditors, Kelly transferred it to his father for the sum of $1,000 with the understanding that, at some future time, perhaps in his will, the patent would be restored to Kelly. Unfortunately, his father died suddenly in 1860. The patent became part of his estate, and the invention was inherited by Kelly's sisters. They did not restore it to Kelly immediately, but kept it in their possession for at least one year. Several reasons have been advanced for their doing so. They may have recognized its value and have decided that in view of their brother's poor business sense it was wiser that they keep it. Or, they may have feared that creditors still could seize the patent if it was in their brother's possession. [40]

Kelly's efforts to interest the established iron makers in his process met with both failure and success. Joseph Butler, Jr. was one of the pioneer iron producers of the Youngstown, Ohio, area. He recalled that around 1854, when he was employed in the store of the Ward family at Niles, Ohio, Kelly came to the Ward home for dinner. James Ward, who was considered an authority on ironmaking and had built the first rolling mill west of Pittsburgh, listened to his guest's description of his process with polite interest. After his guest left, however, Ward's only remark implied that their visitor obviously was crazy. Yet Ward had an exploring mind and he might have been expected to react more positively. An immigrant from England, he had applied to the American iron industry many ideas from the more advanced English practice. Attracted by the high grade coal in Ohio, he had erected a rolling mill and puddling furnaces at Niles. For many years, the Mahoning Valley was second only to Pittsburgh as the greatest puddling center in the United States. [41]

Kelly was more fortunate in his meeting with Daniel J. Morrell, general manager of the Cambria Iron Works. Morrell was an unusual example of the ironmaster of the period. Born in Maine in 1821 in the Quaker settlement of Berwick, he was the seventh son of a large family. He was brought up in the atmosphere of the home spinning wheel and the toil of a subsistence farm. He probably had spent no more than two years in the classroom during his entire life. Morrell left home in 1837 to join an older brother in Philadelphia where, with two associates, he established a dry goods firm. From that time until 1855, young Morrell's entire experience was in the mercantile field. [42]

In that year Morrell was one of a group of Philadelphia businessmen that leased the Cambria Iron Works in Johnstown, Pennsylvania, for a

Daniel J. Morrell. From Herbert N. Casson, *The Romance of Steel* (New York: A. S. Barnes & Co., 1907), opp. p. 24.

five year period. The works had been organized in 1853 but had failed during the following year. Although without experience in the iron business, Daniel Morrell was made general manager, and under his direction Cambria became one of the leading United States iron companies. [43]

Morrell was a public-minded citizen who took a great interest in everything about him. When Kelly sought him out in 1856 to ask for the privilege of demonstrating his process, Morrell not only gave the inventor a corner of the yard in which to carry on his experiment, but also provided the services of a young man to help build a converter for the demonstration.

The assistant, James H. Geer, prepared the patterns for the castings needed for the new vessel. As an old man, Geer related with zest, and probably with certain embellishments, the story of the first demonstration. The workers had not looked with favor on the trial of the converter. The puddlers were aware that if it worked their jobs certainly would be jeopardized. Herbert Casson, who interviewed Geer about the demonstration, understood that it is often fear that makes men scoff and that for that reason, the puddlers were loudest in ridiculing the "Irish crank." [44]

As Casson related the story, Kelly was very nervous when the day for the trial arrived. Fearing that the engineer who was responsible for turning on the blast might not make it strong enough, since that was always crucial, Kelly assured him that he wanted the strongest blast that he could blow. Whether from spite or to oblige the nervous inventor, the engineer hung a weight on the safety valve and blew such a blast that the entire contents of the converter shot from the top in a spectacular display, leaving the vessel empty. [45]

The 200 workers who had gathered to watch hugged themselves and each other in their glee over the outcome of the engineer's great blast of air. For days the shop reverberated to roars of laughter whenever Kelly's fireworks were mentioned. The joke persisted for at least ten years and never seemed to grow stale to those in the iron trade. [46]

Just as Henry Bessemer had regrouped his forces for renewed efforts after failures, so Kelly immediately made preparations for his second trial. He had benefited from the first trial and this time the blast was controlled carefully. Kelly still knew almost nothing about the chemical reaction that brought about the change in the molten metal when the cold air was blown into the converter. Whereas a present-day operator knows when the blast must be turned off by the color and nature of the flame, the only way that Kelly could judge was by a primitive method of his own devising. When the sparks began to fly from the mouth of the converter, Kelly seized a hammer and ran about the yard striking them as they fell to the ground. For almost a half hour, each spark hammered crumbled and disintegrated. Suddenly, he came upon one that did not crumble, but flattened out like a piece of dough. Kelly then was certain that the heat was finished, and he ordered the blast turned off. The converter was tapped, and the molten metal came streaming into a mold. When it was cool enough, Kelly hammered a piece of it into a thin plate. One wonders which the spectators enjoyed more, the volcanic display of the first trial or Kelly's footwork during the second. [47]

Another version of Kelly's sojourn at the Cambria Iron Works was told by John Fry, who in later years was superintendent of the Bessemer plant at Cambria but, when Kelly first appeared in Johnstown, was working as a foundry molder. He, as well as Geer, helped conduct some of the first experiments. Fry recalled that the inventor made his first appearance at Cambria shortly after having been granted his patent for the pneumatic process. According to Fry, Kelly carried on his experiments at Cambria spasmodically. His stays were short, and the intervals between them long. [48] Kelly has stated that in 1857 he moved to New Salisbury, a small town some sixty miles from Pittsburgh. [49] This move coincided with the loss of his iron works during the financial crisis of that year, and the distance may have accounted for his infrequent visits to Cambria.

In his account of the first trials, Fry noted that molten pig iron produced from local ores and fuel was used.

> The apparatus was assembled from scrap heap material and was indescribably primitive. The entire operation consisted of blowing air into the molten iron through a blast-pipe thrust from above into the liquid metal. The only agreeable result was to change this very poor quality of iron for most purposes into a practically worthless one for any. Repeat experiments continuing to give constant results, and the metal being disposed to chill in this small quantity, the next experiment was an attempt to blow air, similarly to the former trials, into the metal in the crucible of old No. 1 blast furnace, when the crucible was full just before tapping, but the blast could not be made to penetrate the iron and the attempt was promptly abandoned.
>
> Mr. Kelly's next effort was to permeate a small quantity of grey foundry iron — crucible-melted — with carbonic acid gas, on the supposition that the metalloids of the iron could thus be instantaneously eliminated. The apparatus, like the first, was crude and inefficient, and no effect was produced upon the iron because of failure to get the insignificant supply of gas through it. But two trials of this character were made, and then the method was abandoned. Mr. Kelly desponding of success at this time.
>
> Next a quite large apparatus was installed in the old mill metal yard, consisting of a cupola to melt pig-iron, extemporization of circular foundry flasks to form a fixed converter, and beds of run-out chills, the entire system connected by claylined troughs. Blast was supplied from the foundry engine near by, and the metal, having been melted in the cupola and been run directly into the converter, was blown with air-blast through a pipe inserted through the converter side into and just below the surface of the liquid iron. Here a fair article of desiliconized iron was produced from grey cast iron, similar to that made in the old-fashioned refinery fire, of which the plant was a poor modification. After several trials with this it was abandoned and relegated to the scrap-heap, its last operation having set fire to the temporary building surrounding it, which was totally consumed. [50]

No more experiments were conducted during the remainder of 1859. Fry claimed that, owing to economic hard times in 1860 and the outbreak of the Civil War during the following year, no further experiments at Johnstown were conducted during this period. [51] A question asked in the columns of *Scientific American* in 1861 awakened interest in the process once again: "Would not some of our enterprising iron manufacturers make a good operation by getting hold of this patent and starting a manufactory of the steel in this country?" Later in the same year, the magazine reprinted an article on Bessemer that had appeared in the London journal *Engineer*. The *Scientific American* noted, "There is a splendid opening for the manufacture of steel by this process in this country. It will be remembered that Bessemer's patent in the United

Kelly's first tilting converter.

States was set aside in favor of W. Kelly of Eddyville, Ky., who proved that he was the first inventor. Can this patent be bought?"[52]

Kelly replied, "In your number of Nov. 16 you ask if the patent for my air-boiling process can be bought? In reply, I would say, that the New England States and New York would be sold at a fair rate. Should any parties wish to negotiate for the purchase I would be pleased to hear from them on the subject. I removed from Kentucky about three years ago and now reside at New Salisbury, on the Cleveland and Pittsburgh Railroad, three miles from Hammondsville (my PO) and sixty miles from Pittsburgh. Accept my thanks for your kind efforts in endeavoring to draw the attention of the community to the advantage of my process."[53]

This interchange of letters may have attracted the attention of Zoheth S. Durfee, who was considered an expert in technical matters concerning iron and steel furnaces and plants. He recently had been advisor to Eber Ward, a shipping and iron industry magnate active in the Middle West, regarding the establishment of a new furnace in Wyandotte, Michigan. Daniel Morrell and Ward were friends of long standing. It is a question whether Durfee or Morrell brought the Kelly patent to Ward's attention but Ward and Durfee secured the rights to the pneumatic process from Kelly. Durfee left very shortly afterwards for England to secure the American rights to the Bessemer apparatus and to study the Bessemer process in operation.[54]

Mr. Fry wrote that, "In 1862, Mr. Kelly returned to Johnstown for a crucial—and as it turned out a final—series of experiments by him with a rotative 'Bessemer Converter' made abroad and imported for his purpose. This converter embodied in its materials and construction several of Mr. Bessemer's patented features, of which, up to the close of Mr.

Kelly's experiments, above noted, he seemed to have no knowledge or conception."[55]

It must have been the experiments with this converter that Geer remembered so vividly, since it was the only tilting vessel Kelly ever used, although Fry made no mention of the dramatic episodes that Geer recalled so lovingly. Fry's description of the Kelly episode was part of a longer article about the Cambria plant. Perhaps lack of space or imagination may have limited the inclusion of such colorful details.

In May 1863, after Durfee's return, the Kelly Pneumatic Process Company was organized with Daniel Morrell of Johnstown, William L. Lyon, and James Park, Jr., of Pittsburgh, becoming partners of Ward and Durfee. Kelly was not included in the organization but was guaranteed a certain amount of the profits. He later returned to Kentucky where he began the manufacture of axes, which he carried on until his death in February 1888.[56]

There are numerous opinions about the amount Kelly received for his patent from the Kelly Pneumatic Process Company. Casson states that by 1870, at which time it was extended for an additional seven years, Kelly had received $30,000 and that subsequently he received another $450,000. Another writer set the figure at only $2,000 or $3,000 by 1870 and at about $25,000 in addition.[57]

When his patent ran out in 1870, Kelly's request for a renewal was granted. The commissioner of patents, in a decision dated 15 June 1871, extended Kelly's rights for another seven years. As his reason for doing so, he wrote that the testimony showed that the patent was very valuable and that Mr. Kelly had been untiring in his efforts to introduce it into use but that the opposition of iron manufacturers and the amount of capital needed prevented him from receiving anything from his patent until the last few years. "He showed expenditures of eleven thousand five hundred dollars and receipts of twenty-four hundred dollars. He has a clear case entitling him to an extension. I do not regard the extension as prejudicial to public use." After listening to the arguments and reading the discussion, the commissioner failed to discover any substantial reason for not extending the patent. "Very few patentees are able to present so strong grounds for extension as the applicant in the case," was his final comment.[58]

Upon the expiration of the Bessemer patent the year before, a request for extension had been denied on the ground that the corresponding British patent had expired at the end of fourteen years, and the commissioner of patents believed that it would be inequitable to protect the patent in the United States when it no longer was in force in England. The commissioner had tempered his decision by observing that "It may be questioned whether [Bessemer] was first to discover the principle [on] which his process was founded. But we owe its reduction to practice to his untiring industry and perserverance, his superior skill and science and his great outlay."[59]

There never has been a satisfactory answer to the Kelly-Bessemer controversy. In 1896, eight years after Kelly's death and two years before Bessemer's, the question, which had lain dormant for some years, was raised again. Joseph D. Weeks, editor of the *American Manufacturer,* fanned the embers by making serious charges against Bessemer in the address given on the occasion of his retirement as president of the American Institute of Mining Engineers. Supporters of both Kelly and Bessemer entered the fray at once, and English and American journals were beseiged by letters in support of one or the other.

Obviously, Week's controversial address was ill-advised. The secretary of the institute commented in the *Transactions* issued in 1897,

> This address having been made the object of much hostile comment, arising, as Mr. Weeks believed, in large part from misunderstanding of its purpose and meaning, was, by his express direction withheld from official publication in the *Transactions* until he should have so modified or added to it as to make such misunderstanding impossible. His illness and death prevented him from executing this intention; and, consequently, the only version of the address now in the Secretary's hands is one of which the author had forbidden the publication. Under the circumstances, therefore, it seems best to publish in this place, only some portions of the address and its appendixes which have a historical value and are beyond controversy.[60]

However, in at least one letter that Weeks wrote to *Engineering,* in answer to an attack by a Bessemer supporter, he did not deviate from the tenor of his original remarks.[61]

Although the controversial speech did not appear in the *Transactions,* the society did print as one of the appendixes various affidavits submitted by Kelly at the time of his original priority claim, as well as statements, including one by Alexander Holley, favorable to Kelly's claim. These had been submitted as evidence in 1870 at the time when Kelly petitioned for an extension of his patent. In answer to the question, "What do you consider the value of the invention of Mr. Kelly in its relation to the pneumatic process as now practiced," Holley replied, "I consider the Kelly invention the first practical development of the pneumatic process and it has been so recognized by the owner of the combined patents covering this process."[62]

In December 1896, a younger organization, the American Society for Mechanical Engineers, elected Bessemer an honorary member. It is impossible to judge whether this was done as compensation for the attack by the president of the older organization or whether it was a coincidence. Since he was unable to make the crossing, Bessemer sent a paper entitled "Historical and Technical Sketch of the Origin of the Bessemer Process," which was read to the members. In the discussion that followed, Robert

Hunt, an outstanding engineer, expressed the general feeling of those who knew the Kelly-Bessemer story best. "The president knows, even more intimately than I, concerning the early experiments and efforts of Kelly," said Hunt.

He was not a chemist and he labored under great disadvantage. In fact chemistry at that time was not generally applied to any of the developments of the iron and steel business and if Kelly had possessed a knowledge of metallurgical chemistry, even as then existed he would have undoubtedly gone much further than he did. But we find that Mr. Bessemer himself was not any too sure of his chemistry, because in his original paper he states that the oxide of iron which was produced eliminated the sulphur from the bath. This was about as direct a chemical mistake or as great a one as a man could make. We know that sulphur is not eliminated, unfortunately, in the process. Again, in some of his early papers, he defined the amount of phosphorus it was possible to use, as 0.02 percent, so that he groped in the dark and took advantage of the developments as they progressed. . . . I have great respect for Mr. Bessemer and it is with great hesitation that in his advanced age I would permit myself to say anything which would seem to take away an atom from his honor and glory, but I wish he were just a little more generous and a little more just."[63]

NOTES

1. *Scientific American* 12 (18 October 1856):43.
2. Kelly received patent number 17,628 on 23 June 1857.
3. The decision, issued by the U.S. Patent Office on 13 April 1857, read as follows:

In the matter of interference between the patent of Henry Bessemer of London and the application of William Kelly of Lyon Co. Kentucky for Improvements in the Manufacture of Iron and Steel, the hearing of which was fixed for the First Monday in April:

It appears, that by the concurrent testimony of numerous witnesses Kelly made this invention and showed it by drawings and experiments as early as 1847 and this testimony appears to be reliable in every respect.

The patent of Bessemer was Sealed in London on the 11th of April 1856 and was dated 11 Oct. 1855.

Priority of invention in this case is awarded to said Kelly and it is ordered that a patent be issued accordingly, unless an appeal be taken within sixty days from this date.

S. T. Shugert
Acting Commiss.

Decision of Commiss.
Kelly vs. Bessemer
     Interference
Decision in favor of Kelly
Filed April 13, 1857
Recorded vol. 2, page 417
(Affidavits filed by U.S. Patent Office in connection with Kelly-Bessemer patent controversy. Reprinted in *Transactions of the American Institute of Mining Engineers* 26 [1897]: 987).

4. Bessemer has nothing to say in his autobiography either about Kelly or about the setting aside of the American patent. It was not until 1896, when he was attacked by J. S. Weeks in his presidential address before the American Institute of Mining Engineers, that Bessemer was aroused enough to send a letter running to many columns to the editor of *Engineering*. In it he denounced the Kelly invention (*Engineering* 6 [20 March 1896]:367-70).

5. Munn and Company were among the leading American patent agents, with offices also in London, Paris, and Brussels. One of their advertisements stated that seven-eighths of all patents obtained by Americans in Europe were secured through their agents. An estimated quarter of all American patent business was transacted by them. In ordinary times the firm petitioned for at least two hundred patents a month.

It should be noted that Martien wrote his letter after Kelly had been awarded his patent, but during the interval of time given Bessemer to make an appeal. It has been alleged that Munn and Company were Kelly's solicitors. If they were acting for him at this time, then their influence in the Kelly award may have been greater than generally has been recognized (William F. Durfee, "The Development of American Industries since Columbus: IX. The Manufacture of Steel," *Popular Science Monthly* 40 [November 1891]:18).

6. James M. Swank, *History of the Manufacture of Iron in All Ages*, 2d ed. (Philadelphia: American Iron and Steel Association, 1892), p. 407.

7. Ibid., p. 399.

8. Ernest F. Lange, "Bessemer, Göransson, and Mushet: A Contribution to Technical History," *Manchester Memoirs* 57, no. 17 (1913):22. In William Durfee's opinion, expressed at a later time, "While there is no reason to believe that Bessemer ever heard of Kelly it is pretty certain that had not Kelly noted the granting of a patent to Bessemer he would never (owing to unfavorable location and pecuniary embarrassment) have been able to procure such attention from the iron trade of this country as would have insured him any reward for his invention" (Durfee, "American Industries since Columbus: IX," p. 26).

9. Editorial, *Scientific American* 14 (31 July 1857).

10. James M. Swank, *History of the Manufacture of Iron in All Ages*, 1st ed. (Philadelphia: Swank, 1884), pp. 304-06.

11. James M. Swank, *Statistics of the Iron and Steel Production of the United States* (Compiled for the superintendent of census by James Swank serving as special agent, 1881), pp. 124-25.

12. John Newton Boucher, *William Kelly: A True History of the So-Called Bessemer Process* (Greensburg, Pa.: Boucher, 1924), p. 1. For reasons unknown, the book was subsidized by a group of ironmasters who never were identified by the author.

13. Swank, *Iron in All Ages*, 2d. ed., pp. 226, 227.

14. Boucher, *William Kelly*, p. 3.

15. Ibid., p. 4.

16. Ibid., pp. 4-5.

17. Ibid., p. 5. William Kelly was to operate the furnace; John Kelly was to manage finances.

18. Ibid., p. 6.

19. Ibid., p. 7.

20. Henry Bessemer, *Sir Henry Bessemer: An Autobiography*, (London: Offices of Engineering, 1905), pp. 143-46.

21. Swank, *Iron in All Ages*, 2d. ed., pp. 397-98.

22. Ibid., pp. 397-98.

23. Ibid., p. 398. Many years later, when a chemical analysis was made of the ore deposits

around Suwannee, it was discovered that they were cold-short, red-short, and neutral ores. In other words, they incuded excessive amounts of Bessemer's old enemies, sulphur and phosphorus. There was a small deposit that contained a large percentage of manganese and, no doubt, it was this ore that produced the better quality of iron that Kelly mentioned (Boucher, *William Kelly*, p. 7).

24. Swank, *Iron in All Ages*, 2d. ed., p. 399.
25. Since only the best, most muscular and intelligent of the laborers succeeded in escaping, the expense of compensating their owners was considerable. For this reason, every section that used slave labor maintained a pack of bloodhounds to track down runaways. When captured black men were returned, the punishment was severe. Boucher relates that Kelly found the process distasteful and unjust (Boucher, *William Kelly*, pp. 24, 25).
26. Ibid., pp. 25, 26.
27. William B. Phillips, "Notes on the Possible Origin of the Pneumatic Process of Making Steel," *Transactions of the American Institute of Mining Engineers* 28 (1898): 745–46. Similarly, John Percy, an eminent English metallurgist, recalled that a paper by Edward Clibborn had been presented before the Royal Irish Academy in 1862 in which Clibborn claimed that the Japanese had used a pneumatic process three hundred years before. Clibborn quoted from *Mandelslo's Travels*, which was printed in London in 1669: "They the Japanese have, among others a particular invention for the melting of iron, without the using of fire, casting it into a tun done about on the inside with about a foot of earth, where they keep it melting with continual blowing, and take it out by ladles full, to give it what form they please, much better and more artificially than the inhabitants of Liege are able to do."
Percy found it difficult to agree with Clibborn. He believed that if the air had been blown through the hot metal the resulting eruption certainly would have been spectacular enough to warrant a description by Mandelslo (John Percy, *Metallurgy: Iron and Steel* [London: John Murray, 1864], p. 816.
As late as 1956, Joseph Needham corroborated this theory in a lecture on the development of iron and steel technology in China, given before the Newcomen Society in England. "The early Chinese ironmasters soon learned how to make steel, by decarburizing their pig iron. This method known as 'the hundred refinings' depended on the use of [an] oxidizing blast of cold air, and was first practiced in the 2nd century B.C.: by the 17th century A.D. it was producing something like cast steel. Parallel with this process was one of cofusion, in which lumps of wrought iron were carburized in a bath of molten pig iron" (Joseph Needham, "The Development of Iron and Steel Technology in China," *Transactions of the Newcomen Society* 30 [1956]: 141–44). A condensed version of the paper was presented by Needham as the Second Biennial Dickinson Memorial Lecture at the Science Museum, London, May 1956. This was published by the Newcomen Society under the same title in 1958.
28. Boucher, *William Kelly*, pp. 15, 18. Gracy began to doubt his son-in-law's sanity, and called on the family doctor to examine him. The doctor was a man deeply interested in scientific matters, who listened attentively as Kelly described the strange behavior of combined oxygen and carbon. The doctor was convinced and put to rest the notion that Kelly was demented (Boucher, *William Kelly*, pp. 17, 18).
29. Herbert N. Casson, *The Romance of Steel* (New York: A. S. Barnes & Company, 1907), p. 6.
30. The expanded story first was told publicly by Kelly's son in 1922 at dedication ceremonies held at the unveiling of a bronze plaque on the site of the Wyandotte plant in Michigan, where the pneumatic process first was practiced. The son said that his father had confided the story to his mother but had made her promise never to repeat

it. After Kelly's death, however, Mrs. Kelly felt herself no longer bound by her promise ("Dedication of Tablet Recalls Bessemer Patent Controversy," *Iron Trade Review* 71 [19 October 1922]:1064.

31. Boucher, *William Kelly*, p. 1, 21.

32. Casson, *Romance of Steel*, p. 7.

33. Boucher, *William Kelly*, p. 23.

34. Ibid., pp. 88–89.

35. Hoping to circumvent his infirmity, in 1870 he began to design a ship on which the salon hung on an elaborate gimbal-like mechanism that was to neutralize the effect of rolling and pitching. One witness commented that Bessemer's new patent was very like stomachs on gimbals and suggested that the same effect could be accomplished in a much easier fashion by sewing passengers into sacks and hanging them from hooks. The vessel, the *Bessemer*, was launched in 1874. It was a total failure, and was abandoned after it had damaged several piers severely in two out of its three channel crossings. Bessemer's loss was approximately £25,000 (W. T. Jeans, *The Creators of the Age of Steel* [New York: Charles Scribner's Sons, 1884], pp. 113–14).

36. "Dedication of the Holley Memorial," *Transactions of the American Institute of Mechanical Engineers* 20 (1891): xx.

37. "The Late William D. Allen," *Engineering* 62 (30 October 1896):562.

38. Michael Polanyi, *Personal Knowledge Towards a Post-Critical Philosophy* (Chicago: University of Chicago Press, 1958), p. 121.

39. Zerah Colburn, "The Origin and Principles of the Bessemer Process," *Engineer* 18 (23 December 1864): 388–89 and (30 December 1864): 405–6.

40. Boucher, *William Kelly*, pp. 77–78. The patent eventually was returned to Kelly to be held in trust for his children.

41. Joseph G. Butler, Jr., "Fifty Years of Iron and Steel," *Yearbook, American Iron and Steel Institute* (1917):287.

42. James H. Swank, *Cambria County Pioneers* (Philadelphia: American Iron and Steel Association, 1910), p. 72.

43. "The Story of the Cambria Steel Company," *Johnstown Tribune*, 5th anniv. ed., 15 September 1928.

44. Casson, *Romance of Steel*, p. 9.

45. Ibid.

46. Ibid., p. 10.

47. Ibid.

48. John E. Fry, "The Bessemer Steel Industry," *Johnstown Daily Democrat*, Souvenir Ed., Autumn 1894.

49. William Kelly to *Scientific American*, n.s. 5, (July–December 1861):343.

50. Fry, *Johnstown Daily Democrat*, Autumn 1894.

51. Ibid.

52. *Scientific American*, n.s. 5 (7 September and 16 November 1861):310.

53. Ibid., p. 343.

54. Fry, *Johnstown Daily Democrat*, Autumn 1894.

55. After Kelly's final round of experiments in 1863, having stood idle for a year or so, the converter was consigned to the scrap heap. Sometime after the Johnstown flood in 1889, the converter was salvaged. In 1892 it was set up on the lawn near the Cambria office. The following year it was exhibited at the Columbian Exposition and later it was displayed in the lobby of the Johnstown plant of the Bethlehem Steel Company. It is now on exhibit at the Smithsonian Institution in Washington, D.C.

56. Casson, *Romance of Steel*, pp. 10–11.

57. Rossiter Raymond, *Engineering News and Railway Journal* 35 (27 February 1896):135.

58. U.S. Patent Office, Decision of Commissioner of Patents, 15 June 1871.

59. U.S. Patent Office, Decision of Commissioner of Patents, 12 February 1871.

60. Joseph D. Weeks, "The Invention of the Bessemer Process," Note by the Secretary of the American Institute of Mining Engineers. *Transactions of the American Institute of Mining Engineers* 26 (1897): 980.

61. *Engineering* 61 (5 June 1896): 755–56.

62. Affidavits filed by U.S. Patent Office in connection with Kelly-Bessemer patent controversy. Reprinted in *Transactions of the American Institute of Mining Engineers* 26 (1897): 983.

63. Henry Bessemer, "Historical and Technical Sketch of the Origin of the Bessemer Process," *Transactions of the American Society of Mechanical Engineers* 18 (1897): 482.

X · Göran Fredrik Göransson

n his autobiography published some years after his death Henry Bessemer failed to mention William Kelly although he did make a comment concerning Kelly's patent in a paper prepared for the meeting of the American Society of Mechanical Engineers held in 1896. However he ignored Göran Fredrik Göransson, who many believe was the first to make the Bessemer process succeed in practice.[1] In 1856, Göransson was the head of the firm of Daniel Elfstrand and Company of Gefle, Sweden, a shipbuilding and ship owning concern that also was widely known as an exporter of lumber and iron. Shortly after that time, the company had purchased a small blast furnace at Edsken, forty miles west of the Baltic port where the Elfstrand company conducted its business. A small forge at Högbo, not far from the furnace, was part of the purchase.

In August 1856 Bessemer had read the paper "The Manufacture of Malleable Iron and Steel Without Fuel" describing his process before the Mechanical Section of the British Association for the Advancement of Science. A month earlier, on July 1, Bessemer had succeeded in obtaining Swedish patent number 35 on the process, and a report of the patent had appeared in July in the official journal *Jernkontorets Litteraturoversikt* in Sweden. But it was only after Bessemer's presentation to the association, an event that was reported widely in the press at home and abroad, that the importance of the process was recognized in Sweden.

Attracted by these accounts and intent on securing at least a part interest in the Swedish rights, Göransson journeyed to England in May 1857. A demonstration of the process at Baxter House convinced him that, in spite of certain adverse reports that already had begun to appear in the journals, the principle on which the process was based was sound. He bought for his company one-fifth of the rights to Swedish patent number 35 from Hoare, Buxton and Company, Elfstrand's London agents, who had bought the Swedish patent from Bessemer for £10,000 a short time before. His company also would have the right to produce 500 tons of steel a year at a royalty of only 2 shillings per ton. Göransson paid one-half of the agreed-upon price of £2,000 on June 7. Because of

Göran F. Göransson. (Courtesy of Sandvik Steel Works Co., Sandviken, Sweden.)

legalities, however, the final contract was not completed until two years later.

Before leaving England, Göransson placed an order with W. and J. Galloway in Manchester for two furnaces, a boiler, a steam blowing engine, and all the necessary apparatus for an installation at Edsken. The machinery eventually arrived at the port of Gefle in August. With great difficulty, it was transported to Edsken. It was necessary first to reload the equipment so that it could go part way by rail, after which it had to be hauled the final thirteen miles by horse and wagon. Because of the bulk and weight of the machinery, sixty men and a large number of horses were needed to bring the machinery through an area of almost impassible roads.

A Mr. Price from Manchester, probably sent by W. and J. Galloway, assembled the converter plant as quickly as he could, but it was November before the trial operations could be started. Göransson estimated that he had lost the equivalent of a year's production at the Högbo forge for the patent and the equipment before it was possible even to begin the experimental working of the converter. The usual difficulties were encountered immediately.[2] The first blows were unsuccessful and in spite of all his care, Göransson was able to produce only a product that was full of slag. A close associate commented that had the Swedish merchant any real knowledge of this field, he soon would have realized the hopelessness of what he was attempting. Nevertheless, Göransson and his assistant persisted in their experiments, keeping careful records of every trial blow.[3]

Per Carlberg in "Early Industrial Production of Bessemer Steel at Edsken" relates that on 12 December 1857, Lindahl, a master mechanic at Gefle, reported to Göransson that he had hammered and welded two ingots sent from the Edsken works. In Lindahl's opinion, the tools made from this steel were as good as any he had made from English steel.[4]

Just as Holley and Kelly saw their businesses swept away in the depression of 1857, so the firm of Daniel Elfstrand was affected and forced into bankruptcy in December of 1857. This was a sad time for Göransson. The experimental work at the small steel plant at Edsken, although far from satisfactory in his eyes, had shown enough progress to convince the administrators of the bankrupt firm that efforts to perfect the process should be continued. They decided that the final payment of £1,000 for the Bessemer rights should be made.

With this encouragement from the administrators, Göransson sent some examples of Lindahl's forgings, as well as a few steel plates that had been rolled by a nearby mill, to the Swedish Ironmaster's Association and asked for their help. Known as the Jernkontoret, the organization had been founded in 1747 to free members of the iron industry from dependence on private loans with excessively high interest rates. It functioned as a bank, technical information clearing house, and a consulting agency for the entire industry. Each member company paid a given sum to the association for each ton of iron or steel produced. This in turn entitled the company to borrow seven-eighths of the value of its product at low interest rates.

Within three weeks, a committee sent by the organization arrived at Edsken to observe the melting of three charges and the subsequent hammering and welding of the ingots at the Högbo forge. They submitted a detailed report on 10 January 1858, and five days later the board of Jernkontoret granted Göransson a loan. With the funds it was possible to install a steam hammer and tilting hammers at the forge. A metallurgist from the Swedish association was assigned to aid Göransson in carrying out his experiments and, after a month's delay for repairing the furnace, operations were resumed in March 1858.[5]

In the early stages of Göransson's experiments Bessemer had sent an engineer, C. J. Leffler, to help. However, Göransson remarked in later years that the man did not know any more than he himself did. After the spring of 1858, Leffler's appearances at Edsken became less and less

Bessemer blow at Edsken, Sweden. (Courtesy of Sandvik Steel Works Co., Sandviken, Sweden.)

frequent although correspondence with Bessemer indicates that he remained in the area and at times visited the plant.

According to Carlberg, Göransson's progress interested Bessemer but the part that he played in the Swedish endeavor is not clear. A letter to Leffler from Bessemer dated 21 June 1858, contained various suggestions that he thought might aid Göransson, but the suggested changes already had been made before the letter arrived. Of particular interest, however, is the mention of the use of manganese. Its presence was vital for the production of good steel in the converter and, as will be seen, was a matter of great concern and interest to Bessemer who wrote: "I am so pleased with the great energy and perserverance Mr. Urenson (sic) has displayed that I shall only feel pleased if I can lend a helping hand to the good work you have so far succeeded in. . . . but ask in return all the information you can give. . . . Have you tried manganese in any way . . .?"[6]

As a substitute for the original tilting Bessemer converter, which had proven difficult to manage, the company installed a fixed standing converter similar to the one Bessemer was using at Baxter House. At this critical stage of experimentation, Göransson evolved the variation in the process that enabled him to make a successful heat of steel on 18 July 1858. He wrote to a friend, Richard Äkerman, a leading Swedish metallurgist:

> We tried every means of augmenting the pressure of the blast by reducing the diameter of the tuyeres and using smaller charges, as we had reached the limit of the pressure which could be produced by the blast engine, but all with less and less success, and, on the point of giving up the experiments, I resolved, in spite of all advisors, to diminish the pressure and instead use a larger quantity of air. For this purpose we put all the twelve tuyeres in one line at the bottom of the converter, and enlarged the diameter of them to seven-eighths of an inch, to see what change they would produce. The result was astonishing. The temperature of the fluid steel was very much raised, the slag came up on the top beautifully, the ingots turned out perfectly even in temper, free from slag, and extremely malleable, more so than iron made on the old process. . . . [7]

Although Bessemer in his autobiography does not credit Göransson with a major role in perfecting the Bessemer process, in Göransson's mind there was no doubt that his efforts, which culminated in the successful heat of July 1858, led to that result. And it will be seen that when Bessemer later was obliged to defend himself against the claims of Robert Mushet, Bessemer argued that the Swedish success justified his ignoring them.

In the letter to Äkerman, Göransson tells of the ultimate success of his efforts.

> I sent fifteen tons of ingots to a firm in Sheffield and fifteen to Messrs. Henry Bessemer and Company's works at the same place and went over to England in September 1858 to have them tested. I found then that Mr. Bessemer had not succeeded in his efforts but had to granulate the steel which he got from the converter in water and afterwards remelt it in crucibles. As such a process could not give any profits the friends who assisted him were losing all hope of success, but they all came down to Sheffield to see my ingots before they finally gave up this business. The Sheffield firm, thinking

it their interest not to forward the Bessemer method, got the whole lot "burnt" at the washwelding, but the fifteen tons hammered and tilted at Messrs. Bessemer and Company's works turned out to full satisfaction after having been tried for knives, scissors, razors, other tools, and plates. . . .[8]

Bessemer must have been impressed by the demonstration. His firm at Sheffield bought one hundred tons of pig iron from the Edsken firm the following month.[9] Carlberg believes that it "gave Bessemer some experience about the qualities required in pig iron for the Bessemer process." However, as Bessemer tells it, he had used the Swedish pig iron before the demonstration.[10]

NOTES

1. Henry Bessemer, "Historical and Technical Sketch of the Origin of the Bessemer Process," *Transactions of the American Society of Mechanical Engineers* 18 (1897): 464.
2. Per Carlberg, "Early Industrial Production of Bessemer Steel at Edsken," *Journal of the Iron and Steel Institute* (London) 189, (July 1958): 201-4.
3. Karl Fredrik Göransson, "The First Successful Bessemer Steel," *Metal Progress* (Cleveland, Ohio) 69 (January 1956): 86.
4. Carlberg, "Early Industrial Production of Bessemer Steel at Edsken," p. 202.
5. Ibid., pp. 202-3.
6. Ibid., pp. 202, 203-4.
7. James M. Swank, *History of the Manufacture of Iron in All Ages*, 2nd ed. (Philadelphia: American Iron and Steel Association, 1892), pp. 404-5.
8. Ibid., p. 405.
9. The Swedish product, made with charcoal and with the remarkably pure Swedish iron ores of Dannemora, was almost entirely free of phosphorus and contained manganese. The use of this type of pig iron in the converter made it unnecessary to add any spiegeleisen.
10. Göransson's statement regarding the method by which Bessemer was producing steel was verified by William Allen, Bessemer's brother-in-law and manager of the Sheffield works. The entry in his diary for 18 June 1859 noted that on this day the Bessemer Company had produced its first steel direct from the converter. This was eleven months after Göransson's first success (Ernest F. Lange, "Bessemer, Göransson, and Mushet: A Contribution to Technical History," *Manchester Memoirs* 57, No. 17 1913): 17).

XI · The Mushet-Bessemer Controversy

illiam Kelly's second experiment at Cambria in 1862 came about both because of heightened interest in the process in England and in America and because of the faith of Daniel Morrell and Zoheth Durfee in its ultimate worth. Morrell not only made the arrangements for the trials, but also continued to maintain an interest in them as they progressed.

One of his good friends was Captain Eber Brock Ward, an aggressive Detroit industrialist who owned numerous enterprises, among them the Eureka Iron Company at Wyandotte and a rail rerolling mill at Chicago. Ward's pioneering instincts were well known, and the Eureka plant was making money. It was working both day and night shifts turning out iron for the war. Morrell brought the Kelly process to the attention of his friend and suggested that they might join forces to promote it.

Eber Brock Ward had been born in Canada in 1811 of American parents who had emigrated from Vermont in search of a better living. But when young Eber was nine, the senior Eber once more decided to seek his fortune elsewhere. With his wife and three small children, Ward set out for Kentucky, where he planned to open a store. Such a trip would have been hazardous and difficult at any season, but a journey in a canvas-covered sleigh in the dead of winter was foolhardy. His wife contracted pleurisy and died after a three-day illness. She was buried under an oak tree in a desolate area of Pennsylvania.[1]

The elder Ward had little heart for continuing and decided to go to New Salem (now Conneaut), Ohio, where his brother Sam lived. Eber Ward and his children remained in Ohio for about four years. During that time his enterprising brother had moved on to Michigan, where he settled on the St. Clair River near Detroit. In the short space of time since he had left New Salem, Sam Ward already had assumed an established place in the small community, now called Ward's Landing, and was the owner of several sailing vessels. The Eber Ward family joined him there.

Sam Ward had no children and may have seen a reflection of his own enterprise in young Eber. Whatever the reason, he soon regarded his nephew as a son. Although the lad started out as a cabin boy on one of his uncle's ships, as soon as he was old enough, he became its captain and eventually its owner.[2]

One of the first American tycoons, the younger Eber Ward, by the time of his death in 1875, had become known as the richest man in the Middle West. His interests were widespread, and he owned or controlled interests in coal, copper, salt, iron ore and silver mines, railroads, banks, vast stretches of timberland, shipyards, a fleet of vessels, newspapers, and a plate glass factory.[3] He was the main force in the building of the Sault Ste. Marie Canal, which made the great Lake Superior iron ore deposits accessible. It was his schooner, the *Columbia*, that had brought the first shipment of 132 tons of iron ore through the canal on 17 August 1855.[4]

In Ward's estimation, the Great Lakes were ideal for cheap and easy shipping of the newly found iron ore deposits in the Marquette Range. He visualized Detroit, with its fine port facilities and its easy access to the Lake Superior ore area, as an ironmaking center. In 1855, with a group of Detroit businessmen, he had organized the Eureka Iron Company and built the Eureka Furnace. Ovens for burning charcoal were built nearby, and it also was bought from farmers who were encouraged to produce it in the off-season to add to the supply. Soon they added a rolling mill. Wyandotte, about ten miles south of Detroit, was considered an ideal location for this new furnace and mill. Marquette, which was nearer the ore supply, was isolated during the winter when the waterways froze.

Two years later, Ward joined a few friends from New England in constructing a mill in Chicago for rerolling iron rails. The presence of six important railroads there convinced Ward that there would be enough business to keep a rerolling mill in full production. The pioneer mill proved to be even more successful than its backers had anticipated.

Although he lived in Detroit, Ward maintained both social and business contacts on the Atlantic seaboard and frequently traveled back and forth. He was an impatient man. If there were any delay in finding suitable local workers for his many enterprises, he customarily sought out experienced men wherever he might be and hired them for his plants. His agent paid regular visits to Ellis Island looking for workmen. He was known to have been responsible for families leaving England or Scotland for Detroit or, at a later time, Chicago or Milwaukee; he also paid their travel expenses.[5]

When he needed a technical advisor for his Chicago and Wyandotte companies he engaged Zoheth Shearman Durfee, who was well known in the East for his ironmaking skill. Born in Fall River, Massachusetts, in 1831, Durfee had received a simple academic education and had learned the blacksmith's trade. Somewhere in the process, he had studied enough about the production of iron to be considered an expert in the field. Durfee became acquainted with the Kelly pneumatic process while he was in New Jersey studying a steelmaking process invented by Joseph Dixon, better known as a producer of the plumbago crucibles that were used in the production of crucible steel. The Bessemer-Kelly controversy aroused Durfee's curiosity and he spent some time investigating its numerous facets.[6] He became convinced of the justice of Kelly's claim and when, at a later time, Ward sought his opinion regarding the merits of the process, Durfee gave it enthusiastic support. Although Ward had plunged into more than one enterprise with little knowledge of its technology, when Daniel Morrell suggested purchasing the Kelly patent Ward was interested but cautious. He was aware that the Bessemer process was gaining a place for itself in England but uncertain whether or not it would be possible to find enough skilled workmen to produce steel if he should build such a plant in Detroit. He also had doubts concerning the exact status of the patents. But carried along by the enthusiasm of both Durfee and Morrell, he joined with them, and early in 1861 the group gained control of the Kelly patent.

Ward now was interested in establishing an experimental works at Wyandotte in order to determine whether or not the Lake Superior iron ores or other American ores could be used to produce a satisfactory steel by the pneumatic process. Durfee believed that the Bessemer machinery would increase the efficiency of the process. With this in mind, as well as to study the Bessemer process as it was practiced in Europe, Durfee went abroad in the fall of the same year.

Traveling about England, the American heard varying opinions about the Bessemer process, many of them unfavorable. The general feeling among the steelmakers, was that the process had yet to prove itself. With typical New England conviction, Durfee refused to be swayed in his belief.

Durfee visited the Göransson works in Sweden, which had been using the Bessemer process for almost four years. He observed that a fair quality of steel was being made from iron taken directly from the blast furnace to the converter. He visited several companies in England that were producing steel from converters. Among them were the Atlas Works of John Brown and the Tudhoe Works, where he saw the process used in the production of axles and rails for certain English railways. He learned that several companies were producing the new steel or building plants for this purpose. Crossing the channel, he visited the first Bessemer installation in France, the works of James Jackson and Son near Bordeaux. He found that the process had been so successful there that an addition to the plant was being considered that would increase production by at least ten times.

It was June before he finished his investigations. Having observed activity that directly contradicted the rumors that the process was a failure, Durfee was even more certain that it had value. He had learned that it was possible to build a complete plant composed of the necessary blast engines, converting vessels, hydraulic cranes, and reverberatory furnace capable of producing twenty tons a day for £5,000. He was amazed to hear that a properly managed Bessemer installation could be operated by only twelve men.[7]

Durfee returned to America and gave Eber Ward a vivid account of the European practice. However, for reasons not known, he returned without the American rights to the Bessemer machinery. Perhaps Bessemer was nettled that the American so brashly suggested the use of the English patents in conjunction with the yet unproven Kelly process. Or perhaps Bessemer was doing so well that he could be indifferent to granting rights in a country where his integrity and honor had been questioned.

As Durfee studied the Bessemer process in Europe, he kept hearing the name of Robert Mushet. From conversations with steelmakers and from articles in technical journals, Durfee learned that it was generally recognized that the ultimate success of the Bessemer process was due in large part to the contributions of this man.

Although Bessemer in his autobiography found it possible to ignore William Kelly and to pay scant attention to Goran Göransson, he could not overlook Robert Mushet. The granting of the priority of invention to Kelly by the American patent office had been a bitter experience, but the Mushet relationship was to be a potion of gall and wormwood that Bessemer was to taste all the days of his life.

The brilliance of the Bessemer achievement remains unchallenged. Industry was transformed beginning with the day on which he read his paper at Cheltenham. Nor can the influence of Kelly be underestimated. Yet, in truth, in the initial stages of their work, both were making iron. They had begun with the intention of producing a superior type of iron and, had it not been for Mushet, their researches might have ended there. It was he, a true metallurgist, who made it possible for both the Kelly pneumatic process and the Bessemer process eventually to produce steel.

Robert Mushet bore the name of a family long identified with the iron business in Scotland. William Mushet, his grandfather, who was born in 1745 near Edinburgh, was the first of the Mushet line to own a foundry. As Edinburgh expanded in the 1760s, so did the foundry, which supplied cast fireplace backs and ranges for the numerous new residences. British naval power also was growing, and the foundry cast cannonballs for the fleet's muzzleloaders. Mushet developed a particularly accurate technique for gauging the missiles and eventually was supplying cast cannonballs to both Lord Nelson and the Duke of Wellington.[8]

David Mushet, born in 1772, was the oldest of William's six sons. Because of his interest in classics and mathematics, he did not enter the family foundry, but became an accountant with the Clyde Iron Works at Tollcross. The works had been founded in 1876 when the Carron Company at Carron was unable singlehandedly to meet the rising demands for iron products during the Napoleonic Wars.[9] The two firms thereafter operated in close collaboration.

Although the business aspects of the Clyde Iron Works initially had attracted David Mushet, he soon became interested in the ironmaking itself. This still was the time of trial and error in the making of iron; yet there were men in the shop who were certain that a knowledge of the nature of the raw materials used in producing iron could help to determine the quality. As the young man mulled over this idea and the talk that he had heard in the shop about iron production, the subject of assaying interested him more and more. However, he had little practical knowledge and no furnace with which to carry on experiments. After some time, Mushet did secure the use of a small furnace and began working intensively on the analysis of iron ores. His methodical mind was useful both in the experiments themselves and in keeping careful records of them.[10]

In about 1800, when he was twenty-eight, Mushet had gained enough confidence in his skill at assaying to leave the Clyde Iron Works. He and two friends, William Dixon and Walter Neilson, pooled their resources to purchase the Calder Iron Works. The plant was in bad repair, and David Mushet, as the new manager, took a leading role in its reconstruction and operation.

An enthusiastic explorer of the countryside, Mushet made a discovery in the following year while hiking that was to be of great benefit to the Scottish iron industry. As he carefully searched for dry footing on stones in the Calder River, he noticed what appeared to be a large stretch of black pavement. Examining it more closely, he discovered that it was interlaminated with globules of a brownish matter, and that its upper edge resembled coal. This newly found material, which would become known as blackband ironstone, was tested and proved to be good quality ore. For years the Scottish ironmasters had imported iron ore from abroad, not knowing that an ore of extremely high quality lay beneath their own soil. For the discovery of this valuable resource the Scottish iron industry was to become immeasurably indebted to the perspicacity of David Mushet.

The Scottish iron producers, however did not at once turn to using the domestic ore. Mushet had leased large tracts of the iron ore veins in that expectation, but to his great disadvantage, only the officials of the Clyde Iron Works were convinced of the value of the iron.

Only a few years later, in 1805, unrecognized in the industry and dissatisfied with his partners, Mushet decided to leave Scotland. He retired from the partnership, relinquished his leases, disposed of his other property, and, with his wife and three children, set out by stagecoach for England. The family settled in Alfreton in Derbyshire, both because Mushet found that the town supported a large iron works and because, having traveled 300 uncomfortable miles, they preferred not to continue.

Mushet soon joined the Alfreton Iron Works and continued writing and experimenting.

Having owned his own shop, Mushet was discontented as an employee at Alfreton. In 1810, he borrowed £500 from a brother in order to buy property in the Forest of Dean. Mushet first bought Forest House in Colford and a tract of land that included the mining rights. With the remainder of the loan, he purchased an interest in the nearby Whitecliff Iron Works.[11]

The Forest of Dean has far-reaching associations with the long history of ironmaking, and it also is of historical interest. It is a forested oval tract of about ten by twenty miles between the Wye and Severn Rivers in Gloucestershire. A royal property maintained by the crown, it has been known since the Roman occupation as a great forest with thousands of holly and beech trees. These, together with large deposits of coal and rich veins of iron ore, provided a bountiful supply of raw materials for smelting. As early as 120 A.D., the Romans set up forges along the Wye, as did the Saxons after them.[12]

Curious customs and laws grew up pertaining to the mines in the forest. "Kings profits," which consisted of "one-third of the value of coal and iron mined," were collected weekly by the "King's gaveller." Only a native of the forest who had qualified by working for one year and a day as a miner, could attain the status of Free Miner. A mine law court, the Court of Verderers, with a jury of forty-eight miners, is said to have been founded in the forest in 1016 by King Canute. The court met at Speech House in the forest's center, and until 1842 the old customs and laws held. The court has continued to meet, and its members still are elected by freeholders of the forest, but it has retained little of its former authority.[13]

The underground workings in the forest are extensive, and successive generations of miners have extended galleries for miles. When the mines were active, it was known as a great sight to see the miners emerging from the diggings with their clothes, faces, and hands dyed red from the oxide of iron.

The ironworkers who lived there became known also for their unusual character traits. An early writer is said to have described them as heathenish in their manners, puffed up with pride, and inflated with worldly prosperity. He tells the story of a visit made by St. Egwin to the smiths of Alcester, to save their "wicked souls." When the saint attempted to preach the holy word, the workers drowned out the sound of his words by angrily beating on their anvils. They are said, also, to have been given to quarreling among themselves over individual rights, to have shunned outsiders, and to have received strangers reluctantly at best.[14]

This was the setting in which David Mushet hoped to recoup his fortunes. He began by buying a share in the Whitecliff Iron Works. The works had been started by Samuel Botham, who, after expending a large amount in constructing the buildings, had to endure a season of un-precedented bad weather. Heavy rains turned the creeks into violent streams that flooded the buildings and carried away the labors of months overnight. Botham was unable to find additional backing for the wrecked works and left the area. His associates, who had refused to help him, took over the works. This was the unfortunate company in which the unsuspecting Mushet had bought a share. He soon realized that it was impossible to work with his partners, and he, too, retired from the plant, which never resumed operations.

In spite of his unfortunate experience in the iron industry, Mushet was not discouraged. He decided to stay at Forest House and to carry on his experiments alone, and he built a small shed nearby for the purpose. Here, in 1811, his son Robert Forester was born. Robert was an old family name and, as a token of the affection he felt for the Forest of Dean, Mushet added Forester to it.

As his reputation as an assayist grew, David Mushet frequently was engaged by other ironworks as a consultant. In 1825, the forest commissioners granted his application for a "gale," which would allow him as a Free Miner to mine a tract of iron ore for the payment of royalty or dead rent. He then built a blast furnace with a capacity of two and one-half tons of pig iron. During the same year, he received the news that, at last, another Scottish ironmaker, the Monkland Company, was using the black-band ironstone with great success. Mushet commented ruefully that even after the twenty-four years it had taken for his discovery to be recognized, there was no financial return to him. "Time," he said, quoting Lord Bacon, "is the greatest of all innovators."

Mushet's hurt pride was soothed on at least one other occasion in later life. In 1843 he was asked to testify as an expert witness in Edinburgh respecting the value of James Neilson's invention, the hot blast. It gave Mushet great satisfaction to return as an outside authority to the scene he had departed thirty-eight years before.

Robert Forester Mushet had been granted the good fortune to have studied and worked with an extraordinary father. For many years he had resided close to Forest House, and when his father retired, Robert and his family moved into the house of his birth. After his father's death, the younger Mushet formed a partnership with T. D. Clare, a Birmingham merchant. Mushet's estate was small and Clare probably furnished most of the financing for the new enterprise to bear the name of R. Mushet and Company, Forest Steel Works.

His father's old shed was maintained for experiments in steelmaking, and the new company installed a scale, a blowing engine, and a small furnace for melting iron in the old building. The new plant was located about a mile from Forest House, close to one of the mines where young Mushet had served his "one year and a day" so that he might become a Free Miner.[15]

The new works consisted of a crucible furnace with ten melting holes and a pair of tilting hammers. Each melting hole, which was square, could contain four crucibles. It is of interest that neither Mushet nor his workmen had seen a steel plant other than his father's. The plant required a fair sized work force: a pot maker to mold the crucibles, two forgemen, an assistant melter, and a head melter, who was the most important member of the team. There were, in addition, two handymen and a blacksmith, who forged the tools needed for the operation of the process. The tongs for lifting the crucibles out of the holes, hammers, chisels, and all other implements that were needed in the nearby mines and quarries, were hammered on his anvil. Robert Mushet himself weighed the materials.

One of David Mushet's closest friends had been Josiah Marshall Heath, another in the long list of men of the iron and steel industry who would never receive their just due. Robert Mushet remembered well the endless hours of metallurgical talk that had taken place between the older man and his father and a similar friendship grew up between Robert Mushet and Heath.

Heath, who had lived for many years in India as a member of the East India Company in the Madras District, had established the Porto Novo Iron Works there. While searching for a small quantity of steel that could be fashioned into spearheads for hunting boar, Heath had been appalled by the crudity of the processes used by many of the steelworks in India. Although not a metallurgist and not previously particularly interested in the subject, Heath's curiosity was aroused and he decided to study metallurgy. He had noticed along the coast of Malabar large areas of magnetic ore that he thought would make a good cheap steel that he could sell in England. But the metallurgists he had consulted had discouraged Heath from pursuing his idea. By this time, Heath was convinced of the value of his discovery and refused to be diverted. He gave up his position, took with him all his money as well as the pension rights he had accumulated, and set out to visit iron and steel plants wherever he could. He studied, experimented, and devoted every possible moment to the pursuit of the idea for which he had given up everything else.[16]

After many years of work, he discovered the importance of the use of what he termed carburet of manganese, a combination of carbon and manganese. Heath often had discussed with the elder Mushet all the phases of his experiments. Mushet had advised Heath to patent the process, but to be certain to include the use of oxide of manganese in his application. This, Heath did not do. When he took out English patent number 8021, dated 5 April 1839, he used instead the phrase carburet of manganese, and for this he was to pay dearly.[17] John Percy, the metallurgist, wrote of the patent, "By this addition, blister-steel, prepared from the comparatively low priced bar-iron of British manufacture, will yield malleable, cast-steel, whereas previously such cast-steel could only be made from high priced descriptions of bar-iron, such as the Swedish or Russian."[18]

In 1840, Heath visited Sheffield and demonstrated his discovery wherever he could find someone who would listen. His enthusiasm convinced many of the steelmakers of the value of his patent. He employed an agent named Unwin to represent him and to introduce the manganese product while he continued his experiments. Heath soon discovered that a mixture of oxide of manganese and carbonaceous matter would achieve substantially the same result and at the same time obviate the need for the separate reduction that the patented process required. Adding the manganese in this fashion was so much more efficient that it was possible to reduce the license charge by two-thirds and the cost of the finished steel by 40 to 50 percent. When Heath forwarded to Unwin packets of the new mixture the agent, more clever than honest, perceived what Heath had done. Unwin proceeded to set up a steel works of his own, using Heath's discovery but refusing to pay royalty for the use of the manganese invention.[19] Heath's patent had referred only to carburet of manganese and Unwin was not using manganese in that form. The predicament and ensuing course of events was described by Thomas Webster, one of the counsel for Heath. "The payment of this or of any sum to Mr. Heath was refused by a section of the steel manufacturers who, relying on the refined distinction just adverted to, created out of their savings a common fund wherewith to contest his rights; the expense of the fifteen years' litigation falling wholly upon himself, fighting single-handed against a common purse, the accumulation of the wealth which he had created."[20]

Heath's struggle affected not only his fortunes but his health as well. The inventor died just before the opening of the Great Exhibition of 1851 at the Crystal Palace, where he had taken space to show some metallurgical specimens produced by his process.

The crucible steel that Robert Mushet was making at his plant did not use Heath's patent. With one difference, his steel was made by the method used by Huntsman so many years before. Whereas Huntsman had added pieces of bar iron and later the necessary carbonaceous material, Mushet melted iron with the appropriate amounts of charcoal and manganese to gain the desired steel. The steel was, of necessity, made in small amounts, and the resulting product was used almost entirely for making tools.

In 1848, seven years after Robert Mushet had begun operations, something happened that not only would change his own method of production, but also would be instrumental in changing the Age of Iron into the Age of Steel. As Mushet later told the story, a lump of white crystallized metal was brought him by Henry Burgess, editor of the *Banker's Circular*. Burgess claimed it was iron found in Rhenish Prussia, and that there was a mountain of it there, he had been told. Mushet recognized that the lump in his hand was not entirely iron ore, but was an alloy of two metals, iron and manganese. Hardly able to contain his excitement, Mushet asked Burgess to inquire further about the material. Mushet learned that the material was called spathose iron and was, as he had suspected, a double carbonate of iron and manganese. He noted that it was selected carefully and smelted in small blast furnaces with charcoal as the only fuel and limestone as the only flux. The molten metal then was run from the furnace into shallow iron troughs and formed into cakes. When cold and broken, these cakes showed large and beautifully bright facets of crystals minutely speckled with uncombined carbon. Because of its brightness, it was known as *spiegel glanz* or *spiegeleisen,* that is, looking glass iron. Burgess found that the average analysis was iron 86.25 percent, manganese 8.50 percent, and carbon 5.25 percent. Mushet now knew the secret of the unusual quality of the steel for which the steelmakers of Rhenish Prussia had become famous. He ordered twelve tons of spiegeleisen.[21]

When the shipment arrived, Mushet began experimenting to determine its qualities and potential. During his years of working with iron, Mushet had seen a certain phenomenon occur again and again. Wrought iron that had been exposed to heat, flame, and drafts of air for long periods of time became valueless. It was termed "burnt iron." In the course of his work Mushet discovered that if he alloyed the burnt iron with his spiegeleisen, the inferior iron returned to its original quality and perhaps was improved. Mushet set to work to discover the reason. After numerous experiments he learned that in some way the burnt iron contained oxygen locked within itself. He knew that carbon had a great affinity for oxygen yet when carbon alone was added to the spoiled iron nothing happened. After further work, he concluded that manganese also had a great affinity for oxygen and withdrew it from the iron in the form of manganese oxide, which passed into the slag. The carbon in the spiegeleisen remained in the molten metal and converted it into steel.

Although Cheltenham was but a short distance from where he lived, Mushet did not hear Bessemer read his paper there. The meeting probably was of no special significance to Mushet, for he could not have foretold the momentous nature of the occasion. A good friend, Thomas Brown of the Ebbw Vale Iron Company in Wales, did attend and came away with a sample of the iron that Bessemer had displayed at the Queen's Hotel.

Robert Forrester Mushet. From *Cassier's Magazine* 34 (August 1908).

Brown stopped off at Forest House on his way back to Wales and showed Mushet the metal. Mushet was interested in hearing about Bessemer's paper and studied the sample carefully. He broke the specimen and examined the two pieces. The fracture presented the telltale marks of a definite crystalline structure and he recognized his old acquaintance, burnt iron. The structure was definite enough for Mushet, as a metallurgist, to discern what Bessemer was not destined to discover for some time and after much travail—the deleterious effects of sulphur and phosphorus. Mushet treated the Bessemer sample as he had other pieces of burnt iron and produced a small ingot that could be forged satisfactorily. Brown was pleased with the outcome.

Mushet suggested that Brown erect a small converter at the works in Wales where a few test heats could be made in the manner described in Bessemer's paper. This Brown did and brought some of the product to Coleford. Mushet described it as resembling nothing as much as "an old fashioned puddle bar from the worst red short iron, only it was far more deeply cracked." When asked if he thought that he could improve on this miserable specimen, Mushet assured Brown that he could.

> I had part of the bar cut into small pieces, and of these I placed 16 oz. in a small clay crucible, and placed the crucible with a lid upon it in a small assay furnace capable of fusing wrought iron. Into another smaller crucible

I put 1 oz. of pure Siegen spiegel, and placed the crucible in the flue of the melting furnace. When the contents of the crucibles were melted I withdrew quickly both crucibles from the furnace, and poured the melted spiegel into the Bessemer metal, and then emptied the mixture into a small ingot-mould. The ingot was smooth and piped and had all the appearance of good cast steel.

I heated this ingot to a fair cast steel heat. Mrs. Mushet held the ingot in a pair of tongs, and with a sledge hammer I drew one half of it into a flat bar. This bar I heated, and twisted in a vice [sic] at a white heat, a red heat, and a low red heat. It remained perfectly sound, and clear in the edges, and not a trace of red-shortness remained. I next doubled the bar, welded it, drew it into a chisel, which I hardened, tempered and tested severely on hard cast iron. It stood well and was in fact cast steel worth forty-two shillings a 100 weight. I saw then that the Bessemer process was perfected, and that with fair play untold wealth would reward Mr. Bessemer and myself.[22]

The spiegeleisen had carried out its functions. This triple compound of iron, carbon, and manganese was the catalyst that would remove the occluded oxygen, pass off the oxide of manganese into the slag, and leave the carbon from the spiegeleisen in the molten metal, thus bestowing upon the charge the qualities of steel.

Although he was confident of his accomplishment, Mushet was not yet willing to leap into print. He asked Brown to send him all of the metal that had been produced in the test converter at Ebbw Vale. Mushet then charged each of the sixteen melting pots at the Forest Works with forty-four pounds of the cut-up metal. While the charge still was molten, he added three pounds of melted spiegel to each crucible and poured their contents into an ingot mold. The resulting ingot was returned to Ebbw Vale where it was rolled into a rail. This rail was placed in a location on the Midland Railway near Derby Station, where iron rails ordinarily had been replaced at least every six months. Two hundred and fifty trains passed over this track daily, not counting the numerous detached engines and tenders that shuttled back and forth. The Mushet rail remained in place some sixteen years before it finally was taken up. It is curious that the first steel rail produced by the Bessemer or pneumatic process was rolled not from steel made by Bessemer, Kelly, or Göransson but by Mushet in his little works in the Forest of Dean.

After he had experimented further with the use of spiegeleisen, Mushet was convinced that he at last had a process that could be patented. He believed that he owed a certain debt of gratitude to Brown, so when the latter suggested that his patent counsel, Mr. W. H. Hindmarch, supervise the application, Mushet could only agree.

During the same period, while Mushet was working in his shed in Coleford, Bessemer was building and tearing down trial converters in London, desperately seeking the fault in his process. Even though his detractors were in full cry, Bessemer went about his experiments with dedication, ignoring the ridicule, the abuse, and even the doggerel that appeared in the daily press. His reputation at that time was probably at its lowest point. Mushet, however, still believed in Bessemer's eventual success, provided that Mushet's contribution became part of the process.[23] As he said in one of his letters, "Bessemer metal without Mushet = Iron; Bessemer metal with Mushet = Steel."[24]

Long before Bessemer's time, there had been numerous researchers in the field of steelmaking, one of whom, Joseph Gilbert Martien, already has been mentioned. In 1855 he had developed and patented a steel-making process that Mushet described in this manner: "His claim was to partially decarbonize melted cast iron by running it from the blast furnace along a cast iron gutter, the bottom of which was perforated with numerous small holes, through which air was forced. The metal thus treated was run into a receptacle — whatever that might mean. But not a syllable was said as to converting the cast iron into steel or malleable iron. The process was indeed a pneumatic process, but one possessing neither value nor utility, unless it was desired to make an exhibition of fireworks at the cost of the iron."[25] Yet Brown's Ebbw Vale Company had thought well enough of Martien's invention to experiment with it, but had found the process impractical.

Thus it happened that when Mushet consulted W. H. Hindmarch, Brown's patent counsel, about applying for a patent on the spiegeleisen process, he suggested that Mushet substitute the name of Martien for that of Bessemer in the specification. Mushet, no doubt impressed by the fact that Hindmarch was known as the outstanding patent counsel of the day, and having little knowledge of the ramifications of patent law, agreed to do so. While Martien no longer is remembered, it was recognized at the time of Mushet's application that Martien's invention, despite its practical failure, contained the germ of the Bessemer process. Soon after the Cheltenham meeting, the authors of articles in several journals had credited Martien and not Bessemer with the development of the pneumatic process. Neither process had been tried publicly, and the two men had lodged provisional specifications within three months of one another so that, for a time, both had provisional protection.

Mushet, in preparing to file for his patent, which was to be the notorious patent number 2219, had made certain arrangements with Brown's Ebbw Vale Iron Company. It was to receive half of any royalties that Mushet might collect. In return, Brown's firm was to bear the cost of the counsel's fee of about £300 and to protect the process from piracy or other infringement. The final specification was to be filed within six months from 22 September 1856, the date of application. The application covered not only England, but France and America as well.

It now was urgent for Mushet to have a converter where he could test his theory that Bessemer metal could be produced directly. At this point, having gone so far, Brown began to hedge, and Mushet sensed that the faith in his process had begun to waver at Ebbw Vale.[26] Frantic lest the time run out before he could complete his tests, Mushet turned to an old friend, S. H. Blackwell, who staked him to an experimental Bessemer plant, complete with a cupola for melting the iron, a hearth, and the necessary blowing apparatus. For this, Mushet assigned Blackwell one-half of his remaining interest. Thus, the inventor was left with a one-quarter share in the process of his devising.

There were some difficulties when the trials began. Because the iron was produced in small quantities, it chilled and hardened too quickly. Mushet succeeded, however, in turning out enough metal to make a few tools that were used successfully for boring rock and iron ore at the nearby Easter Iron Mine.[27]

Mushet, in his pamphlet *The Bessemer-Mushet Process,* says that Bessemer paid him several visits following the Cheltenham meeting. News of what Mushet was doing must have found its way to London. Mushet writes that, although he did not know whether Bessemer thought that he [Mushet] had made a valuable discovery, it was evident that Bessemer did want to know what Mushet was doing. Feeling honor bound to Brown and the Ebbw Vale Iron Company, Mushet disclosed nothing. Bessemer in his autobiography makes no mention of a trip to Coleford or of having talked with Mushet at that time. However, in 1866, after Mushet had attacked him violently at a meeting of the British Association for the Advancement of Science the enraged Bessemer wrote a lengthy letter to the *Engineer* denouncing Mushet's statement.

> . . . I received a letter from Mr. Mushet, who was up to that time entirely unknown to me, in which he announced, in the most inflated language, that he had made a most important discovery which would add enormously to the value of my process, and without which it would be worth nothing. To this letter I immediately replied that I was most happy to hear that he had made so valuable a discovery, and expressed my willingness at once to purchase his patent or take a license to use it, or make such other arrangements as might secure our mutual advantage; and I further said that I would visit him at Coleford, and go further into the matter; I accordingly went to Coleford, but only to find his invention—the nature of which I was perfectly ignorant of, but which was said to be so absolutely essential to the success of my process—had been made over to Mr. Thomas Brown. Thus was I sold to the enemy, and I found myself ruthlessly sacrificed by Mr. Mushet. His invention was offered to me, and I was ready to negotiate for its use or purchase; but without a word of explanation it was parted with, in order to strengthen the hands of my opponent. Was this, may I ask, one of the acts of Mr. Mushet that is calculated to awaken a feeling of profound gratitude on my part?[28]

Blackwell, after having helped with the building of the converter, fell on hard times. When the third year's stamp duty was due on Mushet's patent, Blackwell was in financial difficulties, and Brown had lost interest in the entire project. Neither man paid the fifty pounds that were due, nor did they so notify Mushet. Thus the process came into the public domain. The French patent was forfeited at the same time, but the American patent remained in Mushet's possession. Mushet said of the tragedy, "So my process became public property, and Mr. Bessemer had a perfect right to make use of it and his prosperity dated from that period."[29]

Meanwhile, Bessemer had built his Sheffield plant and was producing steel by the crucible method rather than by Mushet's direct process. Bessemer had much to say as to how he discovered the value of using the manganese in his process. If the idea did not emanate from Mushet, there were other sources that may have put him on the track. Göransson in Sweden believed that he may have supplied a clue. Once Göransson had found out how to use the Swedish ores, which were notoriously free of sulphur and phosphorus but high in manganese, they were an "open sesame" for success with the Bessemer process.

Bessemer remarked in connection with his discovery of the value of manganese, "Events which shape and control lives and fortunes often have no possible connection."[30] Bessemer was an acquaintance of the distinguished Andrew Ure, author of the *Dictionary of Arts, Manufactures, and Mines*. Dr. Ure had been interested in Bessemer's early work with medallions, particularly with copper deposit coating. He had written about the process in the section on electrometallurgy. A supplement to the dictionary had been published in 1846, and in it Bessemer found an article on steel that described at some length the Heath discovery of carburet of manganese.[31] Bessemer also claimed that in an 1852 notebook of Longsdon's he came upon an entry describing their first attempt to use Heath's patent in their experiments. Bessemer took the trouble to list all previous patents in which manganese had been used in iron and steelmaking in an attempt to prove that Mushet was a latecomer on the list of men who patented this idea.

Bessemer made use of the Mushet patent without license from 1857 to 1860 under circumstances that he described in his autobiography.

> About three or four months prior to the date when a further one hundred pound stamp was required to be impressed on them to prevent their forfeiture, I received a letter from a Mr. Clare, of Birmingham, calling himself Mr. Mushet's agent for the sale of steel, and requesting an interview with me and my partner at my office in London on the following morning. On his arrival he explained the object of his visit; it was simply to say that Mr. Mushet was prepared to grant me a license to use his manganese patents for a nominal

sum; he merely wanted his rights acknowledged. I then told Mr. Clare that
we considered Mr. Mushet had acquired no rights under either of his three
manganese patents, and that we entirely repudiated them. I also told him
that we were prepared, on any day to be mutually arranged, to receive Mr.
Mushet and his solicitors and witnesses at the Sheffield Works; that we would
allow them to see the crude iron converted and re-carburized with spiegeleisen,
made into an ingot and forged into a bar, and that I would personally take
that bar to one of my customers and sell it to him in their presence; and the
prosecution of our firm for infringement would be a very simple matter. This
offer resulted in Mr. Clare's retirement from my office and after that interview
we never heard from him, or from Mr. Mushet on the subject.[32]

Shortly after this episode the patents were permitted to lapse. Bessemer had the legal right to use the Mushet process, but there was the
moral right to be considered. Many leading metallurgists were outspoken in their opinion of what Bessemer had done and rose to the
defense of Mushet in the technical journals and the press. As John Percy
wrote, ". . . there is one patent which deserves special consideration,
and to which Bessemer is deeply indebted. I allude to that of Mr. Robert
Mushet, dated September 22, 1856."[33]

Bessemer felt cold rage toward Mushet, and with some justification.
Most stemmed from Mushet's past and present associations with Thomas
Brown and the Ebbw Vale Iron Company. Bessember detested both whole-
heartedly. He never had forgotten Brown's arrogant entrance into his
office seeking to buy the rights to Bessemer's process. When he had read
his second paper before the Society of Civil Engineers in 1859, Brown
had possessed the effrontery to rise and say that he had been hopeful
of the Bessemer process back in 1856 and had spent £7,000 endeavoring
to make the process work, but without Bessemer's knowledge and, of
course, without license or permission. There also had been the occasion
in 1861 when the manager of the Ebbw Vale Iron Company, George
Parry, had taken out a patent that ostensibly was for a steelmaking
process, but that Bessemer was convinced was in fact a ruse to corner
the use of manganese.

The climax to Ebbw Vale's aggression took place in 1864. It was the
last battle that Bessemer had to fight to retain his process. On a business
trip from London to Birmingham, Bessemer was seated in the corner of
a compartment of the train reading his newspaper. Two young men
seated opposite him near the door were conversing in high spirits. Their
conversation was lively and they could hardly control their excitement.
Bessemer ignored their chatter until one suddenly said, "I wonder what
the devil Bessemer will say?" After this Bessemer listened carefully. He
was able to discern that a company was being formed that could cause
him considerable embarrassment, but he could not determine who the

organizers were. When the train came to a stop at Leighton and the young men got off, Bessemer indeed was puzzled, since he had assumed that they were going to Birmingham. Suddenly he remembered that Joseph Robinson, the manager of the Ebbw Vale London office, lived in Leighton. Bessemer rode to Blisworth, the next stop on the line, and left the train in order to take the next one down to London.[34]

Early the next day, he visited the Ebbw Vale office and belligerently confronted the financial agent, David Chadwick. Taken by surprise, Chadwick assumed that Bessemer knew more than he did and admitted that negotiations were under way to reform the Ebbw Vale Iron Company into a joint stock company with a capital of £2 million. The new company would use the pneumatic process in combination with the Parry and Mushet patents.

Although Bessemer later wrote that he knew he could have won any case that might have gone to court in connection with the use of his rights, the amount necessary to fight the suit could have amounted easily to £10,000 and the time consumed would have been disastrous. Bessemer then made a bold move. He told Chadwick that he would apply immediately in the Court of Chancery to restrain the Ebbw Vale men from using certain patents. More than that, he would have 1,000 posters printed in red and blue in large enough type so that they could be read easily by any passerby. He would engage men to carry these posters announcing that Henry Bessemer had applied for four separate injunctions in the Court of Chancery and list the reasons. Inasmuch as the Ebbw Vale company needed a great many investors for such a large scheme, such an action would make the proposed company suspect to potential stockholders.

A settlement was reached as Chadwick, impressed by the words of the compact and powerful figure, was convinced that the inventor would carry out his threat. The Ebbw Vale Iron Company did not have the sole right to the Parry patent, so Bessemer paid £5,000 to procure it and furthermore agreed to deduct £25,000 from the first royalties that the Brown group would pay. In return, Bessemer had full acknowledgement by Ebbw Vale of the validity of his own patents as well as recognition that he now controlled the controversial ones of Martien and Parry.[35]

There remained the figure of Robert Mushet looming on the horizon. Bessemer could face down the large and powerful Ebbw Vale Iron Company, but the proud, dignified figure of Mushet could not be ignored. Furthermore, Mushet was a highly vocal opponent.

Bessemer had taken a stand on the legality of the manganese addition, and he could not or would not alter his position. Why did Mushet not go to court? Mushet has said that at the time he had neither the money nor the good health necessary to fight his case. There also must have been an awareness that the outcomes of long, involved patent suits usually had been disastrous for the plaintiff. Heath's hopeless fight for recogniton of his rights may have influenced Mushet's decision. Zerah Colburn once aptly wrote, "It has sometimes been asked why the entrance from Quality-court to the Great Seal Patent Office, Chancery-lane, should be placed between the Registry in Lunacy and the Registry of Appeal in Bankruptcy."[36]

If Mushet did not go to the court of law for justice, he did go to the court of public opinion. For many years he continued to write long and angry letters to periodicals. If Bessemer's name appeared in print, it was almost certain that a letter to the editor from Mushet soon would be forthcoming. Bessemer usually ignored the letters, but the constant provocation by the aggrieved Mushet occasionally brought forth an answer from the beleaguered Bessemer. Letters then issued from Mushet's pen even more rapidly.

The controversy reached such a pitch during 1861 that the editor of the *Engineer,* which had been printing many of these exchanges, threatened to print no more. The editor accused the frustrated Mushet of being noisy and described him as "that crotchety and irascible Mr. Mushet."[37] Mushet craved merely what he considered his due from Bessemer and from the public for his contribution to the Bessemer process. Zerah Colburn, in the series of articles for the *Engineer* that appeared in December 1864, gave due credit to Bessemer for what he had accomplished. But in the last sentence he said, "Such as it is, however, it is substantially a new manufacture, but its value would appear to be due to the discovery of Mr. Mushet to an extent at least equal to that due to the Bessemer process itself."[38]

Mushet then wrote, "Let it not be supposed that I wish to deny Mr. Bessemer the great merit which his ingenuity and perseverance so well deserve but I wish he had himself taken the manly and honorable course which Mr. Colburn has adopted and given me the just need of praise for having made a perfect success of his invention which neither his own ability nor the united talents of all practical and scientific men could make anything out of except burnt iron."[39]

The continuing controversy may have influenced Bessemer's behavior in 1866, when Mushet's daughter Mary visited him in his office. The only record of the encounter is Bessemer's version in his autobiography. He wrote that his clerk came into his office to say that a young lady was waiting outside and wished to see Mr. Bessemer. She would tell neither her name nor her business. Bessemer instructed the clerk to show her in, and upon entering she introduced herself as Mary Mushet, the daughter of Robert Mushet. Without delay, the girl of sixteen launched into the reason for her call. Her father was ill, and the financial situation of the Mushet family had reached such a low ebb that the bailiff already had threatened to attach the house at Coleford. "They tell me that you use my father's invention and are indebted to him for your success," she concluded. Bessemer stared at her with a penetrating look and assured her that he used what her father had no right to claim. Mary Mushet obviously was frightened at her brashness in opposing this person of such an imposing reputation but, nevertheless, she protested. He went on, "If he had the legal position you seem to suppose, he could stop my business by an injunction tomorrow, and get many thousands of pounds compensation for my infringement of his rights."

Miss Mushet then said that under the present circumstances her father obviously was not in a position to bring suit, that he had lost the patent through no fault of his own, and that the fact still remained that Mr. Bessemer's success was the result of her father's discovery. Without being aware of it, she was facing down a man who through the years had behaved in the same fashion to numerous adversaries. Whether or not this affected Bessemer's attitude, he replied, "The only result which followed from your father taking out his patents was that they pointed out to me some rights which I already possessed, but of which I was not availing myself. I cannot live in a state of indebtedness: so please let me know what sum will render your home secure and I will give it to you."

The amount was approximately £378. Her visit and Bessemer's payment were to have far-reaching consequences.

She thanked me in a flattering voice as I bade her good afternoon. On joining my partner after this interview with Miss Mushet I explained to him what had occurred; he listened to me with surprise and with more impatience than I had ever seen him evince. He thought that what I had done was most unfortunate and imprudent, since from Miss Mushet's words it was evident that the idea was abroad that I had in some way taken advantage of her father. He feared lest my cheque should be considered evidence of my indebtedness.

I was much distressed to find my friend Longsdon so much annoyed, for a more conscientious and just man I never knew; he was, however, somewhat reassured when I told him that I considered it a purely personal matter and had, of course, drawn the cheque on my private bankers. He said he was glad it could never appear as an act of the firm, though he thought it would be long before I should hear the last of it.

Events proved that he was right, for not many months elapsed before a friend—I believe a relation of Mr. Mushet—wrote asking me to make Mushet a small allowance. I objected to do this at first, but afterwards yielded, though I did not care to give my reasons for doing so. There was a strong desire on my part to make him my debtor rather than the reverse, and the payment had other advantages; the press at that time was violently attacking my patent, and there was the chance that if any of my licensees were thus induced to resist my claim all the rest might follow the example, and these large monthly payments might cease for such a period as the contest in the law courts might last. . . .

In the hope that an allowance to Mr. Mushet might have the effect of restraining these attacks on me, I offered to pay him three hundred pounds a year, aiming at abating an intolerable nuisance which I had no other means of preventing. While we were paying over three thousand pounds per annum in the form of income tax, the three hundred pounds was but a small additional tax on my resources, so I allowed it to drag on until Mr. Mushet's decease in 1891, having thus paid him over seven thousand pounds. So naturally, ends this part of the history of my invention, as far as Mr. Mushet is concerned.[40]

It was probably the attack that Mushet had made in his speech before the British Association for the Advancement of Science meeting at Birmingham in 1866 that brought the proud Bessemer to agree to pay the allowance to Mushet. Mushet had said,

I speak no doubt as an interested party, and my opinion is, therefore, open to criticism; but I venture to submit that an invention such as that described in my last patent places all who have benefited by its use under a moral obligation to recognize my claims to remuneration; for although by the accident of the non-payment of the stamp duty my invention became public property, I still think that the accident ought not to debar me from the reward that I am morally entitled to, and could have commanded to so large an extent, but for the oversight of those on whom I relied.[41]

Bessemer was in the audience and at the conclusion of Mushet's remarks was given the opportunity to answer the accusation. "I should like to make a few observations upon the paper to which we have just listened, inasmuch as Mr. Mushet thinks that although he has no legal claim he has a great moral claim upon me," said Bessemer. "Now, I am one of those who regard a moral claim upon a man as possibly greater than any legal one can possibly be, and therefore I wish those who have heard what has been read upon the subject to know the exact state of the case."[42]

Bessemer must have seen an advance copy of the Mushet paper, since he came to the meeting well armed with evidence to dispute Mushet's claims. Bessemer narrated the now-familiar story of the unfortunate Heath, the behavior of Brown, and the trip that Bessemer had made to the Forest of Dean with the intention of purchasing an invention, the nature of which then was unknown to him. The discussion was continued after the meeting when Mushet and Bessemer had one more round of controversy in the columns of the *Engineer* in respect to certain points in Mushet's speech and Bessemer's reply. The moral claim that Mushet insisted he held over Bessemer became a point of issue and Bessemer replied to this accusation.

I therefore utterly repudiate and deny that even a shadow of a moral claim exists against me on the part of Mr. Mushet; but I do think, if moral claims are to be discussed at all in business matters, I had a strong moral claim on Mr. Mushet, whose patents are all founded on mine, and could never have had an existence but for me—a claim that any improvement based on my novel process, and applicable to none other, should have been first offered to me.

It may not be generally known how much I have been pressed by this moral claim in a pecuniary form. One gentleman after another, who appear to have heard a very highly coloured account of my obligations to Mr. Mushet, wait upon me or write to me. One philanthropic gentleman brought me a long list of Mr. Mushet's debts, eleven thousand pounds of which, he said were very pressing, and when these were cleared off he would be happy to discuss with me the amount of annuity I would settle on Mr. Mushet. . . .

I have no ill-will towards Mr. Mushet but I feel neither respect nor friendship for him. The world is wide enough for both of us; and I think it would be well for him to allow a subject to rest, the discussion of which can neither bring him honour nor profit, one that must long since have become a great annoyance to the readers of the scientific journals in which it has already been too frequently discussed.[43]

The sum that Bessemer gave to Mushet over the years was tiny compared to the more than £1 million sterling that Bessemer would have earned from his patent by the time it expired.[44] The payment of the annuity did not, as Bessemer had hoped, quiet the criticism of Mushet's outraged friends. Only Mushet himself was silenced. Now that he was receiving a sum of money annually from Bessemer, Mushet transferred his ire to the Ebbw Vale Iron Company. In a letter written late in 1866, Mushet claimed, "It is difficult to believe that those in whose hands the patent was, should have neglected the duty of fifty pounds and neglected to inform me that they had omitted to pay it. They did not, however, neglect to misrepresent Mr. Bessemer to me, to suit their own sordid views and thus to set me at variance with that gentleman, whom I have since found to be both honorable and kind."[45]

By the year 1868, thanks to the manganese patent, Bessemer was basking in the sun of success, his last battle on behalf of the process over, although the unpleasant attack by Weeks in the United States was still to come. Kelly had made financial arrangements with Daniel Morrell and Eber Ward for his rights in his pneumatic process. His enthusiasm for invention spent, he retired to Kentucky and there established a small business for the manufacture of axes.

The story was different with Robert Mushet, the one true metallurgist of the three. The Forest Steel Works had sapped practically all of the small inheritance that Mushet had received from his father and the loss of the spiegeleisen patent was a harbinger of the future. To the natives of Coleford who had known Mushet from his childhood, it was a continuing worry to know that "everyone was at him for money."[46]

In spite of his difficulties, Mushet had continued his experiments with the use of alloys in steelmaking. Between March 1859 and December 1861, he took out thirteen patents having to do with improvements in the manufacture of iron and steel, and in all of them the use of titanium was indicated.[47]

Thinking that perhaps he could improve his situation by raising capital and bringing in a number of partners, Mushet organized the Titanic Steel & Iron Company in October 1862. The name derived from Mushet's use of titanium in the pig metal rather than from any hopeful reference to the size of the enterprise.[48]

Besides the patents covering the use of titanium, Mushet developed others using tungsten and chromium. L.T.C. Rolt, in his 1967 history of the machine tool industry describes the final important tungsten experiment.

> In the crucial experiment an iron pig cast from an ore rich in manganese was pulverized and to it was added a proportion of wolfram ore (tungsten) finely powdered. The mixture was then placed in a crucible furnace and the molten mixture subsequently poured into an ingot mould. Part of this historic ingot was forged up under a tilthammer into a bar of tool steel by Mushet's faithful assistant George Hancox and it was speedily apparent that they had made a discovery of the first importance, exceeding their most optimistic hopes. Not only was the cutting performance of the new alloy steel far superior to that of carbon steel but, unlike the latter, it proved to be self-hardening, a characteristic Mushet had never conceived possible. Whereas the carbon steel had to be hardened by heating and then quenching in water, the success of the result being dependent on the practical experience and skill of the tool-smith, this new material hardened itself correctly, simply by being left to cool in air after forging.[49]

For several years thereafter Titanic Steel prospered, but in 1871 the company experienced financial difficulties that could not be overcome. When it ceased operation, one of the partners went to Samuel Osborn of the Clyde Steel and Iron Works and offered him sole rights to the new steel in return for which Mushet would receive a royalty on each ton produced. Thereafter, the steel was produced and marketed by the Osborn Company of Sheffield. According to Rolt, "Its superiority as a tool steel was so great than its success was immediate and world wide."[50]

Robert Mushet had battled for years with Bessemer, but in later years the two settled their differences. Mushet spoke gratefully of the money that Bessemer had granted him and looked upon it as an acknowledgement of his claim for recognition. Mushet denied that he felt he was a "re-nowned metallurgist" or an "illustrious inventor." "I merely supplied the rudder to the Bessemer ship," he wrote in a letter, "and a rudder is indispensable, no matter how otherwise complete the ship may be, and in this instance it was truly a magnificent barque—all but the rudder."[51]

In 1873 Bessemer donated the sum of £400 to the British Iron and Steel Institute to establish an annual award for the year's most important contribution to the steel industry. The award, a gold medal, became one of the most coveted in the steel industry. In 1876 Mushet's yearning for recognition was satisfied when he was selected to receive the third Bessemer medal. At the awards ceremony in London the president of the institute, the distinguished William Menelaus, introduced the resolution to honor Robert Mushet.

> Mr. Mushet is the son of a man who was one of the first to apply scientific research to the ordinary operations of iron works, but his great work was the discovery of the Black-band iron-stone, a discovery which founded one of the greatest of our national industries ... the Scottish iron trade. The elder Mushet extended his researches into the manufacture and qualities of steel and I consider that the papers he published, at a time when very little was understood about the metal, threw a great deal of light on the subject.
>
> The son, Robert, walked in the steps of his father, and pursued the same investigations that his father, through a long life had been engaged upon in connection with the manufacture of iron, and more particularly of steel, trying to cheapen its production and improve its quality. It was needless to enquire very particularly what success attended Mr. Mushet's attempt to improve old processes, because they were all overshadowed by the beautiful invention of the spiegeleisen process, as applied to Mr. Bessemer's invention, and it was on that ground that the Council resolved to pay Mr. Mushet the compliment that they then did. ...
>
> It was an invention which was worthy of being associated with the great invention of Mr. Bessemer. The two inventions would go down together, in fact, the one was the complement of the other, and he thought he was right in saying, that no man would be better pleased than his friend Mr. Bessemer, that the Council had resolved to pay that compliment to Mr. Mushet. ... It had made the invention of Mr. Bessemer perfect and probably would be used in England as long as Bessemer metal was made.[52]

The man who seconded the resolution by Menelaus was Bessemer. In addition to seconding it, he stated that he believed that most of the gentlemen knew that in the earliest progress of his invention, differences of an unpleasant character had arisen between Mr. Mushet and himself. These differences were as much a source of regret to Mushet as to himself. There was no doubt that Mushet's invention had supplemented his and that it was a most useful and valuable one. "There were reasons," Bessemer concluded, "why I did not acknowledge the validity of Mr. Mushet's patent. I will not go into the subject. Mr. Mushet has never pressed the question and I may say that for many years past both I and

Mr. Mushet have buried the hatchet and I hope and believe that we entertain the most natural respect for each other. I am happy to know that Mr. Mushet has that medal presented to him and think he very richly deserves it."

Ill health prevented Mushet from attending, and thus he could not savor the pleasure of recognition and the applause of his peers. "It will be impossible for me to be in London on the 19th inst.," he wrote the institute, "in fact, I may probably not be then alive, in which event I trust the Council will present the medal to my widow. I am very grateful to the Council for the honour they have conferred upon me, and I shall be greatly obliged if you will kindly give me the names of the gentlemen forming the Council and the President of the present year."[53]

Mushet was to live another fifteen years, but his health continued poor, and he seldom left Cheltenham. At his death in 1891 in his eight-ieth year, he had lived to see Bessemer become a very rich man, honored by the entire world, and made an honorary citizen of London. Cities were given his name, while the name Mushet scarcely was remembered.

NOTES

1. Bernhard C. Korn, "Eber Brock Ward, Pathfinder of American Industry" (Ph.D. diss., Marquette University, 1942), pp. 17-18.
2. E.M.S. Stewart, "Incidents in the Life of Mr. Eber Ward, Father of Captain E. B. Ward of Steamboat Fame as Related to Mrs. E.M.S. Stewart in the Summer of 1852," *Report of the Pioneer Society of the State of Michigan* 6 (1883): 471-73.
3. Korn, "Eber Brock Ward," p. 136.
4. Harlan Hatcher, *Lake Erie* (New York: Bobbs-Merrill Co., 1945), p. 183.
5. Korn, "Eber Brock Ward," pp. 119, 124-25, 149.
6. *The National Cyclopaedia of American Biography*, 57 vols. to date (New York: James T. White & Co., 1896-), 6:190.
7. *Bulletin of the American Iron and Steel Association* 31 (September 1896): 195.
8. Fred M. Osborn, *The Story of the Mushets* (London: Thomas Nelson and Sons, 1952), p. 4.
9. The Carron Company turned out the short naval guns known as carronades used with such success by the Duke of Wellington.
10. Osborn, *Story of the Mushets*, pp. 7-8. For some years Mushet contributed articles on his various discoveries to the *Philosophical Magazine*. Half a century later, their importance was realized when David Mushet collected these papers and published them as *Papers on Iron and Steel* (London: John Weale, 1840).
11. Osborn, *Story of the Mushets*, pp. 14, 17, 18.
12. Samuel Smiles, *Industrial Biography: Iron Workers and Tool Makers* (London: John Murray, 1863), p. 16.
13. Osborn, *Story of the Mushets*, pp. 19, 20, 21.
14. Smiles, *Industrial Biography*, p. 28.
15. Osborn, *Story of the Mushets*, pp. 24, 27, 33, 123. Mushet was proud that he had served the period of one year and a day, and it was common talk about the forest that

he had been the first "gentleman" to do so in more than a century. This right gave him the privilege of leasing iron mines directly instead of through an intermediary as his father had been obliged to do.

16. Osborn, *Story of the Mushets*, pp. 57-58.

17. Ibid., pp. 33-34.

18. John Percy, *Metallurgy: Iron and Steel*, (London: John Murray, 1864), p. 840.

19. J. S. Jeans, *Steel: Its History, Manufacture, Properties, and Uses* (London: E. and F. N. Spon, 1880), p. 29.

20. Percy, *Metallurgy*, pp. 845-46. Percy, who was close to the situation, quotes the testimony on behalf of Heath given by Charles Atkinson, a leading manufacturer of steel in Sheffield for more than thirty years.

One thing seems undoubted, namely, that the use of carburet of manganese, or of a mixture of oxide of manganese and carbonaceous matter did not come into use at Sheffield until after the publication of Heath's specification; and that is, assuredly, a strong argument in support of its claim to novelty. And to me it appears also clear, that those who professed to have previously employed oxide of manganese and carbon, could only have done so in a very inefficient manner, seeing that intimate admixture is absolutely essential to the success of the process, a condition to which they did not even make a passing allusion. Heath, on the contrary, prepared a most intimate mixture with coal-tar, which was afterwards dried and hardened into lumps by heating it in close vessels. If the mixture be not intimate, rapid corrosion of the pots will be the result. Few persons, if any, have had more experience than myself in the production of metallic manganese or its alloys. .... I have studied the report of Heath's case with great attention, and deeply reflected on the evidence; and the conclusion at which I have arrived is, that if any man ever deserved a patent, Heath was that man. Of course, I pretend to offer no opinion as to the law of the case, but judge simply according to my metallurgical knowledge, and the principles of what I believe to be common sense.

21. Osborn, *Story of the Mushets*, p. 35; pp. 137-38. This material was so little known in England that at the port of Hull it was described by the customs office as crude spelter or zinc.

22. Ibid., p. 41.

23. Ibid., pp. 138-40 passim; p. 144.

24. Ibid., p. 85.

25. Ibid., p. 45.

26. Ibid., pp. 45-46. Before the specification had been filed, the company requested a London chemist's opinion as to the soundness of the idea. The gentleman pontifically announced that, on a large scale, it was quite impossible for the spiegeleisen to mix properly with the Bessemer metal and that Mushet's patent claim was quite worthless.

27. Osborn, *Story of the Mushets*, pp. 46-47. Reprint of Mushet's pamphlet *The Bessemer-Mushet Process*, pp. 136-154.

28. *Engineer* 22 (21 September 1866): 220. It is of interest to know that as early as 3 January 1857, Mushet also had written to George Thomas Clark of the Dowlais Iron Company.

I see by the Newspaper that the purifying process has at Dowlais failed to produce the effect which was I suppose anticipated. The cause of failure is obvious to me, and the remedy simple. If you take an interest in the process I can readily convince you that it may be perfected without trouble or expense [sic]. I have made Iron stronger and tougher than any yet known by my improved processes. My health does not permit me to travel, but if you think it worth while, and could pay me a visit, I have no doubt you would consider that your time had been well employed. The purifying process is a grand discovery, but it wants a further step to make it all it was once supposed to be.

I can supply the want. (Madelaine Elsas, ed., *Iron in the Making: Dowlais Iron Company Letters, 1782-1860* [Published jointly by the County Records Committee on the Glamorgan Quarter Sessions and County Council and Guest Keen Iron and Steel Company Limited, 1960], p. 194. There is no reply to Robert Mushet in the collection).
29. Osborn, *Story of the Mushets*, pp. 47-48.
30. Henry Bessemer, *Sir Henry Bessemer: An Autobiography* (London: Offices of *Engineering*, 1905), p. 281.
31. Andrew Ure, *Supplement, Dictionary of Arts, Manufactures, and Mines* (London: John Murray, 1846).
32. Bessemer, *Autobiography*, pp. 281, 290.
33. Percy, *Metallurgy: Iron and Steel*, p. 814.
34. Bessemer, *Autobiography*, pp. 296-97.
35. Ibid., pp. 300, 302.
36. Zerah Colburn, "The Origin and Principles of the Bessemer Process," *Engineer* 18 (30 December 1864): 405.
37. *Engineer* 12 (4 October 1861): 208.
38. Colburn, "Origin and Principles of the Bessemer Process," p. 406.
39. *Engineer* 19 (6 January 1865): 6.
40. Bessemer, *Autobiography*, pp. 293-95.
41. *Engineer* 22 (31 August 1866): 161-62.
42. Ibid.
43. *Engineer* 22 (21 September 1866): 221. The editor of the *Engineer* concurred with Bessemer's statement when he noted that the question in dispute between Mr. Bessemer and Mr. Mushet was so strictly personal that the discussions between the two disputants no longer would be continued in the magazine.
44. W. T. Jeans, *The Creators of the Age of Steel* (New York: Charles Scribner's Sons, 1884), p. 128.
45. *Bulletin of the American Iron and Steel Association* 3 (2 December 1868): 98.
46. Osborn, *Story of the Mushets*, p. 64.
47. Percy, *Metallurgy*, p. 165.
48. Osborn, *Story of the Mushets*, pp. 64-65. It was the difficulty encountered by Mushet in operating the Titanic Steel & Iron Company that caused his daughter to ask Bessemer for money. Financial chaos was raging around her father and she hoped at least to save their home. The company did survive the crisis and eventually began to produce limited amounts of good steel for English railways.
49. L. T. C. Rolt, *A Short History of Machine Tools* (Cambridge, Mass.: M.I.T. Press, 1967), p. 195.
50. Ibid. "It is merely a coincidence that the Clyde Iron Works to which David Mushet went as a young man in 1792, bears a name similar to that adopted by Samuel Osborn in 1852 for his works when he started in the steel trade" (Osborn, *Story of the Mushets*, p. 80).
51. Osborn, *Story of the Mushets*, p. 146.
52. *Journal of the Iron and Steel Institute* (London) 29 (March 1876): 2-4.
53. Ibid.

XII · Holley's Mission for Edwin Stevens

hen Alexander Holley sailed for Europe in August of 1862 to study ordnance and armament for Edwin Stevens, Bessemer's plant at Sheffield was on its way to becoming a profitable enterprise. Even though his process was being used extensively, Bessemer still was being harried by the Ebbw Vale management and by other detractors. Mushet was in the process of organizing the Titanic Steel and Iron Company Ltd., which would become a reality the following October.

Since the attack on Fort Sumter in April 1861, the United States had been in the bitter throes of the Civil War. Both sides now understood the futilty of their hopes for a war of short duration. The unfinished *Battery* lying in her drydock at Hoboken became to Stevens an increasingly important reason for Holley's mission.

Holley had repaired his relations with his father, and on three days' notice, he made the necessary preparations to sail with his son. The trip was a source of great pleasure to them both. An incident that occurred after they had been at sea several days exemplifies young Holley's thoughtfulness and warmth of feeling for his family and friends. As his father basked in the sun on the ship's deck, Holley approached with a packet of letters. "I believe that you are the only person getting mail this morning," he said as he handed the older man the small package. The elder Holley opened the letters and realized at once that the day was his birthday. Although he had been very busy before their departure, the son had remembered the anniversary and had gathered greetings from family, friends, and neighbors. His father was exceedingly touched by this gesture of thoughtfulness. [1]

The two were a handsome pair, and each had a share of the Holley charm. What was especially important to Holley was that his father was seeing him for the first time through the eyes of friends and professional associates. The immature, debt ridden, irresponsible young man with a tinge of the black sheep had become a poised, confident, and well groomed young man. He was not only accepted but also received eagerly by men of prominence, who were captivated by his wit and conversational gifts. The father observed that the leading engineers of England showed keen interest in what his son had to say and anxiously sought his advice on technical matters. Holley's father always had loved this child of his first marriage but it had required all the patience he could muster to guide him through his difficulties. This was a son he never had seen before, and he was a proud man.

The elder Holley, too, found interested audiences when he pleaded the Northern cause at the numerous dinner parties and social gatherings they attended. Confederate propaganda had been effective in England and the Holleys encountered decided sympathy for the South. The father spoke freely and forthrightly whenever possible, and frequently he came away from gatherings convinced that he had made converts to the Northern cause.[2]

The commission from Stevens proved to be the turning point in young Holley's life. The Stevens name was held in high esteem in Europe, where the *Battery* seemed better known and more valued than in America. Up to this time, Holley had accepted any commission or employment that came his way. None had been of any duration, and his chief purpose in accepting them at all had been his constant need for money. Now, as Stevens's personal emissary, he found himself in a high position such as he never had known before. To execute the commission would not be easy. He would need his sharp reporter's eye to observe; he would use his knowledge of engineering to understand, weigh, and evaluate what he observed; and he would call on his writing skill to explain it all. Not the least useful of his talents would be that ineffable charm and the enormous capacity for friendship that were to open many doors.

With the aid of numerous friends and acquaintances and his own familiarity with Europe, Holley ranged widely in his search for information. Scott Russell, an English friend who was important in naval and military circles, helped by providing introductions to the proper authorities. Holley studied official documents, monographs, and unpublished records and made frequent trips to inspect ironclads under construction. Often he was not welcome officially but was given information privately.

Rossiter W. Raymond, his close friend, relates an incident that took place when Holley wanted to study the lines of a new British ironclad. One of the contractors, who had been first an acquaintance and had become a friend, had said, "If you can manage to get into the yard, I will show you all you want to see. But I am powerless to procure admission for you, and I am sure it will be refused if you ask it."

"Holley hired a fashionable carriage," Raymond relates, "arranged himself in solitary state with folded arms in the back seat, gave the necessary instructions to the coachman and drove straight through the big gate into the yard, acknowledging with a bow the present arms of the guard, as proudly as any Lord of the Admiralty. Once inside he found his friend, and satisfied his curiosity."[3]

Holley's notebook grew as he conscientiously noted everything he saw or read. At the same time, he was sending a stream of material to the *New-York Times* as well as contributing special articles to the *National Almanac* of 1863 and the *Atlantic Monthly*.[4] All this material was to appear three years later in the famous *A Treatise on Ordnance and Armor,* a large work of some 900 pages and 493 illustrations, that would become a sourcebook of steelmaking practice, unique at the time. Holley begins with a description of the standard guns then in use and then describes the process of their manufacture. The first chapter includes a detailed analysis of hooped guns such as the Armstrong, Whitworth, Blakely, and Parrott; solid wrought iron guns; solid steel guns, such as Krupp was making at Essen in Germany; and those of cast iron, such as the Rodman and Dahlgren guns. Then, in orderly progression, sections follow that cover the requirements of guns; strains and structure, taking into consideration their resistance to elastic pressure and the effects of vibration and heat; cannon metals and their fabrication; and rifling, projectiles, and breech loading. The second part of the book deals with accounts of the various experiments that had been conducted on armor plate, behavior of guns while being fired, and the application of gun cotton to purposes of warfare.[5]

James Dredge quotes from a letter written by Mr. G. Canet, one of the ablest of contemporary designers of artillery.

You have had frequent occasion to study the *Ordnance and Armor* of A. L. Holley. You could not refer to a better authority; for it is at once the most complete and the most conscientious work on the history of artillery that has ever been published. The name of Holley recalls to me many souvenirs of the commencement of my career, and I want to tell you that for me no book on the same subject has ever been of so much value. Holley wrote it at a time before artillery had become a science, properly so-called; when even the phenomena attending the combustion of gunpowder were very imperfectly understood. At that time the strains to which guncarriages were subjected were not accurately known, and correct gun design was impossible, because the pressures exerted upon the bore of a cannon could not be measured. When Holley's book appeared there existed only a few theoretical essays, the records of some doubtful experiments, the ill-sustained claims of a few manufacturers, and these were all scattered through different memoirs and reports. Holley was the first to make a systematic summary of all these documents; to pass in able review material, modes of construction, ballistical data, firing experiments, and penetration of plates. This last-named portion of his work had an especial value, for when he wrote the art of constructing ironclad ships was in its infancy. What I find most remarkable in this work of Holley is, that writing at the period from which modern artillery may be said to date, he made all the data he could obtain the object of minute and careful study; that every line bears the stamp of his own work; that it is an earnest study, and in no sense a piece of bookmaking. Every page contains abundance of detail, accuracy in facts, and a sureness of judgment which is truly admirable. His considerations on the resistance and elasticity of material, and on the effects of strains and vibrations, show the large grasp of a truly scientific mind. . . . [6]

Although outdated by the mid 1880s, Holley's book remains an important record of the state of ordnance and armament manufacturing in the 1860s.[7]

It is not possible to reconstruct the itinerary of Holley's European trip. It is certain, however, that he found his way to Sheffield and ultimately to Bessemer's plant. He would have had numerous reasons for doing so. The Bessemer-Mushet controversy, which was being aired in the journals, may have revived his interest in the process about which he himself had written six years earlier. Then, too, he reasonably could have expected to be able to observe the process at first hand since part of the agreement between Bessemer and William D. Allen, his brother-in-law, stipulated that the Bessemer works be available for demonstrations of the process to potential licensees. This was also the year of the Great Exhibition, which displayed numerous iron and steel products ranging from heavy ordnance to fine wire produced both in England and many other countries, and which included a demonstration of the use of Bessemer steel in ordnance.[8]

Whatever reasons led Holley to visit the Bessemer plant, the process in operation impressed him deeply. The converters erupting with their roaring cascades of dazzling sparks always affected spectators, and Holley proved to be no exception.

> The wonderful success and spread of the Bessemer process in England, France, Prussia, Belgium, Sweden and even in India all within three or four years prove that great talent and capital are already concentrated on this subject and promise the most favorable results. . . . The advantage of steel over iron in its more crude forms is that the number and quantity of its ingredients are better known at each stage of its refinement. . . . The new treatment of iron is based on chemical laws. The old treatment was a matter of tradition, trial, failure and guesswork. The Bessemer process is a chemical process—suggested by the study of chemical laws, conducted on chemical principles and prosecuted, modified and improved, according to the results of chemical analyses.[9]

Holley returned to the United States in 1863. His notebooks bulged with information about European practices in the manufacture of armor and ordnance. On this he based his report to Stevens.[10] But he could not forget the erupting converters in Sheffield, and he searched for a company that he could interest in securing the American rights to the process.

In his search, according to H.F.G. Porter, Holley visited Captain John Ericsson, designer of the renowned *Monitor*, to tell him about his trip.[11] Ericsson was associated with John A. Griswold and John Flack Winslow, two iron producers in Troy, New York, in designing and building additional ironclads of various sizes. Holley's visit may have taken place at about the same time that Ericsson received a letter dated 28 February 1863, from Assistant Secretary of the Navy, G. V. Fox. Fox was in direct charge of arrangements for constructing new war vessels and spoke in the letter of the urgent need for a Bessemer plant in the United States.[12] When Holley, in describing his experiences abroad, mentioned visiting the Sheffield Bessemer works, Porter relates, Ericsson decided to send the young man to Troy to talk with Griswold and Winslow. The meeting was so successful that the two men sent Holley back to England to secure the exclusive American rights to the Bessemer patents. "It is a matter of regret to me, on many accounts, that I have to go to England again this summer," Holley wrote in May 1863, "but this is no pleasure trip, but a trip preparatory, I hope, to something settled. Precisely what I am going for, I am not at liberty to say, except that I am going to get information for Corning, Winslow and Company about a new manufacture. If I succeed, they are going to establish the business, which will be in nature allied to, but in a business way separate from their present manufacture. They then wish me to become a partner in the new business . . . and expect, if I succeed, to spend $30,000 in establishing the new manufacture."[13]

Troy at this period was one of the leading iron producing areas of the country. Two important companies there were the Albany Iron Works owned by Erastus Corning and John Winslow, and the Rensselaer Iron Works a mile further up the Hudson River, headed by John A. Griswold and Corning. Winslow and Griswold had been partners with John Ericsson and C. S. Busnell in constructing the *Monitor*.

For this new steelmaking venture, a separate company was formed in order to protect the established businesses of the partners from any risk that might be incurred in the gamble with a new process. Holley had thought at the time he left America that the new company formed to purchase the Bessemer rights would be Corning, Winslow and Company, but by the time he returned from England and started his work at Troy, the company name had become Winslow, Griswold and Holley. It was not a manufacturing company, however, nor did it become one. The Bessemer plant at Troy was owned and operated by the Rensselaer Iron Works. Holley's reference to a partnership offer probably indicates that he had been offered the partnership with the understanding that he would erect and operate the plant. Given the usual state of his finances, it is unlikely that he possessed enough capital to join the partnership, and there is no evidence of a contribution from his father.

NOTES

1. *Alexander Hamilton Holley Memorial Volume* (Privately circulated, 1888).
2. Ibid.

3. *Memorial of Alexander Lyman Holley* (New York: American Institute of Mining Engineers, 1884), p. 13.
4. "Iron Clad Ships and Heavy Ordnance," *Atlantic Monthly* 11 (January 1863).
5. Alexander Lyman Holley, *A Treatise on Ordnance and Armor* (New York: D. Van Nostrand, 1865), p. 396.
6. James Dredge, "Holly Memorial Address," *Transactions of the American Institute of Mining Engineers* 20 (1891):xxx.
7. Rossiter W. Raymond found the work interesting as a specimen of Holley's method of work. "His ability to grasp the leading principles of any mechanical problem; the facility with which he seized the details of an apparatus or an experiment putting them down for future reference in words or in sketches, as might be most convenient and the system with which he indexed everything he had once noted, so that he could at any moment refer to it again without delay" (*Memorial of Alexander Lyman Holley*, pp. 130-31).
8. J. S. Jeans, *Steel: Its History, Manufacture, Properties, and Uses* (London: E. and F. N. Spon, 1880), p. 86.
9. A. L. Holley, *Ordnance and Armor*, p. 396.
10. The *Battery* was destined to remain unfinished. When Edwin Stevens died in 1868, he left a million dollars for her completion but the money was either insufficient or poorly utilized. The ship was cut up and sold for scrap in 1881.
11. H.F.G. Porter, "How Bethlehem Became Armament Maker," *Iron Age*, 22 November 1922. Porter was the son of a man who was well acquainted with both Cornelius Delamater of the Delamater Iron Works, builders of the machinery for the *Monitor*, and with Ericsson.

    The visit to Ericsson may have been suggested by Henry Raymond of the *Times*, who long had been interested in mechanical discoveries. In the early 1850s, Raymond had recognized the importance of Ericsson's invention of a caloric engine designed to propel ships by hot air instead of steam, which was difficult to control and often caused explosions. The two became well acquainted. Later Raymond traveled on the trial runs of the experimental vessel propelled by Ericsson's engine and wrote and lectured extensively about it (Ernest Francis Brown, *Raymond of the Times* [New York: W. W. Norton, 1951], p. 161).
12. G. V. Fox to John Ericsson, 28 February 1863, Ericsson Papers, Library of Congress.
13. *Memorial of Alexander Lyman Holley*, p. 131. Elting E. Morison comments on the anticipated expense.

    It does not seem likely that Holley was referring to the cost of patent rights and, in any case, his figure is quite close to the contemporary cost for the construction of small plants comparable in size with the two experimental Bessemer works. Expenses, either at Troy or Wyandotte, could hardly have exceeded, counting everything, $60,000 or $70,000. And parenthetically it may be added that the only expenses for litigation, besides some travel money and a cask of whiskey, appear to be Blatchford's bill to Winslow. His legal fees from February 1, 1865, to the fall of 1866 amounted to $2594.06 (Elting E. Morison, *Men, Machines, and Modern Times* [Cambridge, Mass.: M.I.T. Press, 1966], p. 146).

XIII · The Wyandotte Endeavor

I n 1862, while Holley was traveling in Europe gathering information for Stevens, plans already were under way for testing the merits of Kelly's pneumatic process at Wyandotte, Michigan. After Zoheth Durfee left for England in the fall of 1861 to secure the American rights to the Bessemer patents, Eber Ward engaged Durfee's cousin William to start construction of a converter. After his graduation from Lawrence Scientific School at Harvard in 1853, William Durfee had been an engineer, architect, and surveyor, as well as a member of the Massachusetts State Legislature before his arrival in Michigan in July.[1]

When he began work on the Wyandotte converter, young Durfee, who was then not thirty years old, was faced almost at once with a multitude of difficulties. Although he had never seen any apparatus for the manufacture of steel as proposed in these controversial patents, it was evident immediately to the engineer that equipment such as Kelly had used in the now abandoned works in Kentucky or at Cambria would be unsuitable for operations on the scale that his cousin and Eber Ward were contemplating. William Durfee had received no detailed drawings from abroad, and not enough had been published in the United States to be of much help. He procured copies of the Bessemer patent from the patent office in Washington. He also obtained a recently published book by William Fairbairn on the manufacture of iron. Although it dealt mainly with iron, it also contained a section on the Bessemer process. Confident that his cousin would secure the rights to the use of the Bessemer designs, Durfee anticipated the purchase and incorporated in his own plans those details that seemed most suitable.

By thus jumping the gun, Durfee infringed Bessemer's rights. When the question was raised later, he honestly admitted using the patents and excused his action by pleading that there had been no doubt in either his cousin's mind or his own that the license was practically signed, sealed, and delivered, an assumption that was to lead to much embarrassment in the future.

When Zoheth Durfee returned from England later in the year, the two and one-half-ton converter was nearly finished. William Durfee was pleased and relieved when his cousin said that it looked very much like

174

the examples he had seen in use abroad. The converter was completed late that year. During the following winter Durfee started construction of the blowing engine, but because of a variety of problems, interruptions caused by the war, strikes, and the fact that he was spending a considerable portion of his time at Ward's plant in Chicago, it remained unfinished until the spring of 1864.[2]

The mechanical difficulties William Durfee had to overcome to build a converter were the least of his troubles. The new project aroused antagonism and open opposition on all sides, and, in addition, Ward did not give it his active support. Twenty years afterward, Durfee told of all that had transpired. The stupidity, ignorance, and eventual sabotage that he had suffered in his efforts to build a converter plant in the United States still rankled.[3]

Durfee was given two assistants who appear to have spent much of their waking hours either concocting inventions on which Ward could spend money or trying to convince him that Durfee was ignorant and inept. The two disapproved of the blowing engine that Durfee considered essential to the efficient operation of the process. They argued that "all that was required to convert cast-iron into steel, was the forcing of abundant oxygen through it when melted, and, as water contained a large proportion of oxygen, the substitution of steam taken directly from the boiler for atmospheric air under pressure would greatly simplify and cheapen the process." They cautioned Ward that if he spoke to Durfee about the matter, he would only condemn their idea. The experiments subsequently failed and Ward lost his investment in the scheme. Durfee considered that his devious assistants had lost their credibility as well.[4] By the end of 1863, Durfee's state of mind was at its lowest. He had not been accepted into the iron making community and, indeed, was regarded as a mild lunatic or as an idiot to be tolerated, but certainly not assisted or encouraged.[5]

Ward's sole aim was to build the experimental plant and to rush it into operation. He was a strange person. To his credit, he rarely interfered with Durfee's operation. He has been described as a man of extremes: self-controlled and passionate, shrewd and credulous, persistent yet changeable.[6] He was not an ironmaster in the true sense of the word and had little real understanding of the details of Durfee's experiments. In his anxiety to make a financial success of the venture, Ward seemed always ready to listen to any suggestion, no matter how ridiculous. He probably had no intention of creating difficulties for Durfee, but seemed unable to resist trying out a persuasive scheme, especially if it were put forward by those unfriendly to Durfee.[7]

The construction of the plant was not Durfee's only problem. Years later, he still could speak with some feeling about the Eureka blast furnace that supplied the metal for the converter. "I never fully understood," he said, "just why the blast furnace . . . was called Eureka; the only theory at all satisfying to my mind being, that some expert in blast furnace history, looking for an example of ancient practice that embodied the most faults in design, contruction and management, and being satisfied when he found this plant that further search was useless, had suggested the word as a most appropriate name for a furnace that more than satisfied his ardent desires for the discovery of the most archaic of metallurgical structures."[8]

The temperament of the furnace manager was similar to that of the monster under his direction, and, from the beginning, he opposed Durfee's ideas. In his exasperation, Durfee labeled this metallurgical authority *Herr Unkunde Unheilschwanger*, much more portentous in the German than in its English equivalent, Mr. Ignorance Pregnant with Disaster. Hot metal for the converter was furnished on contract from the Eureka furnace, and it was weighed before it was charged into the vessel. Herr Unkunde was convinced that, since iron expanded when heated, one hundred pounds of melted iron weighed more than one hundred pounds of cold pig iron. Durfee never was able to convince him otherwise.[9]

Durfee had observed that all of the cinder that remained from the puddling of the Lake Superior pig iron with the Lake Champlain magnetic ore was thrown away. Shocked by this waste, he called it to the attention of Eber Ward. The captain, eager to save money wherever possible, took up the matter with Herr Unkunde, who assured Ward that there was no iron in the cinder. Durfee, convinced that the cinder was more than 50 percent iron, proposed to prove his claim. A few days later, he had the satisfaction of placing in Ward's hand a "button" of iron that represented 55 percent of the cinder from which it had been melted.

Angry at discovering such waste, Ward immediately issued orders that the cinder was to be used in the blast furnace charge. Unkunde was alarmed that its use might result in "bunging up" his furnace and exercised unusual caution. He used 10 pounds of the cinder with each 400 pounds of Lake Superior ore. Since nothing of a catastrophic nature occurred from this homeopathic dose, his confidence gradually increased. With some doubt and fear, during the next months he slowly augmented the cinder charge until he had arrived at what to him seemed the enormous proportion of 40 pounds of cinder to 400 pounds of ore. This was the highest proportion that he could bring himself to use. After Durfee left Wyandotte, Unkunde, with a sigh of relief, returned to his practice of throwing away the cinder.[10]

Zoheth Durfee revisited England in the fall of 1863, charged with two missions. One, as he wrote in a letter of October 6, had to do with the purchase of steel rails for American railroads. "I have now on hand to purchase for the United States Government and several Western railroads orders to the amount of some $200,000. The Pennsylvania Railroad alone gave me an order for steel rails made by the new process amounting to upwards of $100,000 and Mr. Thomson who has seen the rails in use and knows their value, has just given me a letter to the President of the Northern Central Railroad which I think will bring an order from him. I shall have an order from the Fort Wayne and Chicago and probably the Cleveland and Pittsburgh and hope to get some from the Eastern roads."[11]

At that time, the price of British steel rails delivered in the United States was $150 in gold per gross ton. Between 1863 and 1864 imports of iron and steel rails into this country, as reported by the U.S. Customs Service, increased from 20,506 net tons to 142,457 net tons. It is possible that a large part of this increase of approximately 122,000 net tons was accounted for by the Durfee purchase.[12]

Durfee's other and perhaps more important mission was to make another attempt to secure the American rights to both the Bessemer machinery and the Mushet spiegeleisen patent. He found the rights to the Mushet patent clouded. The ownership rested with a group, some of whom were not on speaking terms with others, and, as nearly as he could determine, no person legally could represent the ownership in dealing with him. For months, the patient Durfee endeavored to bring some order into the confused situation. He was able at last to reconcile the group to choose a representative to deal with him. To his shocked amazement, the Mushet representative demanded £20,000. This sum was to be paid in cash before the American rights to the spiegeleisen patent could be transferred to the Wyandotte group. The amount was obviously so preposterous, considering the status of the patent in England, that Durfee concluded he might as well give up the whole idea. On the chance that something still might be realized, however, he continued to negotiate for more than a year. During these months, the amount asked gradually was reduced until, after the year had passed, he obtained the patent for no cash advance whatsoever.[13]

At the same time, by coincidence, Griswold and Winslow had dispatched Alexander Holley to secure the rights to both the Bessemer machinery and process. After lengthy negotiations with Bessemer, Durfee was unsuccessful. Bessemer preferred to do business with Holley, who had the advantage of arriving prepared to make financial arrangements for the whole process. In addition, Bessemer may have preferred to deal with a company as substantial as the one Holley represented and one that had no connection with Kelly.

The Wyandotte partners, believing that plans were far enough along to organize officially, had formed the Kelly Pneumatic Process Company in May 1863. The members of the new company were the Cambria Iron Company; E. B. Ward; Park Brothers and Company; Lyon, Shorb and Company; and Zoheth Durfee. Years later William Durfee described the consortium.

> These gentlemen were selected because of their well-known business ability and their influential association with or ownership of some of the largest and best-appointed iron and steel works of the country, and it was confidently expected that they would take a lively interest in the new process by promptly employing it in the works with which they were identified, and that their example would be very generally followed by the larger iron and steel works of the United States. In this expectation Captain Ward and Z. S. Durfee were greatly disappointed, as neither Mr. Lyon nor Mr. Parke ever adopted the process in their works, and Mr. Morrell only succeeded in overcoming the objections of his associates in the Cambria Iron Company of which he was general manager in such time as to enable him to commence making steel eight years after he was admitted as a member of "The Kelly Process Company."[14]

Later, Charles P. Chouteau, James Harrison, and Felix Valle, all of St. Louis, were admitted to the company, but, according to Durfee, it was in no way strengthened by their connection with it. Kelly, although not a member of the company, retained an interest in any profits that might accrue from the use of the process.[15]

The group that formed the new company was more optimistic than was William Durfee. In a fog of disappointment and discouragement he continued working through 1863 and into early 1864, trying to achieve a successful blow. A miasma of ill will and distrust hung all about him as he endeavored to obtain steel from the monster that seemed a friend one day and an enemy the next. The space allotted him in the casting house of the charcoal blast furnace was so inadequate that if some of the equipment appeared oddly placed or even insufficient, it was only because there had been no place to install it properly.

Zoheth Durfee wrote from England early in 1864 to his cousin that he would not be successful in winning the rights to the use of the Bessemer patents. He suggested new machinery designs that would avoid infringement on Bessemer's mechanical patents. Since the Bessemer plans centered on a movable converter, Durfee sent some rough sketches for a stationary vessel. In Wyandotte, William Durfee immediately set to work constructing a new converter that would remain fast in its moorings. The steel would be tapped from a stationary vessel instead of poured from the tilted converter.[16] Writing of the problem, he said, "Although it would have been possible for me to make steel in a Kelly stationary converter—they were used in Sweden—it was quite evident from the first that the highly original and ingenious apparatus invented by Bessemer (especially the tilting converter and the casting ladle having a tap hole in its bottom) was far superior to anything proposed by Kelly."[17]

These interruptions delayed the completion of the blowing engine until the spring of 1864 at which time Zoheth Durfee returned from Europe, bringing with him Llewellyn Hart, who had worked with the Bessemer converter in France.[18] With Hart's arrival, William Durfee at last had an experienced person to help him. During the summer, with Hart's help, he conducted trial after trial, experimenting with various brands of pig iron. They soon realized that, if they were to continue, a reverberatory furnace would be necessary to provide the liquid metal. Along with their attempts to produce steel, they built the reverberatory furnace.

The heat of the summer matched the temperatures of the cast house where the two men were experimenting. The converter consumed its quota of molten metal repeatedly only to spew it out minutes later in a variety of ways. If the meal offered was palatable, the reaction was normal. But if the molten pig metal was not to the converter's liking, if the temperature was not satisfactory, or if the blast from the blowing engine was feeble, then the converter expressed displeasure. It might vomit the entire charge in a veritable volcanic eruption, as every man ran for cover. Or it could be overcome by a lethargy that chilled the metal into a solid mass that necessitated hacking out the solidified metal from the vessel.[19]

Just as Bessemer, from countless experiments, found reasons for the peculiar behavior of a vessel, so Durfee, laboring with less accumulated knowledge than Bessemer, also found answers. Bessemer had needed more than three years before a metal recognizable as steel had been produced. Durfee had finished the first converter in the fall of 1862, and less than two years later, on 6 September 1864, using Lake Superior charcoal pig iron, he made the first public conversion in the vessel. Daniel Morrell, James Park, and William Lyons, who shared an interest in the Kelly Pneumatic Process Company, came to Wyandotte to witness the important event.[20]

Although Durfee and the workmen had conducted many trial conversions successfully, they all knew that the converter might fail at a critical moment. Thus, all were nervous and tense. The visitors, although they did not share that knowledge, were aware of the importance of the occasion as they studied the converter and watched the preparations being made for the blow. Ward had little to say, but as he waited he slowly shifted silver dollars from one hand to the other — always a certain sign with him of either nervousness or rising temper.

As the group watched, the workmen slowly tipped the converter on

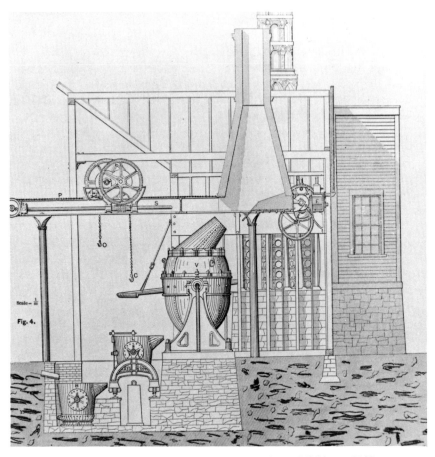

Section of converter house of steel works at Wyandotte, Michigan, 1863.

"This is a view of the machinery as it appeared to a person standing in the pulpit, where the control mechanism was located, and looking toward the converter, V.

"The traveling crane supplied the converter with metal taken direct from the blast furnace. On the transverse shaft nearest the converter were placed two chain drums provided with hook-ended chains of equal length, seen at C. The other transverse shaft was furnished with a single chain drum which occupied a position midway between those on the first transverse shaft; this drum was furnished with a single hook-ended chain, seen at O.

"The traveling carriage was drawn in either direction along the girders by a pair of pitch chains, one of which is shown at P. These passed over rag wheels, R, on transverse shafts which turned in bearings at the end of the girders.

"The whole of the mechanism of the crane was actuated by three small quick working rotary engines, one of which is seen at E. Each of these engines was capable of prompt reversal by means of a rocking shaft, on whose end was a weighted lever, connected by a small wire cord, R, to one of the latch levers of the pulpit. There was a fourth engine of the same description employed for turning the converter, which was controlled by a

William F. Durfee. From *Cassier's Magazine* 8 (May 1895).

its axis and the vessel received its charge of molten metal. Then it was
swung back to its upright position, the blast turned on, and almost
immediately the familiar roar could be heard. At the proper moment,
as the men stood motionless and silent, the brilliant flame shot from
the top, and the process began its familiar cycle. When the flame reached
the proper color and the turmoil within the converter subsided, the
vessel slowly was turned on its trunions, the blast shut off, and the
carburizer run in. A violent reaction quickly took place within the vessel,
and when all was quiet, the molten metal was poured into the waiting

---

lever at the pulpit, as was also the valve which regulated the admission of blast to the
same.

"The lifting of the fluid cast iron delivered by a runner from the blast furnace into the
hoisting ladle *H* placed upon the platform of the scales *A*, [missing from drawing] and
the pouring of it into the converter, was accomplished as follows:

"The carriage of the crane was run to the left-hand end of the girders on which it was
placed, and the hooks on the double chain *C* were attached to the trunnions of the hoisting ladle, and the hook on the single chain was connected with the eye seen on the left-
hand side of its bottom. The converter having been turned into a proper position to
receive the metal, the hoisting gear was put in action and the ladle *H* with its contents
was raised and carried to the right until the end of the spout was over the mouth of the
converter. The raising of the ladle was then stopped, and it was turned upon its trunnions by continuing the hoisting in the chain *O*, until all the metal which it contained
had been discharged into the converter" (William F. Durfee, "An Account of the Experimental Steel Works at Wyandotte, Michigan," *Transactions of the American Society
of Mechanical Engineers* 6 [1885]: foll. pp. 46–48).

molds. Whether or not the watching men realized it, during that half-hour, an historic event had taken place. Under the direction of William Durfee in this experimental converter at Wyandotte, Michigan, they had seen the first steel produced in America by the pneumatic process.[21]

Now that the long period of frustration, discouragement, and hard work was over, Durfee was exultant. In high spirits he bounced from one workman to another, slapping each one on the back in his exuberance. As for the men who had shared the difficult times with Durfee, hardly waiting to gain permission, they made a rush for the door and headed for the nearby town of Sebastopol where the event could be celebrated properly.[22]

To establish the ductility and working qualities of his steel, Durfee had sent some samples of it to a friend in Bridgewater, Massachusetts. There it was rolled into tack plate and then cut into tacks that definitely were superior to the iron variety then in use.[23]

Ward hardly could wait until the steel ingots were rolled into rails. They would be the first all-steel rails produced in America. But the boss roller at the rolling mill was hesitant. He was sure that the steel would prove too hard for the rolls, which had been designed to handle iron rails, and pointed out the difficulties that would be encountered.

Captain Eber B. Ward. From Herbert N. Casson, *The Romance of Steel* (New York: A. S. Barnes & Co., 1907), opp. p. 16.

The superintendent agreed with him. It was a rare person who could muster the courage to question one of Ward's orders, but the roller knew the limitations of his machinery and saw no reason to damage it deliberately.

Ward refused to listen to either the boss roller or the superintendent and, in a voice that could be heard from one end of the mill to the other, roared, "I say, roll the steel or I'll make a goosepatch out of Wyandotte." The men were left with no choice. The ingots were heated, and the first one was started through the rolls and slowly worked down. The rolls were then set slightly closer together and the elongated ingot was started through again. This time, in spite of all precautions, a roll broke. Ward eyed the men and the broken roll for a second and without a word, turned on his heel and stalked off.[24]

Although Zoheth Durfee lost out to Alexander Holley in his bid for the rights to the Bessemer machinery, he played a trump card in seeking out Robert Mushet.[25] Durfee procured a license from the Mushet group on 24 October 1864 to use patent number 17,389, dated 26 May 1857. Mushet's share of the American patent over the years had dwindled to one-sixteenth of one-thirty-second share.[26] The terms of the agreement were such that Mushet, Thomas Clare, and John Brown were admitted to the Kelly Pneumatic Process Company. A year later they were joined by Charles P. Chouteau, James Harrison, and Felix Valle, all owners and operators of iron works in St. Louis. The new group proved of little help in furthering the use of the process either there or anywhere else.[27]

In constructing the converter, Durfee had provided certain safety measures that could be followed if the process were to fail. An elliptical well or reservoir filled with water beneath the rear platform of the structure was to receive any steel that might remain fluid in the teeming ladle in case the taphole chilled. With the reservoir full, the operation was considered safe because, if necessary, two or three tons of fluid steel could be turned into the well. The steel in such instances formed small shot-like balls or masses of a size that could be remelted and reused.

On one occasion the reservoir had not been filled with water; the consequence was an incident that those who were present never would forget. Durfee had gone to Chicago leaving his assistant with instructions to undertake certain lining repairs and not to operate the converter in his absence. Upon his return to Wyandotte, Durfee heard a report that had all the elements of tragicomedy. Ward had brought a group of distinguished men to Wyandotte on one of his steamers. Hoping to impress his guests with a demonstration of the only steelmaking converter in America, he ordered the assistant to operate the converter. Unable to refuse Ward, the unfortunate man undertook the demonstration. He was extremely nervous because of the presence of several senators, judges, bankers, and leading merchants, and not only forgot to check the reservoir but also succeeded in chilling the taphole of the ladle after pouring only two ingots. He ordered the ladle emptied into the reservoir, which contained only a small amount of water. As two tons of molten metal were poured into perhaps a barrel of water, a steam-induced explosion took place. Tough old Senator Zachariah Chandler of Michigan was projected at full length into the pig bed; crotchety "Bluff Ben" Wade, senator from Ohio, was hurled upon a pile of sand in the casting house; while Ward was blown bodily through the open door of the building into the yard and came to rest on a pile of pig iron. Others in the group were hit by flying pieces of metal, although no one was hurt seriously.[28]

When William Durfee arrived in Wyandotte in 1862, he had been faced with the problem of building apparatus he never had seen in order to produce steel. He had scant information concerning either. As he studied the theory of the process, he came to the conclusion that accurate knowledge of the chemical constituents of the metals and raw materials used in the converter was essential. All pig irons did not necessarily make good wrought iron, so it seemed reasonable to Durfee to assume that good steel could not be produced from random pig irons about which little or nothing was known. Durfee believed that by making a chemical analysis of those pig irons that had proved most successful he could establish a guide that in the future would save time and prevent spoiled heats.

More than a year before the Kelly company officially was organized, William Durfee proposed the construction of an analytical laboratory. His cousin, who was acting as secretary of the new group, agreed and approved the plan. In the spring of 1863, Emile Schalk, a German chemist who had been trained in Paris, was hired. Schalk purchased the necessary chemicals and other equipment to set up a laboratory. Since the building had not been completed yet, the chemist was sent with an exploring party to northern Wisconsin in search of suitable iron ore deposits. When he returned the laboratory was ready, and Schalk began analyzing samples of the various iron ores that Durfee intended to use in the converter operations. He also began investigating the effect of nitrogen on steel. Durfee believed that these studies were especially promising and might provide valuable information. However, for reasons that Durfee never explained, except to say that Schalk was in no way responsible, the chemist resigned in December of the same year before he was able to carry out any of his scheduled analyses.

Under Durfee's direction, and against enormous odds, the work continued. The attitude of his fellow workers toward the chemical laboratory, which they called the "apothecary shop," was one of distrust. The sight of Durfee conducting experiments on materials for converter linings, iron ores, or coal quality provoked Durfee's enemies to great lengths in their efforts to discredit his work.

In order to obtain a uniform high temperature for his experiments, Durfee used an oxy-hydrogen blowpipe. One morning Durfee found that after a short period of use, the blowpipe flame went out and could not be relighted. Examining it, he discovered that an auxiliary waterpipe had been disconnected and that a tight-fitting wooden plug had been driven into it. The pipe had been recoupled so that everything appeared to be in working order. Durfee never could decide whether "mischief or murder" had been contemplated.[29]

At the time public attitudes toward chemistry were intensely skeptical. As late as 1872, the general manager of an important iron works wrote to J. C. Bayles, editor of *the Iron Trade Review:*

> The president of our company thinks we ought to follow the fashion and have a chemist. To my mind it is a waste of money. When I want an analysis I can have it made—and that is very seldom; for the furnace manager who needs a chemist to tell him the quality of ore or limestone, or whether his pig-iron is soft or hard, had better resign and go to farming. However, if the president says chemist, chemist it is. My object in writing is to know if you can recommend a young man competent to fit up a laboratory and take charge of it. We have very little society here and it is desirable that he should be a gentleman. My wife plays the piano and I do a little on the flute; and if we can get a chemist who plays the violin, we could have some music evenings. If you can suggest a man who combines these qualifications, I could employ him. I do not know what a chemist would expect; but I should not care to pay more than $10 a week.[30]

In January of 1865, Durfee returned from a short trip to find that the "apothecary shop" was no more. Not a test tube remained in the bare room. In describing the act of vandalism Durfee wrote, "I manifested no surprise; of it I made no complaint; but then and there I mentally resolved that as soon as the first rail was rolled from steel made at Wyandotte, I would leave a community which had afforded me so many painful illustrations of the potential verity of the lines of Gray: 'Where ignorance is bliss Tis folly to be wise.'"[31]

A letter that he received from a trusted friend at about the same time minced no words about Durfee's position at Wyandotte. "I am pleased to hear from you again and yet am sorry: for I know what your feelings would have been upon returning home; as I was unfortunate enough to be at Wyandotte at the time the raid was being made upon your office and laboratory, I saw things at Detroit in which Wyandotte men were concerned that sunk them in my opinion below the most contemptible of our race ... Nothing in your vicinity in writing is safe from the perusal from anyone who wished to read and anything you don't care to have pirated, destroy."[32]

Ward had not forgotten his experience at Wyandotte when he ordered the rolling of the steel rails, but he realized that it would be futile to pursue the matter with the mill there. He was determined that at the semiannual meeting of the American Iron and Steel Association in May 1865, he would have at least one steel rail for the members to admire. What is more, he intended that they should see this rail produced. Accordingly, Ward asked Orrin Potter, who was in charge of the Chicago mill, if his equipment could produce a steel rail. Potter, a relative, was well aware of Ward's temper. He assured Ward that indeed it was possible to roll the steel ingots sent to his mill from Wyandotte. Having made what seemed a rash promise, Potter reinforced and strengthened the rolls to the best of his ability.

Ward, taking no chances, arranged for a preliminary rolling before he invited his guests. Only George Fritz and Zoheth Durfee were on hand. Convinced that the results would be no more successful than before, William Durfee refused to attend. The trial rolling of three rails was so successful that the following day the ironmasters formally were invited to watch the remaining ingots rolled into steel rails. Again Potter's good fortune held, and the steel rails came through the mill sound and well shaped.

Potter regretted William Durfee's absence and wrote him in some detail of the success. The letter obviously meant a great deal to Durfee. Twenty years later he included it in a paper that he read before a session of the American Society for Mechanical Engineers.

> My Dear Durfee:
>
> The meeting of the Iron and Steel men adjourned yesterday to meet in Cleveland the fourth Wednesday in August. I regret very much that you could not have been here, particularly to see how well your steel behaved; and you must allow me to congratulate you upon its entire success. I assure you I was but too proud for your sake that everything we had to do with it proved so very successful. The hammer was altogether too light, of course, and it took more time than it otherwise would to draw the ingots down; yet all the pieces worked beautifully, and they have made six good rails from the ingots sent over, and not one bad one in any respect. The piece you sent over forged is now lying in state in the Tremont House, and is really a beautiful rail, and has been presented to the Sanitary Fair by Capt. Ward. We rolled three rails on Wednesday and three on Thursday. At the first rolling only your cousin and Geo. Fritz were present, at the rolling yesterday were Senator Howe, of Wisconsin, D. F. Jones, of Pittsburgh, R. H. Lamborn, of Philadelphia, Mr. Phillips, of Cincinnati, Mr. Kennedy, of Cincinnati, Mr. Swift, of Cincinnati, Mr. May, of Milwaukee, and three ladies, Mr. Scofield, of Milwaukee, Mr. Fritz, of Johnstown, Mr. Thomas, of Indianapolis, with four

strangers, and everything went so well I really wanted you to see some of the good of your labors for so long a time and under such trying circumstances. You have done what you set out to do, and done it well, and I am glad to congratulate you and rejoice with you, for I can appreciate some of your difficulties, and wanted you to hear some of the praises bestowed upon your labors, as you richly deserve. I know this would make no sort of difference to you, yet we all have vanity enough (especially in such cases as this) to feel gratified at any little compliments we know we are entitled to. But I will not tire you with any more, as your cousin can tell you all and more than I can write, but with kindest regards, allow me to remain, Your most ob't, O. W. POTTER.[33]

Once again the Bessemer process had worked, but only after it had sapped the strength of the man bold enough to make the effort. Durfee demonstrated without a doubt that the process could be carried out. Now, instead of following the original plans, which were to build a substantial plant if the experimental converter proved successful, Ward decided to install a second converter in the makeshift works. This decision, coupled with the generally unpleasant atmosphere, was more than Durfee could accept. Worn and exhausted, he left Wyandotte on 1 June 1865, glad to get away from the meanness and diabolical temperament displayed by what he called the "syndicate of sin."[34]

With his striking personality and intellectual standing, Durfee held numerous positions of responsibility during the remainder of his life and also was a frequent contributor to professional journals. In later years, he often was engaged as an expert witness in patent cases and more than once, when asked to recommend an authority, Holley turned to Durfee as a person to trust for the thoroughness and intelligence that he brought to his work. Durfee never again worked with the Bessemer process and, for that reason, he failed to achieve the recognition that he so amply deserved for his pioneering efforts.

NOTES

1. "Death of W. F. Durfee," *Bulletin of the American Iron and Steel Association, 33* (1 December 1899):203.
2. William Fairbairn, *Iron: Its History, Properties, and Processes of Manufacture* (Edinburgh: Adams and Charles Black, 1861); William F. Durfee, "An Account of the Experimental Steel Works at Wyandotte, Michigan," *Transactions of the American Society of Mechanical Engineers* 6 (1885): 41-42.
3. William F. Durfee, "An Account of a Chemical Laboratory Erected at Wyandotte, Michigan, in the Year 1863," *Transactions of the American Institute of Mining Engineers* 12 (1883-84): 224. Durfee recalled that one influential individual with whom he had to deal literally had sixteenth-century ideas. He was convinced that the world

was flat. "For if it was round," the man insisted, "the Detroit River would be running up hill which it couldn't do, ye know." When he was questioned about the structure of the moon, he replied that "it was a sort of reflector like." When asked what might be holding this reflector up, he answered with many profound shakings of his head, "That is the thing of it" (Durfee, "Experimental Steel Works at Wyandotte," pp. 41–42).

4. Durfee, "Account of a Chemical Laboratory at Wyandotte," p. 235.

5. Durfee, "Experimental Steel Works at Wyandotte," p. 42.

6. Herbert N. Casson, *The Romance of Steel* (New York: A. S. Barnes & Co., 1907), p. 18.

7. Durfee, "Account of a Chemical Laboratory at Wyandotte," p. 236.

   One enthusiast, who firmly believed that the result of an analysis was inevitably a realization of St. Paul's idea of faith, "the substance of things hoped for and the evidence of things not seen," at the particular request of Captain Ward did something which he called an "analytical examination" to a sample of coal. His report was so favorable as to its manifold good qualities, that the Captain purchased the mine from whence it came, only to find, after a large expenditure for pumping machinery, coal-cars, men's houses, and other plant, that there was not enough of this good coal in the mine or on the property to pay for working; hence lawsuits, tribulation and sorrow (Ibid).

8. Durfee, "Experimental Steel Works at Wyandotte," pp. 42–43.

9. Ibid., pp. 42, 48.

10. Durfee, "Account of a Chemical Laboratory at Wyandotte," p. 236.

11. W. F. Durfee, "Z. S. Durfee's Valuable Services to the American Iron and Steel Industry," (letter to the editor) *Bulletin of the* American Iron and Steel Association 30 (1 September 1896): 196.

12. *Annual Report of the Secretary of the American Iron and Steel Association to December 3, 1874*, pp. 42, 53.

13. Durfee, "Z. S. Durfee's Valuable Services," p. 196.

14. William F. Durfee, "The Development of American Industries since Columbus: IX. The Manufacture of Steel," *Popular Science Monthly* 40 (November 1891): 18.

15. James M. Swank, *History of the Manufacture of Iron in All Ages*, 2d. ed. (Philadelphia: American Iron and Steel Association, 1892), p. 409.

16. Durfee, "Experimental Steel Works at Wyandotte," pp. 43, 49. Durfee succeeded in building this new type of converter, but it was not put into use until the fall of 1865, after he had left Wyandotte.

17. Durfee, "American Industries since Columbus: IX," p. 26.

18. Robert W. Hunt, "A History of the Bessemer Manufacture in America," *Transactions of the America Institute of Mining Engineers* 5 (1876–77): 202.

19. Durfee, "Experimental Steel Works at Wyandotte," p. 48.

20. Bernhard C. Korn, "Eber Brock Ward, Pathfinder of American Industry" (Ph.D. diss., Marquette University, 1942), pp. 141–53.

21. Hunt, "History of Bessemer Manufacture in America," p. 202.

22. Korn, "Eber Brock Ward," pp. 119–20, 121, 156. It was not possible to buy a drink of whisky in Wyandotte. When Ward, a teetotaler, had laid out the town, he had forbidden for all time building saloons there. However, his influence did not reach beyond the town limits, and a small crossroads village called Sebastopol had been established about a mile away where the workers from Wyandotte could satisfy their thirst. As John Van Alstyne, superintendent of the Wyandotte mill, once commented, "Every evening the long procession of thirsty souls could be seen going to have their lives saved" (Ibid, p. 121).

23. Swank, *Iron in All Ages*, p. 410. Durfee said in 1892 that he still owned two jackknives

and a razor made from this steel. He admitted that the knives were soft, but that his father had used the razor for fifteen years to his entire satisfaction.

24. Korn, "Eber Brock Ward," p. 157-58. Oddly enough, despite his frontier background and many years as a ship's captain, Ward never used profanity. However, the men with whom he dealt no doubt would have settled for a strong cussing out in preference to the narrowing of the eyes, the setting of the jaw, and the perpetual fingering of the silver dollars—all indicative of his growing anger.

25. Why Holley either overlooked or neglected Mushet is unclear. Perhaps at this stage, Holley failed to realize the importance of Mushet's contribution, and Bessemer may have preferred not to mention it. Holley may not have realized that Mushet's manganese patent still was valid in the United States.

26. Fred M. Osborn, *The Story of the Mushets* (London: Thomas Nelson and Sons, 1952), p. 48.

27. Swank, *Iron in All Ages,* p. 409.

28. Durfee, "Experimental Steel Works at Wyandotte," pp. 43-45.

29. Durfee, "Account of a Chemical Laboratory at Wyandotte," pp. 225-26, 237.

30. J. C. Bayles, "The Study of Iron and Steel," *Transactions of the American Institute of Mining Engineers* 13 (1885):20.

31. Durfee, "Account of a Chemical Laboratory at Wyandotte," p. 237.

32. Durfee, "Experimental Steel Works at Wyandotte," p. 58.

33. Ibid., p. 59.

34. Durfee, "American Industries since Columbus: IX," p. 27.

XIV · Holley at Troy

hile William Durfee was struggling in Wyandotte to complete his vessel, Alexander Holley was beginning his period of servitude to the Bessemer converter. During the fall and winter of 1863, while he negotiated with Bessemer for the American rights, Holley remained in Sheffield studying the various aspects of the process. Although he was an engineer, he had much to learn about metallurgy.

Holley observed repeatedly the operation of the converter: heating the vessel, pouring in the molten metal, beginning the blast, turning the pear-shaped vessel skyward, its becoming a pulsating cauldron of fire as the expanding air roared up through the seething fluid, and the brilliant flame and the sparks that soon escaped from the turmoil inside the vessel. As the flame changed to a dull red and then to pure searing white, the sparks became small hissing points. These gradually gave way to soft floating specks of bluish light as the subsiding upheaval reached the precise moment at which the air must be turned off, and the vessel rotated again to accept the recarburizer and then to disgorge the white frothy stream of metal into the casting ladle.

Holley observed that the senses of the men were vital to the successful completion of a blow. The workmen operating the vessel needed a keen eye to detect minute changes in the color and volume of the flame and to judge the size and nature of the sparks that heralded the end of a blow.

It is interesting to note that the practice of spectral analysis for distinguishing chemical elements from the color of flame began at about the same time as did the Bessemer process. Before 1866 Bessemer himself made some limited use of spectral analysis. So had several companies in France and Germany, but their workers preferred to rely on their own judgment. "There is a natural disinclination on the part of those persons who have to conduct the charges in the Bessemer converters to meddle with a small delicate instrument, which would absorb all their attention, and prevent them from properly overlooking the total series of operations, all equally important, which constitute the Bessemer process," one journalist wrote.[1]

The workmen also needed a good ear attuned at all times to the sound of the pent-up fury within the converter in order to gauge the stages of the blow. The control of temperature was still an uncertain part of the process; with insufficient heat, the metal could not complete its oxidation and was too viscous to handle; with too much, the converter boiled over and spread a sheet of white-hot metal over the floor. Most valuable of all to the workman was the sixth sense that warned him when a charge was about to break loose.

During Holley's stay at Sheffield, William Allen proved to be a good friend, counselor, and teacher. His agreement with Bessemer permitted Allen to use the steelmaking patents without payment of royalties, providing that he demonstrated the process to prospective licensees. To this end he freely exhibited the process and imparted as much information as anyone was disposed to hear. Allen was particularly attracted to Holley and years later clearly recalled Holley's stay at Sheffield. He observed that Holley, then over thirty, looked hardly twenty-one, and he recalled that Holley's eagerness to gather facts and to observe and master details frequently kept him at the plant until late in the evening. Allen declared that "America was never represented by a more perfect gentleman or one more calculated to win respect and esteem of all those he came in contact with in this country."[2]

The months were spent in negotiation as well as in study, and at the end of 1863, Holley returned to Troy with a license to produce steel by the Bessemer process on a royalty basis. The agreement, dated 31 December 1863, also granted a three-year option to the partnership of Griswold and Winslow to buy the Bessemer patent rights for America.[3]

Holley now was ready to build his first experimental Bessemer plant. At thirty-two, he received what was to be the most important assignment of his life. With it he was destined to lay the foundation for an industry whose production would grow from a few tons to millions of tons annually.

On the site of an old flour mill on the banks of the Hudson River, below the city of Troy, New York, Holley started work. Like that of Bessemer and Durfee before him, his way would be tortuous and slow. His relative ignorance of metallurgical chemistry, the lack of proven refractory materials, imperfect machinery, and lack of skilled labor were to be only a few of his problems. As in the locomotive works in Providence where, as laborer, mechanic, and engineer, he had taken off his coat and rolled up his sleeves, in Troy Holley once more found himself down to shirt sleeves.

The converter shell and the handling mechanism were simple to construct using Bessemer's plans. But building the finer details of the apparatus was a different story. The tuyeres and the first stoppers Holley molded himself. No one in the plant knew how to line the converter and ladles with a refractory material that could withstand intense heat, so Holley did this as well. He even dispatched samples of pig iron to Sheffield for trial before attempting to use them at Troy.[4]

Holley labored throughout the summer of 1864 with inadequate knowledge and little available help, repeatedly tearing down and rebuilding the converter. During this period, J. G. Holloway, a Cleveland engineer, stopped at Troy to meet Holley and to investigate the process for the Cleveland Rolling Mill Company. If the Troy experiment succeeded, the company planned to install a converter in its plant at Newburg, then a small town near Cleveland. Holley was working in the plant when Holloway entered the wooden shed that served as a combined office and storeroom. The shed contained a large table and a row of wall shelves that were littered with pieces of steel, pig iron, and iron ore. Holloway surmised that the specimens had come from Sheffield. He observed steel in the form of rail sections and as bars of steel, bent and twisted into various shapes. He saw boiler plate flanged backward, forward, and doubled upon itself, the corners drawn down into small rods that were tied into various fantastic loops and knots. Each sample served as testimony to the ductility of the material from which it had been made.

Holloway long had been familiar with Holley's books on European railways, his *Railroad Advocate*, his contributions to the *New-York Times* as Tubal Cain. He expected to meet a mature man rather than the slender, youthful Holley who came out to the shed to greet him. The astonished visitor was won over by the frankness and open-heartedness with which Holley greeted him. As they toured the works, which at that time were idle, Holley described his past failures and his hopes for the future. Holloway concluded, after making his inspection, that the operation was completely futile. Its only encouraging elements were the steel samples and the confident young man, certain that he yet would make steel in Troy by the Bessemer process. [5]

After working for some months with little success, Holley felt the need of observing a successful Bessemer plant in operation. In October 1864, perhaps as much to restore his fading morale as to increase his knowledge, he returned to Sheffield, taking with him Griswold's son Chester. [6] After a short stay the two returned to Troy to do battle once again with the unpredictable converter. New boilers were procured, the blowing engines altered, another material tried for the converter lining and, what was most important of all, a more suitable pig iron was found.

Because the building that housed the converter was small and inadequate, Holley had to labor under primitive conditions. As had Eber Ward, Winslow and Griswold looked upon this first plant as a trial endeavor. If it proved possible to operate the process under these circumstances, then they would provide the money to build a plant of larger dimensions with a converter of greater capacity than the present two-ton vessel. Consequently it was important that the small converter work efficiently and well.

On 16 February 1865, the vessel received its first charge. Five months after Durfee's first blow at Wyandotte, Holley finally had achieved his first Bessemer blow at Troy.[7] Holley's careful records show that in this first charge he used Number 2 Crown Point charcoal pig iron. The recarburizing metal was Franklinite from the New Jersey Zinc Company.[8] One heat was blown, which used 2,497 pounds of pig iron and 175 pounds of the recarburizer. From this charge he produced three ingots that weighed 482, 491, and 561 pounds. Holley entered in his record book, "Castings 54.4 percent, scrap 26.4 percent, loss 16.2 percent. 1 1/4 inch test piece bent double cold. Blast [of air] 5 to 9 lbs. Blew well and hammered as well as possible. Scrap mostly a large scull due to slow handling. Finer fracture than the charges made from the same brand of iron at Bessemer's works in England on November 30, 1864. Welds pretty well and hardens pretty well."[9]

The second trial, using the same grade of pig iron, was made on February 27. The results were not as satisfactory. This time only 50 percent of the charge ended in the molds as steel, 29.8 percent was scrap, and 20.2 percent was loss. Holley commented, "Blast 10 lbs. Blew 22 minutes. Vessel not hot enough. Ladle nozzle too small—1 1/8 inches."

On the third day, using the same grade of pig iron, Holley profited by his analysis of the previous poor performances and this time succeeded in securing 77 percent of the converter's charge as ingots, 3.8 percent as scrap, and 19.2 percent loss. The notes now read, "Vessel hot and blew well with 9 to 10 pounds of blast. Metal came through bottom by side of tuyere; stopped it with water. Steel all poured out of ladle. Nozzle 1 7/8 inches diameter."

The experiments continued throughout March. At times they went well and at other times poorly. Holley experimented with the dimensions of the ladle nozzle as it became evident how much difference a mere half-inch could make. The amount of the blast was varied and Holley gave his closest attention to the temperature of the molten metal.

On April 4, the converter outdid itself and produced four heats. On April 27, two months after the first charge, Holley produced the first ingot that he considered to be of good enough quality to be marked "Baldwin." Holley entered in the record for that day an enthusiastic, "First tire ever made in America by this process. 'Bullyboy!'" A delegation from the Baldwin Locomotive Works in Philadelphia was on hand for the event. It was a red-letter day for Holley; his converter finally had come through with flying colors.

Holley considered that the operation was still far from perfect and he

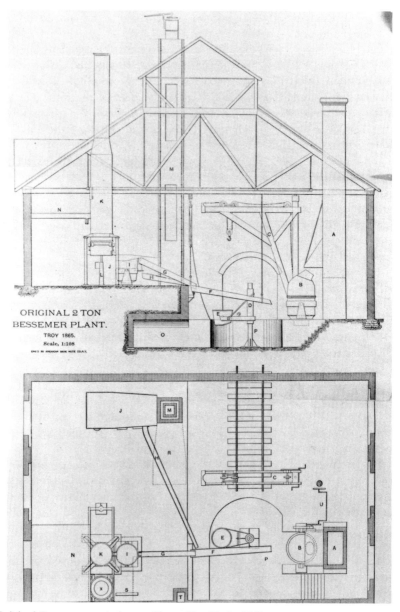

ORIGINAL 2 TON
BESSEMER PLANT.
TROY 1865.
Scale, 1:108

Original Bessemer steel plant at Troy, New York, 1865.

"Two blowing cylinders, 48″ × 48″, were attached to this old wheel, and the rest of the plant in a building 64′ × 41′8″, built for the purpose. The pig iron was melted in the reverberatory furnace *J*, having a bed 7′ long by 4′9″ wide, from which it was run through a gutter in the floor to the wrought-iron runner *F*, and through it into the wrought-iron converter *B*, which of course turned down to receive it. The runner *F* traveled on the rail *Y*, at its higher end and on a corresponding rail, to which its sup-

gave the converter no rest. Throughout May and June he made both major and minor changes. He lengthened the converter by eighteen inches, making it ten and a half feet over all. He enlarged the pit and installed new tuyeres and nozzles. The furnace that was used to melt the Franklinite was raised to permit a better flow of metal into the vessel.

The new cupola was placed in operation in July and the results were much better than before. Using Number 1 Crown Point charcoal pig iron, the iron melted in the cupola in one hour after the taphole had been plugged. The resulting metal was "hot." The coal consumed was 1,226 pounds; 2,997 pounds of iron were melted. The results of the variation in the process were: ingots 85 percent, scrap 2 percent, and loss 14.4 percent. Still not satisfied, Holley continued his work with the process. By the end of July he was entering the results of charges that were numbered 132, 133, and 134 in his journal. [10]

It was not until the middle of July that Winslow and Griswold decided to purchase the Bessemer rights. A letter to them from Bessemer and Longsdon dated 5 August 1865 reads in part,

> We are in receipt of your letter of the 19th ult. We are pleased to learn that you have concluded to purchase our patent interests in America and trust that now peace is restored in your country that the price of gold will soon be more favorable for your remittance to us.
>
> We are very glad to find that you have such good hopes of getting over

---

porting rod was attached, at the lower or converter end. So that after the blow was finished it could be pushed over against the end wall of the building and hence out of the way. A represents a brick stack with a brick hood, which carries off the flame of the conversions. The recarbonizing metal was melted in the furnace R, which is shown by the dotted lines and which worked into the stack M, in common with the furnace J. The resulting metal from the conversion was poured into the ladle E, which was supported by the cast-iron ram or crane D. This ram was controlled by attaching the chain of the wooden hand crane C to it at the hook V, and it was so swung over the moulds set in the pit P. These moulds and their ingots were subsequently taken from the pit by the crane C, and loaded on a car standing on the railroad track shown. The vessel was rotated by decidedly simple apparatus shown at U, which depended upon intelligent labor for its power.

"Such was the original Troy two-ton plant. And it is not now surprising that many difficulties were encountered by the management. Mr. Holley's mind was soon impressed with the advantages of melting in a cupola and one was erected. This was built as shown by K. It was provided with duplicate bottom sections. The extra one being pushed to one side, as shown by X. He also advanced beyond the English method by placing the accumulating ladle I, resting on scales in front of the cupola. G is the wrought-iron runner conveying the metal to the runner F. When the cupola practice was adopted, the spiegel furnace R was torn down and the furnace J converted into one for melting the recarbonizer which was conveyed through the cast-iron gutter H. The cupola bottoms were raised and lowered by the screw L. O was an oven for drying stoppers, and T its chimney" (Robert W. Hunt, "The Original Bessemer Steel Plant at Troy," *Transactions of the American Society of Mechanical Engineers* 6 [1885]: opp. p. 62).

Mushet's interference and also of doing without his plans. We hope that Kelly and Shunk may soon be disposed of satisfactorily to you. We think you will certainly be able to confine those gentlemen in very restricted limits, if you do not succeed in beating them altogether.

Your manufacturing success appears to be quite as good as you can expect for the time and means at your disposal and when you have your larger machinery at work we are sure that you will do well.[11]

In November, Holley finally acknowledged that the water wheel, with its uneven rotation speed, never would give him sufficient blast pressure. Having struggled with the wheel for so long, he abandoned it in disgust. The final charge using the water wheel was blown in December and until work was resumed in the following March, Holley worked on the installation of a suitable steam engine. From the time that the engine was placed in use, the working of the converter became more predictable and efficient.[12]

During this period operators of the plant at Wyandotte also faced problems and anxieties. The question of patent rights never had been settled. In Wyandotte, the plant was operating under the Mushet and Kelly patents and using certain of the Bessemer machinery patents. In Troy, Holley had started out with the Bessemer machinery rights to which he had added many improvements, but he was at the same time infringing on both the Mushet and Kelly patents.

Both a strong rivalry and unfriendly relations existed between the Wyandotte and Troy organizations. Captain Robert W. Hunt, who was to become like a brother to Holley, in later years relished telling of his first meeting with Holley. Hunt had left the Cambria Iron Company in May 1865 to study the pneumatic process at Wyandotte. He was unfamiliar with the steelmaking process when he went to Wyandotte and decided to visit Troy to observe operations there. While Holley had welcomed Holloway and willingly had showed him around the plant, when Hunt presented his letter of introduction, Holley received him graciously, but firmly refused to permit him to enter the plant. Embarrassed, but with the best possible grace, Hunt bowed to Holley and departed.[13]

Meanwhile, Holley was conducting experiments using chromium ore in the vessel, an indication that he still must have hoped to circumvent the Mushet recarburization patent. In reporting his progress with the converter to Bessemer and Longsdon in November, he mentioned that he had an idea for getting around the annoying patent.[14]

Michael Faraday and J. Stodart as early as 1820 had been interested in the effects of adding chromium to steel. During the ensuing years other metallurgists had been interested in expanding their limited knowledge of the effects of chromium and other metals in steel and cast iron.[15] A

French metallurgist, P. Berthier, was particularly attracted by the work of Faraday and Stodart and conducted many experiments using chromium as an alloying element. John Percy's important book *Metallurgy: Iron and Steel* appeared in 1864. Percy quoted at length not only from Berthier's report of his work using chromium as an alloying agent in steel and cast iron but also from the earlier study by Faraday and Stodart.[16] From it Holley may have gotten the idea of trying chromium instead of spiegeleisen in the converter.

In numerous trials using a triple compound of chromium, iron, and carbon he obtained only scrap. The ingots poured from the charge cracked and crumbled badly when worked. In one experiment he evidently threw 120 pounds of red-hot chromium pig metal into the vessel when it was turned down after the blow was finished. The vessel then was turned up for a quarter of a minute, quickly turned back, and the charge poured into the casting ladle. If Holley was looking for a method to circumvent the Mushet patent, he quickly realized that this was not it.[17]

The situation between the two companies now had reached the point where it appeared that the only possible way to solve the patent controversy would be to take the matter to court. Hunt said that things were "shaping up toward a real Kilkenny fight."[18] A circular issued by the Troy company on 15 April 1865, announced that they were ready to accept orders for steel made by the Bessemer process. They were able to furnish steel in the form of rails, axles, piston and connecting rods, crank pins, boiler and ship plates, shafting, cannon and projectiles, bars, rods, spring steel, forgings, ingots, and machinery castings to pattern. The circular set forth the advantages of the new steel and also announced that Winslow, Griswold and Holley were prepared to grant licenses upon reasonable terms, as well as to furnish working drawings for the necessary plant and machinery, to anyone interested. The last paragraphs of the circular contained a warning regarding any infringement of the rights that the company owned.[19]

The two companies prepared for a showdown battle. A court case that would be closely watched on both sides of the Atlantic was in the making. Large sums, both in legal fees and for royalty payments to the winner, would certainly be involved. After both sides had drawn their lines for the fray, the case suddenly was settled out of court. The two groups agreed to join forces and divide the rights with 70 percent of the fees going to the Troy group and 30 percent to Wyandotte. The division of the proceeds from the licensing of the rights has brought about more speculation and wild accusations regarding motives than the situation warranted. Daniel Morrell has been accused of double-dealing. Holley has been labeled a smooth mediator who beguiled the wits out of the Wyandotte group by his oratory and devious maneuvering. The uneven division of the fees at first does appear unusual, but careful scrutiny of the problems involved reveals that the arrangement was not as strange as it looked at first. Morrell, the Quaker who had been given the authority to act for the Kelly Process Company, was unenthusiastic about a court case. Eber Ward, with plants doing well in Chicago and Milwaukee, knew that Troy was producing steel in a much more satisfactory fashion then he was able to do at Wyandotte. He also was aware of Holley's expert knowledge and that Kelly's patent had only four more years to run, with the possibility that an extension might not be granted to Kelly. Thus, he was inclined to agree with Morrell.

According to William Durfee, the prospect of a court battle terrified Ward. Close to the scene of activity in Wyandotte, Durfee never was able to understand why the Kelly group gave in so easily. In later years, Durfee wrote of the situation.

> To the majority of the members of the Kelly company the law was a terror. Lawyers must be paid. Experts would not testify gratuitously. Costs of court would accumulate. Judges were doubtful. Jurors were uncertain. And then if victorious what would they gain? And if defeated utter ruin would overwhelm them. Never before or since has a party of reputable business men been so needlessly alarmed and so utterly oblivious of the first principles of a sound business policy. The various bugabooks and hopgoblins [sic] which their terrified imagination conjured up of the horrors of the life to come among courts, judges, lawyers, experts, witnesses and obstinate jurors in case they ventured to assert in a court their manifest right, at last drove them into making a proposition to Messrs. Winslow, Griswold and Holley looking to a combination of the interests of the two companies and to their final acceptance of an agreement under which they surrendered rights which were of great value to Winslow, Griswold and Holley and obtained practically no rights in return save that of receiving thirty percent of royalties earned by combination and leaving Winslow, Griswold and Holley the remaining seventy percent. In the whole history of business affairs, it would indeed be hard to find a more perfect illustration of "the tail wagling the dog" than this. It is only justice to the late Zoheth Durfee to say that he opposed this compromise and its unjust disposition of the rights of himself and associates with all the energy of which he was capable and the fact that all the royalties the combination ever earned were received under the operation of an extension of the patent of Kelly is quite sufficient to justify his business sagacity and foresight. [20]

Elting Morison, a modern interpreter, explains the settlement in this way.

> In July 1865 Bessemer took out, for the first time since his original patent, five new American patents on machinery, duplicates of British patents taken out from 1860 to 1865. Then Winslow and Griswold, through their counsel, Blatchford, took up their option of all Bessemer's patents for a sum of 8000 pounds on December 7, 1865. In the first week of January 1866 they granted a

license under these patents to "the Pennsylvania group." Thus buttressed with patent rights, a plant in being, and a license to another interest (albeit this license did not include the rights to spiegeleisen), Winslow, Griswold, and Holley sought an agreement with the men at Wyandotte.[21]

Holley, in a letter written to the editor of *Engineering* in 1866, gave his interpretation of the settlement. In his letter he agreed that the pneumatic process first was practiced experimentally at Wyandotte but that the plant was using Bessemer's machinery patents. Holley then said that Winslow, Griswold, and Holley had completed their arrangements with Bessemer and had commenced experiments at Troy before the experiments were started at Wyandotte. Holley either was in error or did not know Zoheth Durfee had been negotiating with Bessemer and that his cousin had begun his work while Holley still was gathering material for Stevens. But Holley was correct in the remainder of his remarks. He acknowledged that the Mushet patent was recognized as valid in this country by the patent office. Whether or not this patent or that of Kelly were absolutely essential to the process or would have been sustained by the courts were questions upon which experts and lawyers were divided. Holley said also that Mr. Bessemer had legal claims for practical details and machinery as well as moral claims as the introducer of a working process, and that neither party disputed this fact. He thought it proper to repeat that in the United States Bessemer had been awarded the chief credit for perfecting and introducing the process that bore his name, and for that reason his assignees properly retained by far the largest share of the royalties arising under the consolidation of interests. "While the lawyers were looking forward to the probable duration of a patent suit, the owners were looking forward to the probable duration of the patents and the business result of their observation was a consolidation of all the patent interests."[22]

The new organization elected Winslow, Griswold, and Morrell as trustees and they, in turn, appointed Zoheth Durfee as their general agent. The wisdom of merging the two competing groups became increasingly evident in the next few years. It was agreed that the Troy works, the Wyandotte plant, and all the Bessemer works to be built in the United States would be licensed under the combined patents at uniform royalties the same as those charged by Bessemer and Longsdon in England, five dollars per ton on all ingots to be rolled into rails and ten dollars on ingots to be rolled into other forms.[23]

Supporters of the cause of William Kelly have regretted that in the amalgamation of the rights, his name should have disappeared so quickly. They have asserted that the evil forces at Troy were responsible, but this was not so. The association formed by the rival groups was a trusteeship termed "The Trustees of the Pneumatic or Bessemer Process of Making

Iron and Steel," later called the Pneumatic Steel Association, then the Bessemer Steel Association, and finally the Bessemer Steel Company. Kelly himself had used the expression *pneumatic process* instead of his own name in referring to his invention. The name Bessemer already was so well known that popular usage soon attached the name of Bessemer to any steel made by this process.[24] Although Kelly was receiving a royalty, he no longer was active, whereas Bessemer had remained in the field. So far as can be determined, Kelly never worked at the Wyandotte plant. He never has been mentioned in connection with the trial heats or the building of converters, nor is there any indication that his advice ever was sought. Had friends of Kelly wished to make an issue of the nomenclature, his name might have been used in combination with those of Mushet and Bessemer. In 1866, however, no one cared whether the steel was called Kelly or Bessemer so long as converters could be built to produce steel. William Durfee probably described the situation best when he said, "Facts warrant calling it the Bessemer-Kelly-Mushet process but as Bessemer by his ingenuity, persistence in methodical endeavor and business sagacity is entitled to first place and if the process is to bear only one name, the popular verdict is justifiable."[25]

Holley's role in this first small endeavor at Troy cannot be overstated. He had become the fountainhead of knowledge about every phase of the Bessemer process, and one of three or four men in America who had the ability to construct and operate a Bessemer plant.[26]

While most of the European countries were developing their steel industries at a more or less steady pace. England had leaped so far ahead that she was threatening to turn the United States into one of her best markets before the American industry could gain a foothold. The Bessemer plant of the Barrow Haematite Iron and Steel Company, built in 1865, already was producing 2,000 tons of steel a week.[27]

Single-handedly Holley had persuaded Griswold and Winslow to purchase the Bessemer rights, and, swept away by Holley's enthusiasm, they had embarked on a venture that would require enormous amounts of capital. Despite his knowledge and enthusiasm, Holley felt the heavy weight of the responsibility he had assumed if the endeavor was to succeed in turning an experimental plant into an example of good steelmaking practice. When the new practice did succeed, his optimism prevailed, but he also was subject to deep discouragement when, as was frequent, his ideas failed in practice. Always in his thoughts was the question of whether or not the project would prosper.

Holley's first child, Gertrude, had been born in 1862, the year he became the enthusiastic missionary of Bessemer. Lucy, his second daughter, was born in 1863, the fateful year when he sailed to obtain the rights to the Bessemer patents for the United States.[28] The responsibility of a family had changed Holley from a free-lance engineer who could afford to gamble on uncertain outcomes to a responsible husband and father. Now every fiber of his mind and body had to be engaged in this, his greatest effort.

Just as in the early days of their marriage, when Mary had accompanied him to the shop and watched the testing of the locomotives and later had helped in the office of the *Railroad Advocate,* she again haunted the shop and waited patiently as he tried one charge after the other in the intractable converter. To a friend who feared that Holley was working himself into a state of nervous and physical exhaustion, Holley, with the same sense of humor that did not desert him in the most trying of times, wryly commented, "Don't be afraid, if I die she can run the concern."[29]

Uppermost in the minds of the industrialists was the idea of applying the new method of steel manufacture to rail production. The major part of iron production at that time was for the needs of the railroads. The steel producers hoped to substitute their steel for the iron. To that end, the first product made from Bessemer steel, both in England and in the United States, was a rail. One of the earliest and most significant tests of the comparative durability of iron and steel rails took place in England. It was similar to the earlier test of the Mushet rail. At Chalk Farm Station near London was a section of track over which every train passed, bound to or from the busy Euston Terminal, and over which extra traffic passed as a result of the many freight trains made up there. In 1865 Bessemer was able to exhibit a rail that had been in use at Chalk Farm Station that had outlasted eleven iron rails similarly placed.[30]

At the same time, longstanding prejudice against the use of steel instead of iron rails rapidly was breaking down. The steel rail producers had become bold enough to guarantee that their rails would outlast iron rails by ten to one. Each company, sure of the quality of its product, began the custom of stamping its name on the rails produced in its mills.

In 1866 the cost of steel rails was still a major problem. The average price of American iron rails was $86.75 a gross ton while that of the British steel rails was $120 in gold per gross ton.[31]

In the United States railways were growing rapidly. Without regard to cost or engineering complexity, each year many more tracks curved through mountains and ran straight across prairies and plains. In 1880 J. S. Jeans projected the future demand for steel rails.

In 1850, only 18,000 miles of railway were open in the entire world. In 1860, this had increased to 63,000 miles; in 1870, to 127,000; and in 1878, to 206,000 miles. Of these 206,000 miles of railway, over 17,000 are in the United Kingdom. In addition, however, to the mileage stated, it is usual to allow an additional 25 percent for sidings, stations, and double lines, thus giving 49,000 miles more and bringing the total mileage of the world up to 255,000 miles, representing an aggregate tonnage of 30,204,000 tons of rails. Calculating the average life of a rail at ten years, it follows that the railway system may be expected to call annually for 3,020,000 tons of new rails for replacement, 1,080,000 tons for the 10,000 miles of new railway system laid down each year, at 108 tons per mile, and 10,000 tons for private lines, mines, and collieries, making the total annual requirements of the world upwards of four million tons of railway material.[32]

Although several American railway companies had begun to use steel rails as early as 1863, it was not until 1867 that the first order for steel rails was rolled in the United States. The order was for only 2,550 gross tons, at an average price of $160 per ton at the mill. The price of British steel rails then was $118 in gold, including duty and all other charges.[33]

As he worked to perfect the plant at Troy, Holley ranged widely in an effort to find new iron ores and pig irons for trial. The company issued a circular offering to process pig iron delivered to the plant at a fee of seventy-five dollars per ton, with the resulting steel to be removed at the owner's expense. For this amount, the Troy company would hammer, roll, and cold bend the material in order to determine its qualities.[34] Robert Hunt later wrote, "It was amusing to think of firms sending a few tons to Wyandotte, Troy or even England to be tried in actual production when a few hours of laboratory work would have settled the entire question."[35]

In spite of the constant irritation from failures in the converter and its operation, failures that often were due to the stupidity or inexperience of workmen, Holley contained his impatience and annoyance. The experimental converter was a dangerous entity. It had to be rotated by hand gearing, and the potential for tragic accidents was a perpetual threat. The molten metal was capable of boiling out of the mold and men then had to scatter as rapidly as they could. At times a bursting mold might catch an unfortunate workman before he could escape. When the hot molds were stripped from the ingots, it was necessary to cool them for reuse. As streams of water were directed against the seething containers, the room filled with hot steam. The rear of the vessel, where the men stood to examine the tuyeres between the blows, was one of the worst locations in the shop. There the hood or chimney, which hung over the converter to carry off the escape gases, also received the slag that was blown from the nose of the vessel. These sloppings hung in great stalactitic masses from the hood and they fell without warning. Falling masses had been responsible for several accidents when inexperienced workers, keeping a close eye on the tuyeres, had failed to watch overhead at the same time.

Some of the undesirable features of the early Bessemer steelmaking

plants were described graphically by Henry Howe, a well known metallurgist.

> In order that the cast-iron might run by gravity to the vessel, the cupolas . . .
> stood close to and higher than the vessels. . . . The cast-iron was tapped from
> the cupolas into stationary tipping ladles. . . . The cupola tappers . . . were
> completely hemmed in by the heat. In front of them were the hot cupolas,
> from whose shells much heat radiated, by their feet were large ladles full of
> molten cast-iron; while behind them rushed in a torrent of hot air, heated by
> the ingots in the pit and by the flames of the vessels. They stood, as it were, in
> a chimney conducting the hot air from the pit and from around the vessels to
> the top of the cupola building. I have often known men to be overcome with
> the heat here, fainting, severe hemorrhage at the nose, etc.[36]

All of these dangerous phases of the new steel process concerned
Holley. He was as sympathetic and cordial with his workmen as he was
with his engineering friends. He could not accept the conditions under
which he had seen the pitmen work in the Bessemer installation at Sheffield. In that plant, the vessels were built so close to the floor that a deep
casting pit was necessary. This allowed the slag to be emptied through
the nose when the vessel was inverted. The converter bottom, however,
was replaced from beneath.

Holley also was concerned about the slow rate at which the small converter was turning out successful heats, although neither Griswold nor
Winslow shared his impatience. Men of the sort who could gamble more
than a quarter of a million dollars in building the *Monitor* were not
discouraged easily. At a conference that lasted late into the night, Griswold and Holley discussed the future of the Bessemer process. Finally,
Griswold said, "Alex how much more do you want to pull us through?"
When Holley replied that it would take at least $100 thousand, Griswold
assured him, "Go ahead, my boy, my faith in you is unshaken; I'll find
the money." The partners at Troy could see that, given a charge of good
metal, the converter consistently turned out good steel at the rate of ten
tons of ingots each twenty-four hours, twice as many casts as Bessemer
was getting at Sheffield. Although his partner's confidence and patience
encouraged Holley during these trying days, their attitude also heightened
his feelings of responsibility.[37]

At the end of 1866, no longer hampered by the patent controversy
and convinced that Holley had proved the value of the process, Griswold
gave the go-ahead for a larger plant. Encouraged by the expression of
confidence in his achievement, Holley set to work. He was given a fifteen-
acre plot of ground well situated above high water, with a good rock
bottom for the necessarily heavy foundations. With the Hudson River
on one side and the Hudson River Railroad on the other, the site was

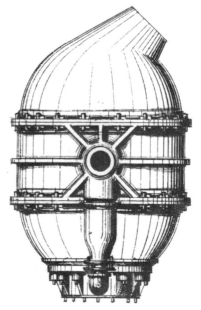

Design of a converter by A. L. Holley. Used on letterhead by John A. Griswold & Co., Troy, N.Y. about 1870. Original size 1 1/2 " by 7/8 ".

ideal for bringing in raw materials and for shipping the ingots to nearby mills for rolling. Iron ore could be brought from the Hudson and Housatonic Valleys, Vermont, Lake Champlain and the promising Adirondack regions.[38]

At last, Holley was able to put into practice all the ideas he had been accumulating since his first involvement with the small experimental converter. Friends who had watched him wrestling with the wayward process gathered around him now and were amazed to see what was taking place.[39] Many of the restrictions of the British practice were to be discarded; the Bessemer machinery would be improved; and space would be utilized more adequately. Day by day, the building progressed with the anxious Holley hovering about as he watched his plans take shape brick by brick.

The building that housed the two five-ton converters was an impressive structure 178 feet long and 65 feet wide. The original experimental converter with all its appurtenances was moved into the building to share the space with the new vessels. With the three converters in place there was still room left for the addition of two three-ton converters, if necessary. Next to the converter shed was another brick building 44 feet long and 24 feet wide housing the cupola that would heat the metal for the neighboring converters. Another building of wood, about the same size, formed a continuation of the converter shed and housed the machinery for working up the ingots. At the opposite end of the converter building was a four-story brick structure 65 feet by 36 feet. In its basement were the Collins turbine used to drive the machinery and blowers on the upper floors, a machine for grinding stone for the converter linings, and a tilt hammer for testing the ingots. The second or ground floor served as a machine shop. The third and fourth levels were designed for tuyeres and other necessary parts of the Bessemer apparatus. The pressure blower and fan and their driving machinery for sending the blast to the cupola and furnaces as well as a mill for grinding finer materials for converter linings also were housed on the third floor. The building even boasted a power lift to make all the areas accessible. The boilers, as well as the blowing and hydraulic engines, were on the river side of the buildings.

The design of this plant became known as the American layout and served as the pattern for the many Bessemer installations that followed. All of the machinery was furnished by American firms. The Delamater Iron Works of *Monitor* fame built the blowing machinery; George Worthington designed the pumping machinery; and George Reynolds was responsible for the steam engine. The building progressed so rapidly that early in 1867 the plant was ready to begin operations.

When the plant was finished, Holley could view the making of steel in a lighter vein. He described it as follows for the local newspaper.

> You will thus observe that to make steel, you take a little iron and a little carbon and a few condiments, apply a hot fire, and there you have it — steel — as simple as the cookery book. But oh! the years of conflict with elements all too subtle and poisons all too willing, before the little niceties that make the difference between success and dead failure were found out at all — machinery, the power — not in a little laboratory upon which you can turn the key, but throughout acres of bricks and mortar, boilers and engines, rollers and hammers, furnaces and crucibles, all surging with pentup energy and fervent heat, day and night, year after year. So that the receipt for making steel would rather be, take great perseverance and equal portions of the knowledge of books and the knowledge of practice — mix them well with no end of money, and then if you know how and have good luck, you can make a grand success of it."[40]

NOTES

1. "Spectral Analysis and the Bessemer Process," *Engineering* 1 (8 June 1866): 381. This journal reported on the use of spectral analysis from time to time. As late as 1868 it reported that the spectroscope was kept in some of the steelworks as a toy to amuse professional and other visitors. The Sheffield workmen had their own method of determining the changes taking place in the flame. "They avoid looking into the flame towards the end of the operation, and select a whitewashed wall or some similar

object illuminated by the flame as the object of observation. The change of light is more clearly visible in that way than by looking at the flame itself, the cause being, of course, the lessened intensity of the reflected light compared with the direct rays" (*Engineering* 5 [17 January 1868]: 49).

2. James Dredge, "Holley Memorial Address," *Transactions of the American Institute of Mining Engineers* 20 (1891): xxxii.

3. Elting E. Morison, *Men, Machines, and Modern Times* (Cambridge, Mass.: M.I.T. Press, 1966), p. 141. According to Morison, who had access to the files of the legal firm representing Griswold and Winslow, a partnership consisting of Griswold, Winslow, and Holley now was formed, but it was never defined on paper.

4. *Memorial of Alexander Lyman Holley* (New York: American Institute of Mining Engineers, 1884), p. 31.

5. Ibid., pp. 84–85.

6. James Dredge, "Henry Bessemer, 1813–1898," *Transactions of the American Society of Mechanical Engineers* 19 (1898): 939.

7. Robert W. Hunt, "The Original Bessemer Steel Plant at Troy," *Transactions of the American Society of Mechanical Engineers* 6 (1885): 61. In "A History of Bessemer Manufacture in the United States," *Transactions of the American Institute of Mining Engineers* 5(1876–77): 203, Robert Hunt says the converter's size was two-and-one-half tons.

8. Franklinite is a zinc-iron-manganese mineral found in quantity only at Franklin Furnace, New Jersey. "The New Jersey Zinc Company treated it as an ore of zinc. The residue was subsequently smelted in a blast furnace and yielded characteristic spiegeleisen" (John Percy, *Metallurgy: Iron and Steel* [London: John Murray, 1864], p. 199).

9. Hunt, "Original Bessemer Steel Plant at Troy," pp. 63–64.

10. Ibid.

11. *Iron Trade Review* 74 (3 January 1924): 124. When Kelly applied for extension of his patent, B. S. Hedrick the examiner in his report of 7 June 1871 mentioned that Christian Shunk had filed an application for a patent in 1854 but showed no method of applying an air blast to molten pig metal except in the hearth of the blast furnace. Opposition to the extension of Kelly's patent was entered by S. W. Kirk and Shunk but they were overruled (From Interference File of U.S. Patent Office, Kelly vs. Bessemer. Parts appeared as appendices to Joseph D. Weeks, "The Invention of the Bessemer Process," *Transactions of the American Institute of Mining Engineers* 26 [1897]: 980–91).

12. Hunt, "Original Bessemer Steel Plant at Troy," p. 64.

13. Robert W. Hunt, "A History of the Bessemer Manufacture in America," *Transactions of the American Institute of Mining Engineers* 5 (1876–77): 207.

14. *Iron Trade Review* 74 (3 January 1924): 124.

15. Carl A. Zapfe, *Stainless Steels* (Cleveland: American Society for Metals, 1949), p. 6.

16. Percy, *Metallurgy*, p. 199.

17. Hunt, "Original Bessemer Steel Plant at Troy," p. 64.

18. Hunt, "History of Bessemer Manufacture in America," p. 204.

19. Circular issued by Winslow, Griswold and Holley, 15 April 1865, Scrapbook. Author's Collection.

20. William F. Durfee, "The Development of American Industries since Columbus: IX. The Manufacture of Steel," *Popular Science Monthly* 40 (November 1891): 28–29.

21. Morison, *Men, Machines and Modern Times*, p. 142.

22. Alexander Lyman Holley, "The Bessemer Process in America," *Engineering* 2 (19 October 1866): 308. There may have been another reason why Eber Ward was willing to settle the case out of court, a reason Ward's competitors in Troy would have had

difficulty understanding or even believing. During the middle 1800s Spiritualism had become popular in the United States and many serious people had become its converts. Shortly before the threatened lawsuit, Ward had become deeply involved with Spiritualism and frequently was accused by his associates of seeking the advice of a medium before transacting business affairs. His youngest son Frederick is said to have committed suicide some years later because Ward believed a Spiritualist's report that the son was a bastard. If this common talk had any basis in fact it is possible to believe that Ward gave in to the Troy interests on the advice of a medium (Charles William Shinn, "Captain Eber Brock Ward, Iron Master of the West" [Master's thesis, University of Wisconsin, 1952], pp. 71, 93).

Bernhard C. Korn in his Ph.D. dissertation "Eber Brock Ward, Pathfinder of American Industry" (Marquette University, 1942) tries to prove that Ward was not a Spiritualist, although he admits that Ward did "evince a keen interest" in the subject. Korn states that, even though Ward was a Unitarian, he made liberal donations to churches of all denominations. After Ward's death there were estate suits that lasted many years and all of his affairs were gone into thoroughly. Ward was very close to his sister Emily, and usually asked her advice in many of his business transactions. Ward's Spiritualism came up in the questioning of Emily Ward during one such case. She reported that he once said, "A spirit in the body is better than one out of the body in all business transactions. Anyone who relies upon the representation of spirits will be ruined financially. There are as many liars on the other side as there are here" (Michigan State Supreme Court, Owen v. Potter, Emily Ward Testimony, 1886).

23. Alexander Lyman Holley, "Bessemer Process in America," p. 308.

24. The following advertisement appeared in the trade journals after the merger.

The Pneumatic or Bessemer Process
To Manufacturers of Iron and Steel
The Proprietors and Assignees of the several Letters-Patent granted by the United States to Henry Bessemer and Robert Mushet, of England, and to William Kelly, of this country, for certain Improvements in the Manufacture of Iron and Steel, having, for the purpose of avoiding all conflict of claim thereunder, consolidated their interests in said Patents under a Trusteeship, styled, "The Trustees of the Pneumatic or Bessemer Process of Making Iron and Steel," the undersigned, the Trustees aforesaid, invite attention to the advantages of the above process, which can be fully and freely examined at the Works of Messrs. Winslow and Griswold, at Troy, N.Y.

The great Strength, Toughness, perfect Homogeneity and comparative Cheapness of Production, of the Pneumatic or Bessemer Cast Steel, as well as the enormous extent of its manufacture and use abroad, are too well known to require comment: and the undersigned are pleased to state that works have been, and others are now being erected in different sections of the country, to supply, in part, the demand so rapidly increasing here for STEEL RAILS, TIRES, BOILER PLATES, FORGINGS, etc. made by their process.

Licenses will now be issued on reasonable terms to all who desire to engage in the new manufacture and the undersigned hereby give notice that they have appointed Mr. Z. S. Durfee of 418 Walnut Street, Phila. their General Agent, for the business, and that all communications respecting, Licenses, and inquiries concerning the process and cost of the apparatus therefor etc. should be addressed to him as above.

John F. Winslow, Troy, N.Y.
John A. Griswold, Troy, N.Y.
Daniel J. Morrell, Johnstown, Pa.
Trustees
Bulletin of the American Iron and Steel Association 5 (7 September 1870): 8.

25. Durfee, "American Industries since Colombus: IX," *Popular Science Monthly* 40 (November 1891): 26.

26. Hunt, "Original Bessemer Plant at Troy," p. 61.

27. J. S. Jeans, *Steel: Its History, Manufacture, Properties, and Uses* (London: E. and F. N. Spon, 1880), p. 86. In England, each year since 1859, when Bessemer had taken the bold step of building the plant at Sheffield, additional companies had signed with him as licensees. By 1866, there were fifty-two converters, producing 6,000 tons of steel a week operating in Britain.

On the Continent, the converters were being installed almost as rapidly. In Germany, the Krupp plant at Essen was among the first to take up the process and had installed thirteen vessels, some with a capacity of as much as seven tons. The Friedrich Krupp firm originally had proposed to pay Bessemer £1,000 per year for all his plans for mechanical appliances as well as information about ongoing improvements. After the first year Krupp failed to carry out the agreement inasmuch as Bessemer never had been able to obtain a German patent. The Prussian government considered that the Bessemer process was the same as the Nasmyth patent of puddling iron with steam. As a result, anyone in Prussia was free to use Bessemer's process without the payment of royalty.

This ruling directly affected the fortunes of the process in Belgium, where the John Cockerill plant located at Seraing had introduced it in 1864. Soon after Krupp began to produce Bessemer steel, the Belgian company felt the pinch of the competition from the royalty-free steel that the German company was shipping all over the Continent. Cockerill unsuccessfully complained to Bessemer, who tried to investigate the allegation. The Belgian company then resorted to the ancient ruse of dispatching two of their workmen to Essen to seek employment. Representing themselves as experienced in the operation of the process, the two were hired. When the workmen reported that indeed Krupp was freely using the process for which the Belgian company was paying a handsome royalty, Cockerill discontinued royalty payments after the spring of 1870. Bessemer now had lost the income from rapidly increasing industries in Belgium and in Germany.

Following the reading of the Bessemer paper in 1856, William Jackson, a well known iron manufacturer in France had built a trial converter at St. Seurin. After five years of experimentation he had erected a permanent Bessemer plant, the first in France. Despite Jackson's success, other French steel producers did not adopt the process, so that, for many years French steel production lagged. In 1863, a son of the Schneider family, owners of the great works at Le Creusôt, went to Sheffield to negotiate for a Bessemer license. Arrangements were concluded and he returned to France with detailed drawings for the building of a Bessemer plant. The managers of Le Creusôt then observed the process in operation at Sheffield and even sent a sample load of their pig iron to Bessemer to assure themselves that the material was suitable for use in the converter. For two years the company built the machinery for the works, undoubtedly the finest in Europe; they were completed only a few weeks before Bessemer's French patents were due to expire. Even though finished, the plant was not placed in operation until the patents had expired and thus Bessemer received no royalties from Schneider. Terre Noire, the other important French company, followed their example and also refused payment.

Sweden, under the guidance of Göransson, first had produced Bessemer steel at least ten years before any other country, but production there was less extensive than in other countries. Even though the country possessed exceptionally pure iron ore, it had virtually no coal deposits. Charcoal was necessary to remelt the pig iron. Its cost discouraged companies from installing large-scale plants. The iron produced in Sweden always had been considered the best in Europe. When producers turned to steel manufacture, economics dictated that they forego large items such as rails in favor of quality steel for tools and cutlery.

Styria in lower Austria had been a celebrated center for steel production for several generations before the development of Bessemer steel. Austrian steel scythes had become as famous as the earlier Toledo and Damascus blades. Other types of steel had been produced in small hearths at the works of Prince Schwartzenberg at Turrach before November 1863, when the Bessemer process was introduced there. Three more installations, in Carinthia and at Neuberg and Gratz, soon followed. Austrian iron ore was of the highest purity, ideal for the Bessemer process. Then, as now, two mountains, the Erzberg of Eisenerz in Styria and the Huttenberg Erzberg in Carinthia, supplied the ores. Operations were limited to a certain degree because of the poor quality of Austria's coal and the necessity of relying on charcoal for blast furnace fuel. Although Bessemer lost his Austrian rights because he failed to operate his process there within the time limits of Austrian patent law, he did supply the Neuberg management with detailed drawings of the converter apparatus and descriptions of the process and he conducted a trial of their pig iron at Sheffield. The Austrians paid him a fee of £1,000 for the work.

In India, the Beypore Iron Company in Madras had been one of the first to obtain a license for the Bessemer process, and it developed a variation of the process similar to that used in Sweden. By 1862, it was turning out steel of good enough quality for display at the exhibition in London. Visitors to the hall could see an interesting exhibit of jungle knives hammered out of Bessemer steel by native smiths.

Other countries lagged for various reasons. Spain, rich in resources, exported large quantities of iron ore but had neglected to develop a domestic steel industry. Italy had extensive reserves of iron ore but lacked coal. J. S. Jeans commented at the time on Japan's position. "Having regard to the natural resources of this empire and its advancement in many arts and sciences, it is rather remarkable that the metallurgy of iron and steel should have been in so backward a state as it is at the present day," he wrote (Ibid., pp. 303-14).

28. Ancestry of the Olmsted and Holley families compiled by Francis Howard Olmsted.
29. *Memorial of Alexander Lyman Holley* (New York: American Institute of Mining Engineers, 1884), p. 115.
30. J. S. Jeans, *Steel*, p. 675.
31. "Production of Rails in the United States," *Proceedings of the American Iron and Steel Association* (Philadelphia: AISA, 1873), p. 32.
32. J. S. Jeans, *Steel*, p. 673.
33. "Production of Rails in the United States," p. 32.
34. *Bulletin of the American Iron and Steel Association* (5 December 1866): 97. The owners at Troy referred to serious derangement of the works and interruption of business resulting from the use of a very bad iron that the offer produced, but claimed that their serious concern with increasing Bessemer production in the United States was the reason for the offer. The last paragraph of the circular is of interest because it indicates the state of technical knowledge of the time; the company declared that it would not publish results of the tests since they deemed them the property of the company paying for them. They admitted that they did not know what kind of steel an untried iron would produce, but believed that opinions based on analysis could be formed as accurately by others as by themselves.
35. Hunt, "Original Bessemer Steel Plant at Troy," p. 205.
36. Henry Marion Howe, *The Metallurgy of Steel* (New York: Scientific Publishing Co., 1890), pp. 318, 328, 345. Howe also had observed the Bessemer process in operation at Sheffield and commented, "In this confined, unventilated and comparatively inaccessible indeed infernal abyss, hemmed in by red hot ingots and molds, bespattered by

the vessel's white-hot spittings as it turned up or down, scorched by the slag which it dropped between heats and threatened by the floods of molten steel which now and again broke through its nether parts, the salamandrine pit-men intolerably reeked and wrought."

37. *Memorial of Alexander Lyman Holley*, p. 36.
38. *Troy Daily Times*, 27 July 1868.
39. Hunt, "History of Bessemer Manufacture in America," p. 204.
40. *Troy Daily Times*, 27 July 1868.

XV · Pennsylvania Steel Company

hile Holley was completing the plant at Troy, plans were going forward for the construction of another Bessemer plant, to be called the Pennsylvania Steel Company. The installation, to be located near Harrisburg, Pennsylvania, was financed principally by a group of railroad executives who decided to produce their own steel instead of importing expensive English rails. Prime movers in the enterprise were Edgar Thomson, president of the Pennsylvania Railroad, and Samuel Felton, former president of the Philadelphia, Wilmington, and Baltimore Railroad. They were joined by Nathaniel Thayer of the Baldwin Locomotive Works and William Sellers, the outstanding machine-tool maker of Philadelphia.[1]

The company engaged William Butcher, an engineer from Sheffield. Butcher had settled in the Philadelphia area where, in 1864, with the aid of local capital, he had built a steel works. However, owing to poor management and the fact that steelmaking conditions were different in the United States from those he had known in Sheffield, the company was not a success and Butcher left.[2]

The backers of the new Pennsylvania Steel Company were aware of Butcher's capabilities and engaged him to carry out the plans for the new works. Late in 1865 the group made overtures to Winslow and Griswold for a Bessemer license and in January 1866 the final agreement between the Troy group and the Pennsylvania company was signed. The Bessemer plans were to be used along with all the later improvements and changes made by Holley. Elting Morison writes:

> By its principal terms, the Pennsylvania Steel Company was to receive complete and correct working drawings of the best 'Plant' and machinery known to said parties of the first part. The company was also to have the right to send not more than two men to work for not more than two years in the Troy plant, without pay, to learn the necessary techniques for steelmaking. In addition, anyone identified in writing as a member of the Pennsylvania Steel Company could obtain free access for a limited period of time to inspect the works of the Troy plant. In return for these rights and privileges, the company paid $5,000 for working drawings, a royalty of $5.00 a ton on all steel used for rails, a royalty of $10.00 a ton on all steel used for other purposes, and covenanted to keep accurate production records available at all times for inspection by the party of the first part.

211

By a separate and intricate series of negotiations the steel company also obtained the services of Alexander Holley to design and construct the new plant. ... In return for Holley's services, the owners agreed to build the new superintendent a house similar in size and design to one he already possessed and to liquidate the debt of about $20,000 Holley owed to Griswold, on whom he apparently had a drawing account. In return for this liberal arrangement, the company was assigned through a third person, so as not to prejudice the formal and public royalty contract with Griswold and Winslow, the right to manufacture about one blow a day royalty free.[3]

Ground was broken for the new plant in May 1866. For some time extensive work had been going on planning the necessary layout of the works. The construction would involve railroad facilities within the plant, heavy foundations to support the converters and equipment, the converter house, a machine shop, a smith shop, machinery for air fans and water pumps as well as water lines, cranes, and ladles to handle the steel in its various stages from raw materials to ingot. Boarding houses and houses for the workers also were part of the huge steel plant.

In December 1866, the steamer *Indus* carrying the converter and machinery intended for the Pennsylvania Steel Company was wrecked off the coast of Ireland and sank. As a result American companies, principally those located in Philadelphia, now provided the machinery and equipment.[4]

Although Holley had supervised the work and had become affiliated with Pennsylvania Steel Company, he did not move to Harrisburg until January 1867. To some extent Holley had the disposition of an adventurer. He felt a need to explore fully his capabilities and this feeling goaded him on. After his trials and worries had been brought to an end by the visible symbol of the working Bessemer plant at Troy, Holley could have relaxed and managed the plant with little stress. Instead, moved by his need to struggle once more with the "grim sphinx," as he once called the converter, and with the well nigh "inscrutable process," he went to Harrisburg to start anew.

With the move to Harrisburg a new field of experience opened for Holley. For the first time he was brought into intimate relationship with an unusual type of person, the Pennsylvania ironworker who had grown up near blast furnaces and had breathed the smoke-laden air from birth. His childhood had been spent in a community devoted primarily to iron founding and at night from his window he had seen the sky glowing and alive with the reflections of fires from the furnaces. Many of the ironworkers came from the second and third generations of families that had known only life around a furnace. They had lived close to catastrophe from childhood, their ears attuned to the shrill sound of a mill whistle that might mean either a change in shift or the warning of a blast furnace out of control.

As Holley had lived and fought with the converter and given it the faults and attributes of humans, so these men of the iron mills both loved and hated the furnaces but were never content away from the roaring cauldrons. Because of their shared way of life, which had demanded very little formal education, the ironmakers understood each other and consequently formed a close-knit group to which a newcomer could not gain admission quickly. This was especially true for a newcomer with a college education who brought new ideas that threatened change.

The plant of the new Pennsylvania Steel Company was similar in design to the one at Troy. However, instead of a fifteen-acre plot on the banks of the Hudson, Holley now had a one hundred-acre site near the Susquehanna on which to place his converters and the necessary buildings. With a canal and the Pennsylvania Railroad close by, good transportation was insured. Situated in the heart of the iron and coal region in a suburb of Harrisburg called Baldwin, this new works eventually would include a rail mill as well as the Bessemer department. Consistent with his belief that provision always should be made for growth, Holley provided for it in his plans, terming it "systematic growth" and laying out his buildings accordingly. [5]

Holley's new position was that of chief engineer. Building the Harrisburg plant carried with it the usual frustrations, but Holley was wiser and more experienced than he had been at Troy. As a result, the new plant went into operation faster than had the one at Troy. Because there would be some delay in constructing the rail mill and the Pennsylvania Railroad management was most eager to lay American rails, arrangements were made to ship steel ingots halfway across Pennsylvania to the Cambria Iron Company at Johnstown for rolling.

It was this arrangement that brought Holley together with George Fritz, the chief engineer at Cambria. Fritz was a huge man who until he was eighteen had been a farm laborer and then had served apprenticeships to a carpenter and to a patternmaker. At some time he had lost parts of all the fingers on his right hand. His knowledge of ironmaking had been obtained from hard day-to-day battle with the furnaces. There was scarcely a place in the mill where he had not labored. Self-educated, he had little patience with book learning, and, like other ironworkers, he distrusted what he called "assumed scientific knowledge." [6]

Fritz's first impression of Holley can only be imagined. Fritz must have had a definite reaction when he laid eyes on this polished outsider with the manners of a New York gentleman. The same simplicity and directness of approach that had enabled the tall, sturdy Holley to move easily in distinguished company disarmed Fritz. The Cambria engineer must have observed more than these surface characteristics, for the brilliant, outspoken Fritz and the newcomer quickly established a deeply trusting, close, mutual friendship.

Fritz was close to Holley's own age, but while Holley already had ranged the United States east of the Mississippi as well as a good portion of Europe, his new friend seldom had left the state of his birth. Fortunately, Fritz possessed an inquiring mind and an observant eye and shared with Holley a consuming interest in all phases of the Bessemer process.

When the first lot of steel from Harrisburg arrived at Cambria to be rolled into rails, certain difficulties became apparent at once. The mill had been designed and used to produce iron rails. Although it had been strongly built, the machinery could not cope with steel. Broken coupling boxes, split rolls, and engines that refused to drive the machinery would bring the rolling process to a dead halt before the ingot had gone further than its first pass. "There is an inherent cussedness about rolls, which, so far, no man has been able to find," Holley muttered as he and Fritz continued their efforts to coax the rails through the uncooperative mill. [7]

A group of five men, four from the world of iron and one from steel, along with unskilled helpers, eased the recalcitrant mill through these difficult days. All of them would be remembered with respect and affection among steel men of future generations, and their exploits and accomplishments would be told and retold. No one of them can be considered more important than another. Each selflessly contributed the best of his experience to the general fund of knowledge that was needed to solve the new problems of producing steel rails.

Another member of the group was Robert Hunt. He had returned to Cambria from Wyandotte in May 1866, at the time that the Cambria management was considering building a Bessemer plant. Born in Falsington, Pennsylvania, in 1838, Hunt was the only child of a family whose forebears had come to the state with William Penn. His father, a graduate of both Princeton and the University of Pennsylvania, had been a successful doctor, but ill health had forced him to discontinue medical practice while he was still comparatively young. With his wife and child, the doctor went to Covington, Kentucky, where he opened a drug business, and where he died in 1855, shortly after the move. Although only seventeen, Robert Hunt endeavored to carry on the business. After two years, his health impaired by the strain, the boy and his mother returned to Pennsylvania, where they settled in Pottstown.

Young Hunt's cousin was a senior partner in the iron rolling mill of John Burnish and Company where, after a short period of recuperation, young Hunt found work. For several years he labored at jobs that ranged from the puddling of iron to the heating and rolling of the iron rails. His eager intellect led him to learn more about the material with which he worked. In 1859, he took the bold step of entering the laboratory of

Booth, Garrett and Reese in Philadelphia to study analytical inorganic chemistry.

This was the way Hunt gained employment at the Cambria Iron Works. He became their chemist at a salary of twenty dollars per month, and on 1 August 1860, he established the analytical laboratory, the first of its kind to have been maintained by an iron and steel company as part of its organization.

In the spring of 1861, Hunt left Cambria to become night foreman at the Elmira Rolling Mill, which had been organized by his cousin at Elmira, New York. In April Hunt volunteered for duty and served throughout the Civil War. He was mustered out with the rank of captain, a title of which he always was proud. He returned to Cambria as chemist, but in May 1865, the company sent him to Wyandotte to study the pneumatic process so that, whenever the Cambria management decided to make steel, he would be prepared to supervise the installation.[8]

One month after Hunt arrived at Wyandotte William Durfee, whose disillusionment and discouragement there already has been described, left the plant. Llewellyn Hart had resigned early in 1865 to enter the employ of the works at Troy, and Ignatius Hahn, from the Krupp works in Essen, had stayed for only a few months. Thus it fell to Hunt, with only a month's apprenticeship, to take over steelmaking at Wyandotte.

Hunt wrote in later years of this experience. "As Mr. Hahn's retirement left the company without any practical steelmaker, and the works had thus far been conducted on an experimental basis, the proprietors determined upon making the most hazardous experiment of all and put them in charge of the writer, who had gone there a few weeks before, in the interest of the Cambria Iron Company. In accordance with this arrangement the writer made his first 'blow' and by some strange fatality happened to 'turn down' at just the right time."[9]

Hunt served his steel apprenticeship at Wyandotte for a little over a year. During this time he tried out various American pig irons in the converter. In October 1865, the first heat of steel was made from pig iron that had been produced from the famous Missouri Iron Mountain ore and the results were good. While Hunt had his problems at Wyandotte, they were never considered to be in the same category as those that had beset both Durfee and Holley. Hunt, always generous in giving credit where it was due, said of the pneumatic enterprise:

> While at the Wyandotte works, steel was made at an earlier date, the Troy establishment was the first to bring the process to a commercial success. Not having been personally connected with these works during the early days, I cannot so fully realize the doubts and difficulties through which they passed, but I do know from Wyandotte, to say nothing of any later experience, that it has required faith made perfect, to carry one through the sea which seemed

Robert W. Hunt. (Courtesy of H. H. Morgan, Robert W. Hunt Co.)

to be bounded by no shores. As I have often expressed it, if we, knowing there was a way through all our troubles, felt so hopeless, what must have been Bessemer's pluck, to enable him to persevere through his difficulties, when the desired end was known only through faith. [10]

When Holley arrived at Harrisburg, Robert Hunt was the second man of the group of five whom he met. Both remembered their previous meeting with some embarrassment. This was the occasion on which Hunt had been refused admission to the Troy plant by the polite but firm Holley. Now the two met again, for Hunt was in charge of the steel department at Cambria and the honor of rolling the first commercial order of steel rails would fall to him and his mill. [11]

That the men who then were producing steel still had much to learn is borne out by Hunt's description.

The word Steel then conveyed a very definite conception to the minds of all men. The character and peculiarities of its behavior under treatment, either hot or cold were expected to be like those of tool-steel, as that had been observed while under the manipulations of the maker, or afterwards of the smith. And the one thing most earnestly impressed upon every one's mind, and emphasized as the very essence of good workmanship was, that steel would bear only the most moderate heat. In the converting-house, all the possible practices of crucible-steel teeming were introduced. The ingot-moulds were carefully brushed out, heated and smoked before being used. When the steel was teemed, all doors and windows of the casting-house were closed, etc., and time was not spared on any of the details. This was the English practice; and we in America followed.[12]

Such practice proved to be expensive. It was not possible to produce cheap steel rails on mills that had been built to turn out iron rails. Handling each heat separately made the entire practice of producing steel time-consuming and costly. In addition, the men who produced steel rails were paid twice as much as those who rolled the iron product, since it was expected that they could roll only half as many steel as iron rails. Since Holley was providing the steel ingots from Pennsylvania Steel Company converters, and Fritz and Hunt were producing the rails, the three men soon were working as one. As they labored with their problems, they pooled their ideas. First they had to determine the future of steel rails. Initially, no one thought that steel would displace iron. It was estimated that only 50 percent as many steel rails as iron would be produced annually, and that the steel ones would be used only in locations of especially heavy traffic.

The producers could not yet foresee how they might produce cheaper steel. After the ingots were delivered at Johnstown, they were rolled into blooms and then allowed to cool. It was believed that the steel would be inferior if it retained any portion of the initial heat. If time permitted, the ideal method was to stack the blooms in the yard to weather for six months. In the next step, they were carefully cold-chipped before being charged into the reheating furnaces. After one heat of blooms had been removed, the furnace was allowed to cool. Then, the next heat was charged and the furnace temperature slowly raised, giving the steel plenty of time to soak. The truth of the matter was that not one of the three—Fritz, Hunt, or Holley—was certain of the best way to treat steel in order to produce superior rails. Their uncertainty resulted in a cautious approach to the entire process.

In making iron rails, the established practice was to work the ingot down to a bloom with a steam hammer. Then the bloom was rolled into a rail. This same procedure was used on the initial shipments of steel sent to Johnstown from Harrisburg. The three were dissatisfied with the method, because the rails were coming off the mill in unsatisfactory condition. As they studied the problem, George Fritz questioned the need for using the hammer at all. Acting on his suggestion, special rolls were designed and hammering was eliminated. The ingots now could be rolled into blooms fresh from Harrisburg, and one costly step eliminated. The suggestion then was made that the great caution about the rolling temperature of the blooms might be unnecessary. Gradually the heat was raised until eventually the blooms were run into the mill with the cinder still dripping from their seething sides.

Another person brought into the small circle at Harrisburg and Cambria was George Fritz's older brother John, a mechanical genius like his brother. John Fritz was chief engineer and superintendent at the Bethlehem Iron Company. Although the companies themselves were rivals, their engineers were as close as only brothers can be. Hunt said of the relationship, "In all my experience, I have never met with such mutual admiration, trust and love as existed between those two brothers. The brains of one were at the command of the other."[13]

Both brothers had begun as laborers learning the manufacture of iron. When John Fritz had entered Cambria, he had brought George with him. The older Fritz had risen through a succession of jobs in the iron industry before he achieved the position of chief engineer and general superintendent at Cambria in 1854. He then had moved on to design and take charge of building the Bethlehem Iron Company plant, which included forges, blast furnaces, and rolling mills. The Bethlehem interests were planning to erect a Bessemer plant in the near future and the group, the center of Bessemer knowledge in America, worked together to plan the installation. John Fritz enjoyed the long hours they spent discussing their mutual problems. The group also shared Holley's elation the day he visited Johnstown with the exciting news that his dream had been realized, and the Bessemer plant at Harrisburg was making four conversions at each turn, or eight per day, and was producing forty tons of ingots.

That year, in addition to the trials and worries of his job, Holley had known personal sorrow. A son, Alexander Lyman Holley, Jr., had been born to the Holleys but had lived only a short time. The men at Cambria had felt Holley's loss keenly. As sensitive to Holley's troubles as they were to those of a blast furnace, each in his own way, tried to make Holley's year easier.

Mary Holley now was kept busy at home with young Gertrude and little Lucy as Holley worked the clock around getting the rail mill finished and into operation. George Fritz and he shared many hours planning the new mill, which would be the first in America specifically designed to roll steel rails. As Holley worked, he felt the weighty responsibility of knowing that a skeptical industry awaited the results. It was an enormous undertaking for a man of his age, and at that time Holley was grateful for the support of his understanding friends.

The Fritz brothers, both skilled mechanics, had no use for a machine that did not operate at top efficiency. John Fritz often said, "It matters not how well you may be skilled in all other branches, if your machinery is imperfect you will surely come to grief and the only possible way to attain success is to obtain a thorough knowledge of both the engineering and the mechanical construction of all the machinery used in the art."[14] His brother George was intrigued with the problems Holley faced in building the new mill. Inasmuch as the two had planned it together, the Pennsylvania mill was like the Cambria mill in general plan and equipment. The furnaces, settings, and the rail train were all George Fritz's design.

The rail mill went into operation in May 1868 and Cambria saw no more steel from the Harrisburg plant. The rail train was a miracle of performance. The heaviest in America, it consisted of four sets of three-high rolls, twenty-three inches in diameter. One set was used to roll ingots into ten-inch blooms, the next roughed the blooms on their way through to their destiny as rails. The third set finished the rail, and the fourth was used for rolling beams or if a rail of a different pattern were needed.[15]

As the Fritz brothers, Hunt, and Holley worked out their problems at Cambria and Pennsylvania, another young man, a latecomer to Cambria, entered the circle. His baptismal name, William Richard Jones, would be recognized by few, but the name Captain Bill Jones is that of a man who was to become a legend in the iron and steel industry and about whom numerous fantastic stories are told.

Notwithstanding the vocabulary that he is known to have developed in his mature years, Jones was the son of the religious and intellectual leader of his family's Pennsylvania Welsh community. Bill had been born in 1839 in the squalid village of Catasauqua. The father earned a poor living as a patternmaker, and it was necessary for the boy also to earn. When only ten years old, he was apprenticed to the Crane Iron Company.[16]

A slight, roistering, blue-eyed boy, young Jones had a temper that already had evidenced itself when he nearly had wrecked the Catasauqua schoolhouse because he believed that the teacher had whipped one of his friends unjustly. His curiosity was of the sort that impelled him to cut open his fingernail to find out what was underneath. Jones at ten possessed a mind that engaged itself with whatever was at hand. His father owned a collection of books. Jones learned to read from works such as those by Plutarch and Shakespeare, and he never lost his ability to quote from them. The smoothness and versatility, if not the substance, of his profanity probably stemmed from his early readings in the classics.

Anyone who enters an iron plant at the age of ten can start only at the bottom. Young Jones worked first in the foundry and later moved on to the machine shop. By the time he was sixteen he was receiving the wages of a journeyman machinist. He had left the Crane Iron Company for another company, and by 1856 was working as a machinist at I. P. Morris & Company in Philadelphia. Out of work during the panic of 1857, he tried his hand as a lumberman, a farmer, and finally as a raftsman on the Ohio River, not an occupation for the weak or sensitive. As he made his way up and down the river, it is probable that the awesome vocabulary that had begun to develop at the ironworks reached the maturity that would be the envy, despair, and joy of men who later worked with Jones.

After almost two years as a raftsman, in 1859 he returned to his craft of machinist at the Cambria Iron Company. In a very short time he was off again, this time to Chattanooga, where he helped build a blast furnace and married Harriet Lloyd after an intense courtship. At the outbreak of the Civil War, he returned to Pennsylvania with his wife and once again secured a position at Cambria at the rate of two dollars per day. But in 1862, when President Lincoln issued the call for nine-months men, Jones enlisted and marched off with the 133rd Pennsylvania volunteers. [17]

After the war, his taste for wandering at least temporarily satisfied, Jones returned to Cambria as assistant to George Fritz, who had taken his older brother's place as general superintendent after he moved to Bethlehem. By this circuitous route, Jones arrived upon the Cambria scene in time to become one of the group of American authorities on the production of Bessemer steel who were pooling their aptitudes, knowledge, and experience to advance the new technology.

The Bessemer converters and the rail mill at the Pensylvania Steel Company were now in fair operating condition but certain inefficiencies there still bothered Holley. Not content that the plant had been built and was in operation, he continually searched for the inadequacies in the converter or the rail mill that prevented the perfect blow or the finest rail. He later remarked, "These manufactures can neither stand still nor be suddenly metamorphosed. . . . We must feel our way into larger development; we must work gradually into better practice; we must improve a little at a time." Economies might be realized by small improvements, he thought, but it was ridiculous to mend an outmoded machine. A necessary change should be made at once rather than nursing along machinery until its collapse. [18]

Holley soon had the chance to put his theories into practice. On 19 October 1868, the roof of the plant he had built at Troy caught fire, and before the fire apparatus could quench the flames, the major part of the roof collapsed. The collapse, combined with the intense heat, damaged most of the Bessemer machinery. The works were in ruins, and Holley returned to Troy as manager to rebuild them. He had been away from Troy for almost two years; during this time two managers had come and gone and John Griswold had bought Winslow's interest in the Bessemer steel works. Holley's relations with Griswold always had been close and marked by feelings of mutual trust and confidence.[19] At Harrisburg, in contrast, Holley for some time had been feeling that the owners were exerting unwarranted pressures on him to increase production of steel rails. Thus, he left Harrisburg with a feeling of relief from constant anxiety, a feeling tinged with regret at the destruction of his cherished works at Troy.

During his Harrisburg years Holley had gained not only the friendship of the men in the circle of five but also that of the laborers who worked under him in the plant. They mourned his departure.[20]

At Harrisburg, as at Troy, Holley had tried to hire men with no previous steel experience to work in the converter department. He found that untrained workmen had fewer preconceived notions and prejudices. He knew that workmen always were unwilling to concede the usefulness of improvements, and if they were not directed carefully, they would follow their own inclinations rather than the directions for a new process. For this reason, Holley may have been more closely associated with his workers than was customary. Holley's charm and habit of digging in and working alongside them had won them over completely. Once convinced that what he had planned was an improvement that could lighten their labors, the men would turn in with a will to make amends for their obstinacy and prejudice.[21]

From poor backgrounds, with little education, and unused to displays of emotion, the laborers nevertheless felt a need to let Holley know in some tangible way of their regard and their regret at his departure. They collected $500 among themselves for a suitable gift. C. T. Arnberg, a draftsman at the works, designed a perfect miniature of a Bessemer converter, and A. A. Moore, a machinist, drew a base for the model that would contain a specimen from the first steel rail made at the Pennsylvania Steel Works. A Harrisburg jeweler fashioned the converter out of solid silver.[22]

The renovation of the Bessemer plant at Troy was to occupy Holley during all of 1869. That year also marked the first failure of a Bessemer plant in the United States. The Freedom Iron and Steel Works, located near Lewiston, Pennsylvania, had been the fourth Bessemer works built in the United States and the only one built during Holley's lifetime with which he had no connection.[23]

In 1866 the company was reorganized and named the Freedom Iron and Steel Company with the intention of converting the plant into a Bessemer steelworks. John Wright, an old friend of Andrew Carnegie and president of the company, went to England and purchased the most complete assemblage of Bessemer machinery then available. The only

Silver converter given to Alexander Lyman Holley by the workers when he left the Pennsylvania Steel Company. (Courtesy of Frances Olmsted.)

part of the mechanism that was not English was the blowing engine constructed by I. P. Morris and Towne of Philadelphia. According to Hunt the company intended to make boiler plate and tyres and when that failed, they switched to the great American product, rails. But Carnegie's biographer relates that before the building of the Bessemer plant, Carnegie already had been an enthusiastic supporter of a steel-faced rail pattern invented by Thomas Dodd in England and had invested in the project while still in England. Despite the fact that he never had clear rights to the process, he did produce these rails and went to great lengths to induce his railroad friends to buy them. But J. Edgar Thomson, who had purchased some of the rails, told Carnegie frankly that they were so poor that the patent might as well be abandoned. Subsequently at the Freedom plant Carnegie became involved with another rail patent, for the Webb steel-headed rail. Unlike the Dodd patent, the Webb patent covered the rerolling of old iron rails with steel heads. Again,

Carnegie's experience was the same: confusion over patent rights and the failure to produce a satisfactory product.[24] In his autobiography, Carnegie makes no mention of the Webb patent, but maintains that if the Bessemer process had not been successfully developed, "I verily believe that we should ultimately have been able to improve the Dodd process sufficiently to make its adoption general."[25]

Carnegie claims that he had not failed to notice the growth of the Bessemer process. He writes that his friend John A. Wright had visited England especially to study the process. "He was one of our best and most experienced manufacturers, and his decision was so strongly in its favor that he induced his company to erect Bessemer works. He was quite right, but just a little in advance of his time. The capital required was greater than he estimated. More than this, it was not to be expected that a process which was even then in somewhat of an experimental stage in Britain could be transplanted to the new country successfully from the start. The experiment was certain to be long and costly, and for this my friend had not made sufficient allowance."[26] Although here Carnegie does not mention his own involvement, other letters show that he intended to use the Webb process at Freedom, rerolling iron rails and using Bessemer steel produced at the plant for the heads.[27]

After the first blow in May 1868, the new company tried for a year to produce satisfactory steel. But their iron supply contained too much phosphorus, and within a year the company was out of business.[28]

The idea of steel-headed iron rails was not new. The first ones manufactured in the United States were rolled on 1 March 1866, at the Trenton Iron Company. Carnegie must certainly have been aware of the process in use there, especially since it was successful. The first large order went to the Pennsylvania Railroad for trial. Other railroads followed, and by 1869 the Trenton Iron Company, using both puddled steel and Siemens Martin steel, was producing 24,000 tons of these rails a month.

As Abraham S. Hewitt's biographer points out, the transition from iron to steel rails was not immediate and steel-headed rails were important in the changeover.

> Steel-topped rails were to prove a mere temporary bridge from the iron rail to the all-steel rail; but they were a useful bridge, and the Trenton mills attained an easy leadership in their American manufacture. Indeed, they became one of the principal manufacturers of this peculiar product in the world, for the similar rails from England and Westphalia were not so good. Users of the Trenton rails bore testimony to their durability, cheapness, and safety. Various historians have written as though the all-steel rail made a swift and easy conquest of our important railroads after the war, but in reality the part-steel rails were an important factor in the situation. When, in 1866, an American named Hart applied for a patent on the steel-topping

process, Hewitt and his brother Charles easily proved that they had been the first to employ it.[29]

Another Bessemer works, the fifth in the United States, had been constructed by the Cleveland Rolling Mill Company at Newburg, Ohio. It was completed in September 1868, from plans that had been furnished by Holley. The design was similar in most details to that of the Pennsylvania Steel Works. The machinery, with the exception of one converter, was built by the Cuyahoga Steam Furnace Company headed by J. W. Holloway. The construction at Newburg was directed by H. Gmelin, an engineer who had traveled widely due to his positions with the various Bessemer steel-producing companies. He started his career with the Bessemer works in Styria and in Carinthia in southern Austria and moved on to Sheffield where he remained for some time. He had accepted the position only temporarily because he already was committed to build a new blast furnace in Austria. His successor was to be John Thompson, who had been superintendent at Troy under Holley's watchful eye. When Thompson finally arrived to take over, Gmelin returned to Austria. The fifth Bessemer plant produced its first conversion of steel on 5 October 1868, five months after the opening of the ill-fated Freedom plant. It went on to become one of the most successful works in America. Its advantage was its situation regarding raw materials. As Holley described it, "Cleveland is the principal outlet of the iron and ores of the Lake Superior region, most of which are well adapted to Bessemerising. The best raw coal and coke irons in this country, are made within a circuit of seventy-five miles from the works. Railway supplies for the West and Northwest, and lake and manufacturing machinery, are already furnished largely by the various iron works in the region, and there are few, if any, better points for the manufacture and supply of steel rails and products."[30]

Although Holley at the time was busy in Troy rebuilding the Bessemer works, he also kept a watchful eye on both the Pennsylvania and Newburg plants for which he continued to feel a responsibility. In January 1869, he received an offer to work again in the field of technical journalism. Looking back at his pleasant experience with the *Railroad Advocate,* he could not refuse. Holley's experience had given him outstanding training in reporting and writing on technical subjects. His ability to organize material was recognized as so unusual that Van Nostrand, the New York publishing house, sought him out to edit *Eclectic Engineering Magazine,* a new publication. He was able to keep the position for about a year until it became evident that being both editor and engineer demanded more time than he had at his disposal. Yet despite all that he had accomplished, Holley seemed to feel that he had not yet proved himself when he wrote from Troy in August of 1869, "I have not got

Alexander Lyman Holley about 1869. (Courtesy of Mrs. Ashbel Wall.)

Mary Holley about 1869. (Courtesy of Mrs. Ashbel Wall.)

along far enough in life to look back on much work or much fruit from it; but I have lived long enough to conclude with certainty, that leisure is the hardest thing in life to get along with. I try to have as little of it as possible."[31]

The rebuilding of the Troy plant occupied Holley for the remainder of 1869. The original converter was maintained in constant operation turning out some 300 tons of ingots a month, while the walls of the new plant were going up and the machinery was being installed. Availing himself of all the knowledge acquired from his Harrisburg experience, Holley extensively remodeled the works, made changes in the melting or cupola house, and made the blowing engines the largest in the country. He often must have thought of the difficult and harrowing experiences connected with the construction of the first Bessemer works. Referring to it in later years he said, "It was a hand-to-hand fight, involving mechanical details, refractory linings, celerity of operations, regularity of melting and conversion and economy of labor. With every fact written in his book, the closeted scientist could no more adequately prescribe the practical conditions of improvement, than could the student in optics specify in work and formulae the glory of an Italian sunset."[32]

Although Holley now was the master instead of the servant, never for one second did he underestimate the potential of the converter. The men, too, still stood in awe as the tilting monster threw up its flame with a threatening roar, and they still could be astonished that the giant had been tamed by the quiet engineer standing near them. One rolling mill engineer summed up the situation when he watched the flame shooting from the nose of the vessel. "A puddle ball or a rail pile will lie still on the floor if anything breaks down but if five tons of fluid steel gets the upper hand of you, there's no telling where it will stop."[33]

The Wyandotte plant also was abandoned during 1869, the year in which Freedom Iron and Steel Company failed. With the passing of these two companies, Holley was left as the dominant engineer in the Bessemer field, a circumstance in which his well-known modesty, friendliness, and warmth of sympathy might have given way to conceit. This never was to happen. Jarvis Edson, an engineer, told of paying a visit to Troy in order to observe what the engineers had been doing during his extended absence abroad.

> Knocking at the door with nothing to contribute and no claim upon their attention, I asked if I might see the process of manufacturing Bessemer steel. After remaining seated a few minutes, a messenger came and took me to the end of the works where I was introduced to Mr. Holley. My case was stated plainly to him—I was simply a trespasser upon his time and hospitality. But for two or three hours it was impossible for me to get away from him. I was shown the same attention which probably I would have been shown had

I been the president of a railroad or the president of some scientific society. I found out enough to appreciate what he was doing. I was shown the innermost parts of the interesting process as it was carried on. I was not only shown the success but I was led into the secret of failure, of mistakes and the hard road over which he had traveled, and it was with great difficulty that I left him rather than increase the debt of gratitude which I owed him. [34]

Holley had traveled far since the days of the *Railroad Advocate* and his associations with Zerah Colburn, whom he had seldom seen since. They had been fellow passengers on the maiden voyage of the *Great Eastern,* and they must have seen one another occasionally in London where Colburn was editor first of the *Engineer* and then of *Engineering,* a journal that his restless and feverish energy had led him to start. In Holley's estimation, as well as that of others, Colburn had developed the best serial publication in the field. But his well-known unstable temperament, coupled with overwork, finally had their effect. In 1867 Colburn again began to drink heavily. Impulsively, he left *Engineering* early in 1870, went first to Paris, and then sailed for America. In New York, avoiding friends, he wandered the streets or spent his time in cafes. Finally, he journeyed to Belmont, Massachusetts, where, at the age of thirty-eight on 20 April 1870, he committed suicide, a tragic conclusion to a brilliant career. Remembering Colburn's fine qualities and the happy associations of their youth, Holley attended Colburn's funeral and wrote a sensitive obituary for the *New-York Times.* [35]

On 12 January 1870, the first blow was made at the rebuilt works at Troy. The rolling facilities had not been installed yet and ingots were shipped a mile up the river to the Rensselaer Rail Mill owned by Griswold and Corning where they were rolled into blooms and rails. Following the completion of the new Bessemer plant, Holley turned his attention to the design of a blooming mill. With the cooperation of James Moore and William George, who shared the patent of the new mill with him, the thirty-inch blooming train was ready by the beginning of 1871. The top and bottom rolls were stationary while the middle roll moved up and down to obtain the necessary compression. It was equipped with front and back lifting tables that were raised by hydraulic power. By means of these tables, which carried small rollers, a twelve-inch ingot big enough to make two rail blooms was placed and then pushed into the rolls of the mill. With this improvement only eight men were necessary to operate the mill. The mill was used until 1872, when Holley added improvements developed by George Fritz, among them driven table rollers and a pusher. The number of men required to operate the mill and the new tables now was reduced to three men and a boy. [36]

The Cambria Iron Company at Johnstown, Pennsylvania, became the sixth to install a Bessemer plant. In 1856 Daniel Morrell had been interested in the pneumatic process and, as an early supporter of William Kelly, had permitted him to build a converter in the yard at Johnstown to carry on experiments. In this way, Cambria gained the distinction of having been the first plant to harbor a converter. Whether or not it ever produced steel is not certain. Robert Hunt said of it, "It was calculated to convert about a half a ton of metal, and received its blast from the foundry blowing engine. But I never heard even a tradition of a perfect conversion made in this vessel."

Morrell subsequently became one of the first partners in the Kelly Pneumatic Process Company and played a leading role in the amalgamation of the rival patents. At that time, Morrell proposed to the board of directors of Cambria that if he were paid the value of his Cambria stock, he would put the money into building a Bessemer converter. This suggestion met with silence, and Morrell temporarily gave up the idea. But by the time five Bessemer plants were in operation, and steel rails were being laid in large numbers, even the most conservative mind had been convinced. In 1870, George Fritz, the chief engineer, was given the signal to go ahead. Holley, of course, was called upon to design the plant.[37]

Holley and Fritz eagerly entered upon the new venture. Assembling the labor force was Hunt's job. At that time, rolling mill men preferred to work in new plants, and, even though the projected Bessemer plant was far from being a certainty, many rollers wished to switch over. In spite of his enthusiasm for the new department, George Fritz was unwilling to have the fine organization of other mill departments disturbed and gave strict orders that Hunt was not to raid any of them to obtain skilled men. This presented a problem for Hunt. The order eliminated even his trusted associate, Jacob Dunlap, who had gone to Wyandotte with Hunt as a puddler and had been promoted to puddle mill foreman on his return to Cambria. He was keenly disappointed when he learned that he could not join the Bessemer department. John Rinard, who had been in the army with Hunt and had advanced to heater in the rail mill, actually resigned and reapplied for work as a laborer in his eagerness to join the new venture.

In the Bessemer department, Holley was the leader and George Fritz the associate, but in the installation of the blooming mill, their roles were reversed. Fritz introduced two features in the blooming mill that were different from those at Troy: driven rollers in the tables and a hydraulic pusher for turning over and moving the ingots on the tables. The number of men required to operate the mill dropped from eight, as at Troy, to three men and a boy. The design of the mill became the standard for Bessemer works in America.[38]

The first blow at Cambria was made on 10 July 1871, under the direction of Robert Hunt, while Holley and George Fritz looked on with proprietary interest. The rails that were rolled from the blow sold for $104 per ton. Of the more than seventy men who had labored to build and bring the plant into operation, Holley, Fritz, and Hunt probably were the only ones who ever before had seen the dazzling spectacle of a converter in action. These men believed they had shared an experience that bound them together, and forty years later those who were able met in Johnstown to honor those of their number who were no longer alive. The men gathered to reminisce with Hunt, the only survivor of the three who years before had guided the building of the original plant.[39]

Cambria became a symbol in the history of the iron and steel industry. For a generation or more, it was the training ground for the countless young men who entered the iron industry in America. The pattern for that training had been established by George Fritz's brother John, first at Cambria and later at the Bethlehem Iron Company, and for years a man could present no finer reference than to say that he had been one of Uncle John Fritz's boys.

NOTES

1. Elting E. Morison, *Men, Machines, and Modern Times* (Cambridge, Mass.: M.I.T. Press, 1966), p. 163.
2. Alexander Lyman Holley and Lenox Smith, "American Iron and Steel Works: III. The Midvale Steel Works," *Engineering* 23 (30 March 1877): 239.
3. Morison, *Men, Machines, and Modern Times,* pp. 164-65. Morison had access to the papers of the law firm of Blatchford, Seward, and Griswold.
4. *Engineering* 2 (28 December 1866): 505.
5. *Troy Daily Times,* 27 July 1868.
6. *Engineering* 16 (29 August 1873): 168-69.
7. Robert W. Hunt, "Steel Rails, and Specifications for Their Manufacture," *Transactions of the American Institute of Mining Engineers* 17 (1888): 227-28.
8. Information supplied by H. H. Morgan, president of the Robert W. Hunt Company.
9. Robert W. Hunt, "A History of the Bessemer Manufacture in America," *Transactions of the American Institute of Mining Engineers* 5 (1876-77): 203-204.
10. Ibid.
11. R. W. Hunt, "Steel Rails," p. 228. The first steel rails rolled in the United States were those that Ward had produced at the North Chicago Rolling Mill in 1865 from steel supplied by the Wyandotte works, but that operation could only be considered the experimental rolling of six rails. Not until August 1867, was it possible to say that steel rails were being produced on a commercial scale in America.
12. Ibid., pp. 227-28.
13. Ibid.
14. John Fritz, *Autobiography of John Fritz* (New York: American Society of Mechanical Engineers, 1912), p. 47.

15. *Troy Daily Times,* 27 July 1868.
16. Herbert N. Casson, *The Romance of Steel* (New York: A. S. Barnes & Co., 1907), pp. 21, 22, 23. The Crane Iron Company operated the first furnace in America to use anthracite coal successfully as fuel.
17. Ibid., p. 23. The 133rd Pennsylvania regiment saw service in the Army of the Potomac and participated in the battles of Fredericksburg and Chancellorsville. Jones soon earned a reputation for his impetuous and daring ideas. When a pontoon bridge was being constructed across a river, Jones did not wait until it was finished but shouted, "Come on boys let's swim." In setting the example, however, he split his nose diving into the water, which was only a few feet deep. When he was badly wounded at the crossing of the Rapidan, characteristically he refused to leave the ranks. When it became evident even to him that he was in no condition to continue, he reluctantly left the company and returned to Cambria. Still hopeful of getting back into the war, he organized another group of men known as F Company of the 194th Pennsylvania Volunteers, of which he was made captain. Jones was well enough in 1864 to go to Baltimore to serve as provost-guard.
18. Alexander Lyman Holley, "Some Pressing Needs of Our Iron and Steel Manufactures," *Transactions of the American Institute of Mining Engineers* 4 (1876): 77.
19. Hunt, "History of Bessemer Manufacture in America," p. 206.
20. Casson, *Romance of Steel,* p. 19. At about this time Holley began what was to become a habit, when confronted with a difficult problem, of packing his bag and journeying around the country to visit his friends. Often Fritz or Hunt or Jones or some one of the others might look up at any time from his work to see Holley coming through the gate. Just as in the old days at Harrisburg or Johnstown, the friends would sit down and talk, sometimes for hours, until the problem was solved or the worry concerning it eased.
21. A letter from William F. Durfee to the *Bulletin of the American Iron and Steel Association* 28 (5 September 1894): 194 reads as follows:

    Notwithstanding their manifest and manifold advantages the triumph of the inventions of the Messrs. Fritz was not without opposition for in 1882 the pseudo practical men who mismanaged one of the western mills advised its owner that it was utterly impossible to roll rails in a three high mill as it was difficult enough to roll rods and small bars in such a mill but as for rolling rails the idea was absurd. And when these parties were finally compelled to yield to more progressive ideas and submit to having a mill in their charge made to conform somewhat to the best practice of its competitors such was their animosity against the new three high train and the engineer who had advised and erected it that they surreptitiously removed all the keys in the holding down bolts of the roll train and part of those under the engines with the intent of wrecking the machinery when it was started but fortunately the engineer in charge of the improvements discovered this treason against property and progress in time to avert disaster.
22. The document that accompanied the miniature converter occupied an honored place in Holley's scrapbook and bears the marks of the many fingerprints it acquired as it was passed from worker to worker for signature. Holley was deeply touched by this expression of their respect (Document in Scrapbook, Author's Collection).

    Again five years later, he was pleased by the receipt of a letter from Joseph Potts, written for the workmen at the Pennsylvania Steel Company that told him that the workers in the plant had decided to organize a library, which they named in his honor.

    Dear Mr. Holley: Our men have at last succeeded in organizing a Library Association which they have chosen to call the A. L. Holley Library Association of Baldwin, Pa. in remembrance of you and your interest in the subject, as also in acknowledgement of your kind gift of a number of volumes to form the nucleus of a Library for their use. All of which I was directed, as Secretary to inform you of, with their grateful thanks. I thought you would be gratified to hear of this, as denoting the kindly remembrance we have of your association with us, altho some years, eventful to us all, have elapsed since we had much intercourse.

The movement originated in the Rail Mill, through Jno. A. McClure who was subsequently made President, John W. Grove, Treasurer, C. W. Babb, the Nebingers and others.

We have a little over $200 subscribed. The members pay $1 admission fee, and fifty cents per month, dues. Have rented a room on the Turnpike between our house and the Canal bridge, and propose at once to secure some papers, magazines, etc., and have a reading room open every evening for the members.

I am glad to hear of your safe return and with respect and kind remembrance remains as always, Your friend, Joseph Potts

23. Hunt, "History of Bessemer Manufacture in America," p. 209. Carnegie manuscripts, vol. 3, Manuscript Division, Library of Congress, Washington, D. C. During the Civil War, a tax was levied on income and Carnegie's report of income for 1863, among other items, shows an amount of $250 from the Freedom Iron and Steel Company.

24. Hunt, "History of Bessemer Manufacture in America," p. 209; Joseph Frasier Wall, *Andrew Carnegie* (New York: Oxford University Press, 1970), pp. 255-57 passim, 258.

25. Andrew Carnegie, *Autobiography of Andrew Carnegie* (Boston: Houghton Mifflin Co., 1920), p. 186.

26. Ibid., p. 185.

27. Wall, *Andrew Carnegie*, p. 260.

28. Hunt, "History of Bessemer Manufacture in America," p. 209. William Durfee wrote of the misadventure: "As late as 1868 a large establishment for the manufacture of steel (in which over a million dollars was invested) commenced operations in western Pennsylvania, and at the end of one year it was abandoned and dismantled, the whole of the investment having been utterly lost in consequence of attempting to use material which an analysis costing not over fifty dollars would have shown to be absolutely unfit for the purpose intended" (William F. Durfee, "The Development of American Industries since Columbus: IX. The Manufacture of Steel," *Popular Science Monthly* 40 [November 1891]: 23).

29. Allan Nevins, *Abram S. Hewitt, with Some Account of Peter Cooper* (New York: Harper & Brothers, 1935), p. 236.

30. Hunt, "History of Bessemer Manufacture in America," p. 209; *Troy Daily Times,* 27 July 1868. The wonder of actually making steel in the curiously shaped Bessemer container still was a novelty to many men, as is illustrated by a story told by one old timer who had served his apprenticeship at Newburg. The Newburg mill had imported a group of Yorkshiremen to run the new plant. In their own country, when some extraordinary feat was accomplished, the company distributed rounds of ale to the workers in recognition of their achievement. The group of Englishmen at Newburg were so impressed each time they brought off a successful blow that they felt an allowance of ale was in order (William Hainsworth, "Reminiscences of an Old Iron-master," *Bulletin of the American Iron and Steel Association,* 26 (23 and 30 March 1892): 82.

31. *Memorial of Alexander Lyman Holley* (New York: American Institute of Mining Engineers, 1884), p. 135.

32. Alexander Lyman Holley, "The Inadequate Union of Engineering Science and Art," *Transactions of the American Institute of Mining Engineers* 4 (1876): 193.

33. *Engineering* 14 (22 November 1872): 357.
34. F. R. Hutton, "F. Holloway" [Remarks by Jarvis B. Edson], *Transactions of the American Society of Mechanical Engineers* 18 (1897): 641.
35. "Obituary of Zerah Colburn," *Van Nostrand's Eclectic Engineering Magazine* 2 (June 1870): 654–55; *Engineering* 9 (20 May 1870): 361.
36. Hunt, "History of Bessemer Manufacture in America," pp. 206, 207, 210.
37. Robert W. Hunt, "Cambria Steel Workers' Reunion," *Iron Age* 88 (5 October 1911): 740.
38. Hunt, "History of Bessemer Manufacture in America," pp. 210–11.
39. Hunt, "Cambria Steel Workers' Reunion," p. 740.

XVI · John Fritz

ohn Fritz was born in 1822 at Londonderry, a small town in Chester County, Pennsylvania. His father had immigrated from Germany when he was only ten. His mother, of Scotch-Irish Presbyterian stock, was born in Londonderry and lived there all her life. Fritz and his younger brothers, George and William, all worked on their father's farm from childhood. At six, John began planting corn and carrying rye whiskey and fresh water to the men working in the fields. He spent a few months during the winter seasons going to school but his education never went beyond that.

At sixteen, he was apprenticed to a blacksmith to learn to shoe horses and repair wagons. He also learned many small machine shop jobs such as repairing threshing machines and making parts for the local grist-mills, sawmills, blast furnaces, and forges. After laboring with primitive tools for almost six years, he decided that his aim was to learn the iron business and, in 1844, he struck out to search for a job in that field. Having unsuccessfully tried the Phoenix Iron Company, he set out for Trenton, where the largest ironworks in the area was located. On his way, he stopped at the Norristown Iron Company, which was building a new mill; it was there that he found his first job in the iron business.

At first young Fritz felt strange among the older and more experienced men, but he performed his tasks with such enthusiasm and efficiency that he soon was placed in charge of all the machinery. The cogwheels were in constant need of repair. Either the cogs were out of mesh or the wheels had to be replaced, and such repairs might close the plant for as much as a week.[1]

Fritz was so involved with machinery problems during the day that there was no time then for him to learn the true nature of the iron business. Norristown operated in two shifts, and Fritz soon began to return to the mill after supper to observe the entire process. After the heat was charged into the furnace and as they waited for its completion, the founders, all of them carrying the weight and brawn required for their jobs, would sit down on a pile of pig iron and haul out their pipes for a smoke. Fritz finally gained the confidence of the picturesque group, and was invited to join. Most of them had served their apprenticeships in the ironworks of England and Wales, and they spun endless yarns

about work in the mills in the old country. As they described the construction of the furnaces, the operation, and puddling practices, the
young man stored away every fragment of information.

It was the puddling process that Fritz watched first, and before long
he began to work shoulder to shoulder with the puddlers. The dramatic
process of iron being brought "to nature" and then being formed into
glowing fiery balls dripping slag on their way to the squeezer never
failed to fascinate him.

The rolling and finishing departments next claimed his attention.
The labor in the rolling mill was less onerous than in the puddling
department, but different problems presented themselves. The Welsh
and English rollers were reluctant to divulge the secrets of their skill.
They kept their templets, those precious brass patterns that represented
the required dimensions of the piece to be rolled, in their pockets and
watched the observer suspiciously. But Fritz's eagerness and inquisitiveness
prevailed, and in a short time he was made night superintendent. Shortly
thereafter he took charge of the day shift.[2]

Fritz stayed at Norristown for about five years. A new blast furnace
was planned at Safe Harbor, a small town about ten miles from Lancaster
on the Susquehanna River. In order to learn about this phase of iron
production, Fritz left to help in the construction of the new plant. In
the midst of a mutually satisfactory experience there, he contracted
malaria and eventually had to leave the unhealthy setting for home.
At Londonderry, he tried unsuccessfully to shake off the recurring chills
and fever. For several years he rested at home and at last consulted a
doctor in Philadelphia who cured the disease.

He returned to Norristown, but stayed for only a short time before
moving on to work on the construction of a new furnace designed to
use anthracite coal. Later he and his brother George built a machine
shop in Norristown with the intention of going into business for themselves. But a chance meeting with an old friend, David Reeves, changed
their plans when Reeves offered John Fritz the position of superintendent
at the Cambria Iron Company, located across the state at Johnstown.
By accepting, John Fritz at last had selected the place where his talents
would begin to flower.

Fritz landed in Johnstown late one night in June 1854. There was
little to be seen at that hour except the glow from the beehive coke
ovens reflected in the dark water of the Conemaugh River. Next morning,
when the new superintendent left the hotel to survey the surroundings,
he saw unpaved muddy streets stretched out in the distance with only
a few board sidewalks visible in the business section of the town. "Cows,
hogs and dogs all ran at large," he said. "The dogs would get after the

pigs, they would squeal, the cows would bawl, the dogs would bark and fight and I should have been much amused if I had not been there to stay."[3]

Cambria at the time was struggling to establish itself on a solid financial basis. But Fritz discovered that the mill was already under way, so that it was no longer possible for him to make any of the improvements that he considered necessary. When the mill began to produce rails, the iron from which they were rolled presented immediate problems. Changing the rolls failed to effect improvement, and the rails continued to come from the mill with their flanges badly torn and the heads cracked.

The low iron content of the ore that the Cambria plant was obliged to use was responsible for the repeated failures. For the small forges, which for many years had worked in the area, there always had been a plentiful supply of high iron content ore. But these small forges needed much smaller quantities of raw materials than were needed for a large plant like Cambria, which had been built to roll rails.[4]

Because of these inferior ores, Fritz faced a continuing problem of how to produce rails of the quality that his sense of craftsmanship and honesty demanded. Inspection did not exist in 1854, and the cracks in those rails that escaped the scrap pile were patched and puttied.

George Fritz followed his brother to Cambria to serve as his assistant. The two stood day after day watching the iron bar, or the "pile" as it was called, start its way through the pair of rolls until the procedure was fixed indelibly in their minds. The two rolls, mounted one above the other, revolved only towards each other. The bars thus could be rolled in only one direction. One roller started the pile through one groove in the rolls; the catcher on the other side stood ready with tongs to grasp it and return it over the top of the roll to his partner on the other side. As it returned through the next groove in the rolls, it was caught swiftly and sent over the upper roll, which was revolving in the proper direction.

Although he only could admire the extraordinary skill required for this rhythmic performance, John Fritz was dissatisfied that so much hand labor and heat were being wasted and so much production was being lost. The fact that the rail was cooling as it slammed its way through the rolls and back over the top must in some way be responsible for the defects, he reasoned.

When the flanges began to crack, the defects in the rails tended to become larger and longer with each pass. Sometimes the flange would tear off, wind itself around the roll, and break the roll or damage some portion of the assemblage. The pile might split as it entered the roll and open out like the mouth of an alligator, causing endless delays and more damage. It seemed impossible that Fritz would succeed in his efforts to roll perfect rails. Except for the putty and patches, rail production indeed would have been minuscule.

As he watched the fumbling process, Fritz became increasingly dissatisfied. He confessed in later years that he often had to overcome the urge to flee. The men doing the work did not share his discontent. The roller and catcher teams were proud of their rhythmic performance. Fritz's talk about rolling inefficiency injured their self-esteem and at the same time made them fear they might be supplanted by newfangled machinery.

Cambria's finances at this time were at an ebb, and, along with his engineering duties, Fritz took on the task of finding additional money to keep the plant in operation. He finally found six men who agreed to take equal shares valued at $30,000. The refinanced company took the name of Wood, Morrell and Company. Daniel J. Morrell was selected by the group as their general manager and this was how Morrell, the merchant from Philadelphia, found himself in the iron business. The financial support for Cambria came none too soon, for Fritz later remarked that the sheriff could have sold the property at auction the next day.

Although the new management may have been satisfied with the operation of the plant as they found it, Fritz was not. Finally, unable to control his impatience, Fritz informed the shareholders that, in his opinion, Cambria's situation soon would decline to the point of disaster. Because the mill had been built so recently the men were skeptical, but they listened as Fritz outlined his idea for a new mill that would have three twenty-inch rolls instead of the customary two. The rails would travel back and forth alternately through the bottom and top pair of rolls. This method would force the bar into its final shape before it had had time to cool. The duration of the rolling process from bar to rail would be shortened enough to increase noticeably Cambria's rail production. Otherwise, he said, it would be necessary to build both a new train of rolls and a new engine to turn the rolls. Fritz was demanding what was tantamount to a completely new rail mill.

A few of the stockholders were convinced, and they agreed to discuss the matter at the next meeting of the board. Fritz was amazed when he was told after the meeting that the board had decided to build only an eighteen-inch two-high geared mill to replace the one that was so troublesome. Shocked that his argument could have been so unpersuasive, Fritz refused to consider their decision as final. The board, in turn, believed that as a mere employee Fritz was behaving highhandedly by refusing to build the mill they had approved. For several days during the impasse that followed, Fritz tried to convince individual board members of the correctness of his point of view. At last the board decided to go along with Fritz's three-high rolling mill, but, because of cost, insisted on paring down the size of the rolls from twenty to eighteen inches.

Before they could change their minds, Fritz arranged for the roll patterns and used all possible speed to get the mill under way. Despite his activity, a minority within the board, which never had approved of this radical departure, served notice in the form of a legal document that it would hold the management personally responsible for the results if the new mill were built. This action had the desired effect, and the managing directors again endeavored to persuade Fritz to forget his three-high mill and to build a conventional one. Many ironmasters had assured Morrell that the whole idea was foolhardy as well as unsound. Everyone knew that the rolling practice was the result of years of experience, and they questioned forcing rollers to learn a new method when they had spent their lives perfecting the old. The workers joined the controversy and sent a delegation to the managers to plead the case for the type of mill they always had known. The harried Fritz went to see Edward Townsend, vice president of Cambria, and outlined the deficiencies of the old style mill and described the benefits of the new. Although Townsend was by nature modest and retiring, once he decided upon a course of action, he became forceful and efficient. Convinced by Fritz's argument, Townsend authorized him to proceed and he himself assumed the responsibility of convincing those who were opposed.

Fritz's friends, fearing that he was staking his reputation on the outcome of this new design, did their best to dissuade him. Fritz agreed that indeed he was staking his future on what he was about to do but that he could only go ahead.[5]

Fritz was ten years older than Holley and his labors with this first mill of his own design occurred about ten years before Holley had begun his involvement with the Bessemer process at Troy. As Holley would solve the problems of a new process, so Fritz had to work out the difficulties in bringing his new mill to completion. Since it was customary for the standard rolling mills to break rather regularly, it had become the practice to design for a relatively desirable type of breakdown. Breaking pieces were built into the mills at points of unusual pressure in the train so that the pieces rather than the expensive roll would give way. For the same reason, coupling boxes and spindles were lightly constructed and a special breaking box even was installed above the rolls to hold them in place. If the rolls were subjected to heavy pressure, the box would be crushed, saving the rolls from harm. Consequently, the so-called safety devices broke continually, causing delays in production while they were either repaired or replaced.

Fritz decided to break away from this design method. He determined that the machinery should be built strongly and well and that there would be no more crutches and coddling. When the mill manager thought the patternmaker had erred in making a solid instead of a hollow box to hold the rolls, the patternmaker explained that "the old man" had designed it that way. Although still in his early thirties, Fritz already had earned the title "the old man," which he carried into his old age, when for an unknown reason it was changed to Uncle John.

Much perturbed, the manager went to Fritz and suggested that the mill would be smashed when the rolls encountered strain, to which Fritz heatedly replied that he would rather have a grand old smash-up once in a while than the constant annoyance of broken leading spindles, couplings, and breaking boxes. The manager walked away muttering, "By God, you'll get it." When the news of this departure in mill design had circulated, all the diehards and bearers of ill omen gathered around to forecast doom.

The controversy did not delay construction, and the new installation was ready at the beginning of July 1857. On July 3, the old mill was stopped and immediately torn down. The new engine was put in place and by July 29, the mill of Fritz's dream was ready to roll. Because of their opposition, the heaters were not invited to be present when the rolls were started. One heater who had been more friendly to the new endeavor than the others was asked to heat the piles for the first rails. At ten in the morning, the first pile was sent on its way through the rolls, carrying with it the fortunes of a man who had staked his reputation on the trip it was taking.

The pile emerged a perfect rail. Two more rails also made the trip successfully, but the onlookers, fascinated by the demonstration, had neglected to monitor the newly installed engine. As a result, the bearings of the eccentric heated and caused the eccentric rod to bend. It then was necessary to shut down the mill for the day. Hearing the exhaust of the overheated engine, the heaters, who were nearby, came to investigate the trouble. Spying unrolled material on the ground, they immediately assumed that the new mill was a failure and all had suitable comments to offer. Alexander Hamilton, the mill superintendent, assured them in language that any workman could understand that there was no failure but that rails had been rolled that were superior to any these Welsh workmen had seen rolled in Wales. Thomas Lapsley, who had charge of the rail department, William Canam, and, of course, George Fritz, who had been an important member of the Fritz team, all shared the pleasure of Fritz's success.[6]

The following day the mill again was placed in operation. Rails came from the rolls all that day and night. The day after, which was Saturday, the day shift finished rolling at noon. But Fritz and Hamilton did not leave the plant until six that evening, starting homeward together, content that the long siege with the building of the mill at last was over and that they now could turn their minds to making other improvements.

Fritz had barely reached home when he heard the dreaded fire alarm whistle at the mill. Racing back to the plant, Fritz was shocked to see his precious mill a mass of flames and the surrounding shops threatened by the inferno. Since their roofs were of wooden shingles it seemed barely likely that they could be saved. Praying that the walls of the rolling mill would fall in rather than out, everybody worked frantically to save the machine shop, pattern shop, and foundry. When the walls collapsed inward, Fritz saw that the rest of the plant was safe but that his rail mill was a complete ruin. Gazing hopelessly at the smoldering rubble, he asked himself where the money might be found to rebuild the plant.

That same weekend Fritz made plans to start reconstruction. Since the mill had proved its worth before the catastrophe, the stockholders proved willing to advance more money. The workmen, anxious to contribute their share, offered a day's labor without pay. Fritz decided not to erect a building, that instead a temporary structure could serve. He hastened to recondition and replace the damaged machinery. To the amazement of everyone, Fritz had the mill ready for operation within twenty-eight days. It is true that there was barely a roof over the rolls, but the mill was in place and operating. Based on the experience of the fire, Fritz decided that the new plant building would not be constructed of wood. While the mill turned out rails, he proceeded to erect a stone building over and around the working men.[7]

Until 1860 John Fritz continued to perfect the Cambria plant, improving the machinery and physical structure and molding the labor force into a loyal, hard-working group. This was not easy. The men did not accept supervision or new ideas readily and could be handled only with the utmost tact. Fritz once ruefully remarked, "There was so much objection coming from the workmen to anything that was new that I once told them if I got up anything new and they said it was all right, I should look over my drawings again, thinking there must be some mistake." But, by the time Fritz decided to move on to the Bethlehem Iron Company, he knew that he had developed a group of foremen who not only were loyal to the company, but also were capable of working in harmony with each other.

When Fritz left Cambria for Bethlehem, he left a plant that was considered one of the outstanding companies of its time. According to Robert Hunt, who arrived a month after Fritz's departure, he left also a house of mourning. Even though George Fritz, who had worked closely with his brother and who was one of the mechanical geniuses of his time,

John Fritz. (Courtesy of Bethlehem Steel Corporation.)

succeeded to the position of superintendent, the men talked constantly of "the old man" who had left them for Bethlehem.

John Fritz stood on the great divide between iron and steel. When he began work in the iron industry, most of the machinery was made of wood, and the important members of the labor force were the carpenters and millwrights who turned out the shafts and wheels. By 1844, when the production of iron rails commenced, the machinist had begun to appear and the millwrights to disappear. The machinist, in turn, gave way to the mechanical engineer as methods of manufacture became more involved, and exact design became the criterion for fine production.[8]

John Fritz arrived in Bethlehem as general superintendent and chief engineer of a nonexistent plant. Within two weeks, ground was broken, and Fritz was on his way to building the great plant that would serve as his monument. Once more, the way was strewn with obstacles. The Civil War interrupted construction and floods washed out the foundations. Despite everything, the plant was in operation by September 1863, a plant that was to serve as a model of fine construction and efficient operation for many years. All the buildings were constructed of hard gray sandstone from the neighboring hills and the roofs, as Fritz had resolved that plant roofs ought to be, were covered with slate.

The Bessemer process was still controversial and had not yet been brought to America. Fritz watched its development in England and

Sweden with interest. After he learned that the patents in the United States were owned by Winslow, Griswold and Holley, he was curious enough to travel to Troy to investigate the process. Somewhere Fritz had learned of the role of phosphorous in Bessemer steel, and at Troy he discussed the subject with Griswold. In a circular issued by Bessemer, Griswold showed Fritz a statement that the permitted phosphorous limit in finished steel was 0.02 percent. Fritz was disappointed because analyses that had been made at Bethlehem indicated that all the available iron ore and coal contained a larger amount of phosphorous than that. He told Griswold that, in view of the low phosphorous limit, it was foolish to discuss the matter further and that it was evident that the process could be of little value in this country. "As a consequence," said Fritz, "the Bessemer fever which I had when I met Mr. Griswold was changed to a Bessemer chill after the interview." Fritz decided not to experiment with the new steel process and, instead, to concentrate on improving the quality of iron rails.[9]

By 1868, developments in the technology of rail manufacture made it obvious to John Fritz that, if Bethlehem was to hold its own in that field, a Bessemer installation was essential. He did not know where he would secure the necessary low phosphorous iron ore and the coal for such a plant but he hoped that either they might be imported or additional deposits found. Mindful of these problems, he began to investigate Bessemer installations both in the United States and abroad. His investigation came none too soon. Five Bessemer plants already were operating in the United States and four others were being planned.[10] With the exception of Freedom, Holley had provided the plans and served as consultant for all of them and he was engaged as the consulting engineer for Bethlehem. It was inconceivable that a Bessemer plant could be built without his counsel, encouragement, and ability to produce a satisfactory blow from a difficult converter.

Much of the Bessemer machinery used in the other plants was produced by the Delamater Iron Works in New York, and Holley took John Fritz there to see a blowing engine designed by George Reynolds, chief engineer of the company. But Fritz had his own ideas and he determined to design and build his own engines and machinery at Bethlehem. For this reason, it took a relatively long time to complete the plant, but as a result it was possible to construct high quality machinery at a relatively low cost. When Holley suggested that perhaps Fritz had made some of the machinery unnecessarily strong, he laughed and said, "Well if I have, it will never be found out." After studying earlier installations, Fritz decided not to build the traditional separate structures to house the converters, melting house, engine room, blooming and rail mills, but instead to place everything in one large building shaped like a Greek cross, with transepts but with no partitions whatsoever.

Although Bethlehem was the tenth Bessemer plant, Fritz served his apprenticeship as had those before him. He discovered from bitter experience what could result from high heat and hot metal. He learned that accidents were most likely at the time of greatest heat. He once said that when a charge of metal was poured into the vessel, the blast turned on, and the vessel turned up, his anxiety increased in a corresponding ratio, until both became intense. There still was no suitable refractory material that could withstand the intense heat. Linings were apt to burn through and the bottom could fail, sometimes under the first heat, allowing the hot metal to stream out. An explosion was certain when the metal met the water, some of which always was present near the converter.

Even with the watchful Holley checking and making certain that everything was in readiness for a converter to begin operations, things could go awry. Even when he had become an internationally known engineer, Holley still would enter a pit to help the workmen clear out the debris from a heat that had not waited to be poured into molds, but had gushed through the bottom of the converter. Little wonder then, that when an ingot toppled off the buggy hauling it to the rail mill, Holley, thinking of many wild scrambles to escape from a flood of rampaging seething metal, could grin and say to Fritz, "Boss, that will lie still."

In the summer of 1873, the group of friends lost its first member. George Fritz of Cambria died at the age of forty-four. He had made notable contributions to the infant steel industry. The blooming mill that he developed had been adopted by the works at Troy, North Chicago, Joliet, and Bethlehem. Every piece of machinery in the mill, furnaces, shops, or mines of Cambria bore witness to his mechanical skill. Although his brother is credited with the invention of the three-high rolling mill, which revolutionized the rolling procedure in the steel industry, George Fritz is known to have helped with its development. His death was a blow to the men who had worked under his direction, and on the day of his funeral, the mill was shut down. Many stores and offices in Johnstown closed, as thousands of his friends walked silently in the procession.[11]

The Bethlehem Iron Company made its first Bessemer blow on 4 October 1873. The entire town turned out to gaze in wonder at this latest addition to the growing group of Holley-designed Bessemer plants. The designs for each of them had profited from what Holley had learned from designing the previous one. In his mind a plant never was finished, and he roamed incessantly from plant to plant both in the United States and Europe checking on operating efficiency and suggesting improvements, large and small. He often remarked that "Small and cheap plants make dear steel." Holley favored operating cost reductions through increased efficiency, but not through the use of cheaper machinery. He was equally unwilling to patch up old engines; he once remarked that the greatest profit from a bad type engine comes from melting it down in the cupola.[12]

NOTES

1. John Fritz, *Autobiography of John Fritz* (New York: American Society of Mechanical Engineers, 1912), pp. 1, 8, 9, 32.
2. Ibid., pp. 40, 45, 47, 48, 50, 53.
3. Ibid., pp. 56, 59, 67, 75, 78, 90, 91.
4. John Fritz, "The Development of Iron Manufacture in the United States in the Past Seventy-Five Years," *Journal of the Franklin Institute*, 4 December 1899, p. 443.
5. Fritz, *Autobiography*, pp. 102-7 passim. With one exception, no important change in rolling procedure had taken place for at least eleven years. This one improvement was the introduction of a rail-straightening machine that had taken the place of a man with a sixty-pound sledge. As Fritz mused over this improvement, "When he wanted a rest, the works had to come to a standstill until such time as he was completely rested, sobered up or restored to health."
6. Ibid., pp. 111-15 passim.
7. Ibid., pp. 116-20 passim.
8. Fritz, "Development of Iron Manufacture in the United States in the Past Seventy-Five Years," p. 442.
9. Fritz, *Autobiography*, pp. 150-51.
10. Ibid., pp. 152-53. The Cleveland Rolling Mill Company, the fifth plant, had made its first blow in October 1868. Cambria, Union Steel Company, the North Chicago Rolling Mill, and the Joliet Steel Works in Joliet, Illinois would come on line before the Bethlehem plant. The Union Steel Company at South Chicago and the Cleveland Rolling Mill, owned by the same group, had almost identical Bessemer installations. The first blow at Union was made on 26 July 1871. A little over two weeks previously Cambria had gone into production. Number eight was constructed at the North Chicago Rolling Mill. The first blow there was made on 10 April 1872, just one year after the ground breaking ceremony. Although Eber Ward had abandoned his early intention of making steel at Wyandotte, he still had enough faith in the process to invest in a new installation at North Chicago. O. W. Potter, the old friend of William Durfee, was in charge of erecting the plant; and Robert Forsyth, who had served his apprenticeship under Holley at Troy, became the first manager.

   Most of the machinery in the ninth plant, the Joliet Steel Works, came from the defunct Freedom Iron and Steel Company, which had been the fourth installation. The Joliet plant made its first blow on 26 January 1873, and rolled its first rail on March 15 of that year (Robert W. Hunt, "A History of the Bessemer Manufacture in America," *Transactions of the American Institute of Mining Engineers* 5 [1876-77]: 211-12).
11. "Death of George Fritz," *Bulletin of the American Iron and Steel Association*, 7 (13 August 1873): 396.
12. Alexander Lyman Holley, "Some Pressing Needs of Our Iron and Steel Manufactures," *Transactions of the American Institute of Mining Engineers* 4 (1876): 81.

XVII · The Edgar Thomson Steel Works

I n the year 1858, when Alexander Holley published his first article in the *New-York Times,* when Holley and Colburn published their *Permanent Way,* when Bessemer started up in Sheffield, and when Mushet was struggling to make a living in the Forest of Dean, a small forge began operating in a place called Girty's Run in the outskirts of Pittsburgh.

Owned by two brothers, Anthony and Andrew Kloman, the small plant, which forged axles for the railroads, was the beginning of what would develop into the formidable Carnegie Steel Company of the future. Anthony, the older, was easygoing and convivial, while the younger, Andrew, was the epitome of the thrifty, hard-working German craftsman who took pride in the superior quality of his product. In contrast to his older brother however, he was inclined to be suspicious. Among their customers was the Ohio and Pennsylvania Railroad, whose purchasing agent was Thomas N. Miller. Because of their business relationship and because Miller often had brought other customers to Andrew, the two became friends. [1]

Within a year the growth of the business necessitated expansion of the plant. It was natural for Andrew Kloman to turn to Miller as a possible business partner. Miller, however, believed that because he was Kloman's customer, he should not at the same time become his financial backer. Instead, Miller introduced Kloman to Henry Phipps, a young man whom he thought might be interested in investing the necessary $1,600. When Phipps was unable to raise the money, Miller came to his aid and became a silent partner with his share listed in Phipps's name. Phipps and Miller were to share equally in one-third of the profits from the business. [2]

The times were fortunate, and the men worked hard. Miller was able to direct new business to the young company. Phipps kept his day job and took care of the Kloman books at night. Kloman sweated with his workmen keeping the forge busy. Before long, because of demand for material for the Civil War, the price of railway equipment soared, and government orders soon formed the major part of the business. Andrew Kloman, believing that his older brother hindered more than he helped, suggested that Miller acquire Anthony's interest. Miller did so in April 1863 at a price of $20,000 for the one-third share. After the sale Andrew Kloman had the agonizing realization that he now owned only a one-third interest in the business that he and his brother had started.[3]

To assuage Kloman's distress, Miller sold him enough of his interest to bring Andrew's one-third share up to one-half, but before long his suspicious nature made him worry about the one-half that he did not own. He tried to buy out first Phipps and then Miller. For weeks the unhappy man went from one partner to the other hoping that he could prevail. Neither was willing to surrender his share in a growing and prospering endeavor. Soon an atmosphere of ill will, recrimination, and suspicion enveloped the partners. Kloman's ire was directed particularly towards Miller. Mediation seemed to be the only feasible move, and Phipps and Miller, even though also at odds with one another, agreed upon a mutual friend as mediator. By this curious route, Andrew Carnegie entered the steel industry.[4]

Kloman earlier had rejected Miller's suggestion that Carnegie be taken into the plant as a partner. Kloman considered Carnegie too important and well-to-do. For the same reasons, Kloman now was willing to accept Carnegie as a mediator.

The story of Carnegie's rise is one of the best known Horatio Alger legends in the annals of great American fortunes. Carnegie was brought to America in 1848 by his father, a Scottish weaver. The family, including a younger brother Tom, settled in Allegheny, Pennsylvania, where as a young boy Andrew began work in a cotton mill. After a year he became a messenger boy in nearby Pittsburgh for the Eastern Telegraph Line, where he seized the opportunity to learn telegraphy.

The telegraph office handled all of the Pennsylvania Railroad Company's telegraph traffic for its Philadelphia to Pittsburgh line, in 1853 only a single track. The superintendent of the western division of the Pennsylvania Railroad, Thomas Scott, frequently dealt with Carnegie on railroad business. Scott was impressed by Carnegie's enterprising nature, and, when the railroad installed its own telegraph lines, hired him as both telegraph operator and private secretary.[5] Soon Carnegie became known among the railroad people as Scott's Andy, a title that pleased them both. Scott kept an eye out for opportunities for this protégé. As Scott rose through the system, his protégé shared his good fortune, and, in 1859, when Scott became vice president of the Pennsylvania Railroad Company, Carnegie, at the age of twenty-three, immediately was appointed superintendent of the western division.[6] On Scott's advice, young Carnegie in 1856 bought his first stock, and he soon followed that purchase with others.[7]

Although Carnegie never acquired technical knowledge, he possessed an uncanny facility for allying himself with technically competent men. Carnegie made the acquaintance of John L. Piper in 1856, when he moved to Altoona with Scott, who had been promoted to general superintendent of the Pennsylvania Railroad Company. Piper was a master mechanic, known as the best mechanic in the employ of the Pennsylvania. He had two interests, horses and bridges. Most railway bridges in America still were constructed of wood and often burned or were washed away by floods. Piper had rebuilt many wooden bridges and his thoughts naturally had turned often to iron bridges, which were beginning to be built both in America and Europe. Two colleagues, Jacob H. Linville, a bridge engineer, and Aaron Shiffler, in charge of bridges for the Pennsylvania, shared Piper's interest. Piper and Linville had designed a small iron bridge that Carnegie saw in the shop at Altoona. Piper's enthusiastic description of iron bridges in America and the tangible evidence he showed Carnegie sold him on the idea.[8]

At Carnegie's suggestion, the three bridge enthusiasts went to Pittsburgh and, in 1862, formed the Piper and Shiffler Company, using Linville's patents. Carnegie purchased a one-fifth interest for $1,250 and assumed a leading role in the management of the new company.[9] It was an immediate success. A fortunate combination of circumstances was responsible for its astounding rise. All three knew their business well. They received powerful backing from the Pennsylvania Railroad and, what was more important, the Civil War brought all the orders the plant could handle. By the time Carnegie became the mediator in the Kloman quarrel, the bridge company was providing a substantial part of his income.

For several weeks, in his role as mediator, Carnegie trotted from one to the other of the three dissidents, listening to their charges. The outcome was curious, since everyone apparently profited by the mediator's intercession except Carnegie's best friend, Miller. Carnegie suggested that a new company called Kloman and Phipps be organized, with Miller holding only a one-sixth interest and remaining as a special partner. An odd clause was inserted in the proposed agreement that would give Kloman and Phipps the right to buy out Miller's interest and eject him from the partnership upon sixty days' notice if the two men believed that he had acted in any way prejudicial to the best interests of the company. Carnegie, having come this far with Kloman and Phipps, then persuaded his brother Tom to try to obtain Miller's signature. Tom Carnegie accomplished his assignment by telling Miller that his brother Andrew's efforts would be placed in a very bad light if Miller refused to sign. The signature, however, had not been gained without protest. Not long after, during another quarrel, Miller was ousted, and his capital was held in the name of Tom Carnegie as trustee.[10]

Miller may have sensed that his days with Kloman were numbered, and, while the negotiations still were going on, Miller began to think of building another ironworks. He bought a five-acre cabbage patch only four blocks from the Kloman forge. Miller evidently felt no ill will toward Carnegie, since Miller invited him to become a principal in the new organization, which was to roll beams and other shapes for the bridge company. In spite of his brother's large interest in the Kloman company, Andrew Carnegie accepted Miller's offer and, along with Shiffler, became a partner in the new company, the Cyclops Steel Company. John C. Matthews, the member of the new group with knowledge of the iron business, constructed the plant, and during the entire winter of 1864 and 1865, work went forward in great haste. However, in the spring, when Cyclops Steel was about to begin operating, it was discovered that the building foundations were too weak for safety.[11] Recriminations flew back and forth. Matthews, who was the object of most of the criticism, remarked bitterly that he had been expected to build a $400,000 plant for $100,000. When Andrew Carnegie told his brother that a great deal more money would be necessary to bring the Cyclops plant up to acceptable operating condition, Tom Carnegie was outspoken in his reply. In his opinion, he told his brother bluntly, they were all amateurs. Kloman was the only person in either company who knew how to make iron, so why not join forces? Anxious to get off the hook, the older Carnegie agreed that if his brother could persuade Kloman and Phipps, he would talk to Miller. Both Carnegies were so persuasive that a merger took place that resulted in the formation of the Union Iron Mills Company. Kloman's Iron City Forge became the Lower Union Mill, and the Cyclops plant, the Upper Union Mill. Kloman was named manager of the entire company, and Miller, for the time being, was the largest stockholder.

Kloman's special aptitudes were needed almost immediately. The new mill proved to be of even worse construction than the partners had believed. Moreover, the company had been established at a low point in business activity. At the end of the war the government had ceased purchases of war materials, and a peacetime economy had not been reestablished yet.

Carnegie, using his own special skills, proceeded to reorganize the bridge company as the Keystone Bridge Company and brought in additions to the roster of partners, including men such as Edgar Thomson and Thomas Scott, president and vice president of the Pennsylvania Railroad, as well as other rail officials. Carnegie now could count on Piper's skill in the art of bridge building, the Linville patents, and Kloman's skill in iron manufacture at Union Iron Mills, plus the interest of the railroad men in Keystone. Their business, he was sure, would keep the rolls turning at the Union Iron Mills. With matters thus arranged, and despite threatening uncertainties, in May of 1865 Carnegie, Phipps, and a friend, John Vandervoort, set off on a European vacation.

Kloman was left in charge of operating the two mills with Matthews under him as superintendent of the Upper Union Mill. Tom Carnegie and Miller were to manage the business end of the enterprise. A fortunate combination of circumstances pulled the unseasoned company through these difficult times. Foremost was Kloman's excellent management. Moreover, in Andrew Carnegie's absence, his younger brother Tom became known and was found to be completely trustworthy. He was a quiet man who was less ambitious than his brother. His integrity helped a great deal to stabilize the new company. Finally, Miller, who by then had amassed a fortune from his oil venture, contributed financial support at critical times. By the fall of 1865, more than one-third of the iron and steel mills in Pittsburgh had shut down. The struggling company held on by a slim margin so that when the tide turned, the company was prepared to run with it. The railroads in the South needed rebuilding, and the way to the West was opening. The bridge company came in for its share of the new prosperity. Just as Andrew Carnegie had calculated, it sent its iron business to the Union Iron Mills, and the two companies worked together closely. [12]

Scarcity of money continued to be a problem, however. When Carnegie and his friends returned from their European trip in the spring of 1866, they found that, although business definitely was improving, payrolls were difficult to meet. Phipps resumed financial management of the new company. [13]

Miller still blamed Phipps for the outcome of the Kloman and Phipps mediation, and when his old antagonist was returned to the board of directors of Union Iron Mills in 1867, Miller no longer wished to stay. After extensive discussion with Miller about its value, Andrew Carnegie succeeded in acquiring Miller's shares. Carnegie now owned 39 percent of the company.

Under the hard-working team of Tom Carnegie, Phipps, and Kloman, the Union Iron Mills Company became a profitable endeavor. Andrew Carnegie, however, yearned for something more. The trip to Europe had reawakened in him an old desire for culture. Moreover, he wished a broader field in which to exercise all of his talents, and, in 1867, leaving his brother to supervise the business, he moved to New York. Andrew Carnegie and his mother took up residence at the fashionable St. Nicholas Hotel on Broadway, and Carnegie opened an office in Broad Street for the purpose of handling investments. New York City was the American center of finance as well as culture, and he found it to be a maelstrom of activity. Money was plentiful. Carnegie soon discovered that European capital was enormously attracted to American railroad bonds. At the same time, the railroad companies found that he had a singular ability to sell them, a success that undoubtedly was helped by his close relationship with the executives of the Pennsylvania Railroad. Between 1867 and 1873, Carnegie sold approximately $30,000,000 worth of bonds on which he received commissions. In 1868 his annual income was $56,000, which included $15,000

Andrew Carnegie. From Herbert N. Casson, *The Romance of Steel* (New York: A. S. Barnes & Co., 1907), frontispiece.

from Keystone, $20,000 from the Union Iron Mills, and $6,000 from his interest in a rail mill. He was then thirty-five years old.[14]

Carnegie never lost sight of his reasons for having left Pittsburgh. Despite his strenuous activity, he spent a portion of each day with tutors in literature, economics, history, and government. He managed, through clubs such as the Nineteenth Century Club, to associate with the people of culture and social background with whom he identified. He also learned how to transact business on social occasions and became a charming and witty guest at dinner parties and literary gatherings.[15]

During these years, Carnegie had developed a clever business strategy. He learned how to profit in numerous ways from a single venture. The building of the Eads Bridge, which spanned the Mississippi, is an example. After procuring the contract for Keystone to build the superstructure, he marketed the bonds to finance the construction. Keystone, in turn, had its material rolled at the Union Iron Mills from both of which Carnegie derived income. He not only received a block of shares for his services, but eventually became a director of the bridge company. When the bridge was completed, the enterprising Carnegie received a $30,000 bonus for early completion.[16]

In 1869, at the height of the railroad boom, the companies of the

Carnegie group all were operating at full capacity rolling beams, rails, and other products. They were not, however, producing their own pig iron. Only seven small blast furnaces in Pittsburgh then produced pig iron, and since the demand exceeded the supply, it was expensive. When a delegation from a group of Pittsburgh plants approached Phipps with the idea of pooling resources and erecting two blast furnaces to meet their mutual needs, Phipps was attracted by the plan. For advice he went to Tom Carnegie's father-in-law, William Coleman, the pioneer Pittsburgh ironmaster, who pointed out that, in his judgment, it was more expedient to own all of one furnace rather than one-seventh of two.

Thus it followed that in December 1870, the Union Iron Mills partners organized a new company called Kloman, Carnegie and Company. The following spring the first Lucy furnace, named after Tom Carnegie's wife, was started. The other group, whose invitation to join the Carnegie partners had been refused, likewise built their own furnace, Isabella Number One, and both went into blast at about the same time in 1872. The new blast furnaces were considerably larger than any that had been built previously. Since they were similar in size, their comparative production records was a matter of interest for years.[17]

After building the first Bessemer plant at Troy, Alexander Holley had been involved successively with other plants in other locations. But Pittsburgh, although rapidly growing as an iron manufacturing center, had not entered the Bessemer picture yet. The nearest installation that used the new process was the Cambria plant at Johnstown about fifty miles away. William Coleman, a lifelong pioneer in the iron industry, had watched with interest as the various Bessemer works were built. Despite the fact that he already was sixty-five years old, Coleman decided to visit all the Bessemer plants in operation and to see for himself what the new process had to offer. He was sufficiently impressed to persuade his son-in-law, Tom Carnegie, to join him in building a Bessemer plant.[18]

The partners obtained 107 acres of farmland about twelve miles from Pittsburgh; the area had been known as Braddock's Field since the defeat of General Braddock there in 1755. Much more important to Coleman than the land's history was the setting. The Pennsylvania Railroad ran along the north boundary, the Baltimore and Ohio through the center, and the Monongahela River flowed along the southern edge. Easy transportation for both raw materials and finished products thus was assured. In Coleman's view, Braddock's Field was ideal for a Bessemer plant. David Stewart and John Scott, both railroad men and close neighbors of Coleman, were brought into the planning of the new company. Although Andrew Carnegie still remained outside, Pittsburgh merchant and banker David McCandless was interested in joining provided that his good friend,

William P. Shinn, long connected with railroads, likewise would come into the new partnership.[19]

Carnegie disliked the thought that his brother and his friends were going into Bessemer steel without him. On his next business trip to England, he again went to Sheffield to investigate the Bessemer process. The failure of the Freedom plant in 1869 may have colored Carnegie's initial attitude toward its worth but, after he had witnessed a demonstration of the process at Sheffield, Carnegie's canny good sense prevailed. Since by nature Carnegie could not bear to take part in an enterprise in which he did not play a leading role, he subscribed $250,000 to the new company. Coleman put in $100,000, and each of the other men contributed $50,000.[20]

The new company, hoping to benefit from his experience with Bessemer works, engaged Holley to furnish the plans and to supervise the construction. No one else would have been capable of designing the new plant. Despite his prestige, Holley's fees had remained modest. He furnished all the necessary working drawings for the plant and served as supervising engineer for a fee of $5,000 and a yearly salary of $2,500.[21]

Holley began designs for the plant in September 1872. On November 5 of that year the firm of Carnegie, McCandless and Company formally was organized for the purpose of building the eleventh Bessemer plant in America.

Of the Bessemer plants with which Holley had been associated, only two originally had been designed and built to produce steel. The others had been built for established ironmaking concerns. Ingenious planning had been needed to fit converters and auxiliary equipment into existing spaces. For this plant Holley could begin with a spacious bare plot of ground on which to build the finest steel plant he could devise. Holley set to work with his usual enthusiasm, and ground was broken on 13 April 1873. Phineas Barnes, who had served with him as resident engineer at some of the other mills, left the Joliet plant to work once again with his friend.

By a curious stroke of fortune, the new works also secured Captain Bill Jones, Holley's close friend since his time at Cambria. After the death of George Fritz in August 1873, the directors at Cambria had chosen Daniel Jones to succeed Fritz. In this choice they probably were influenced by Daniel Morrell. Captain Bill Jones had been at Johnstown for sixteen years and, as Fritz's assistant, seemed the logical choice. But Captain Jones had been very close to Fritz and was a staunch supporter of his wish to maintain high wages, a policy with which Morrell did not agree.[22] Moreover, Jones's bold and unorthodox ways of working frequently shocked the more conventional Morrell. Although Captain Jones could get more work from his men than any other boss in the plant, such antics as shutting his department to take his men to a baseball game or to a horse race never made sense to Morrell. Little wonder that Morrell passed over Captain Jones and appointed Daniel Jones, a steady and even tempered man, as superintendent of Cambria.

Daniel and Bill Jones were close friends and had worked together for many years; thus the promotion was a shared embarrassment. Daniel Jones believed that the job belonged by right to the Captain and offered to refuse it, but Bill Jones would not hear of such a thing. He recognized that the management had spoken and decided that he had better look elsewhere for a position. [23]

Holley may have spoken a good word to the directors in Jones's favor since he quickly was engaged by general manager William Shinn as superintendent of the new steelworks at Braddock. [24] Shortly thereafter, when a labor dispute developed at the Cambria plant, Morrell served notice that the foremen would have to end the trouble or forfeit their jobs. Smarting under the threat, many of the foremen and workmen quit Cambria and went to Braddock under Bill Jones, the man with whom they most enjoyed working. [25] The Carnegie plant thus gained all at one time a group of experienced hands skilled in the art of operating a Bessemer converter.

Captain Jones's generous nature is demonstrated by the facts of an incident that occurred at about the same time as the exodus from Cambria. When Robert Hunt wired Jones to say that he, too, was out at Cambria and was on his way to Troy to see if there might be a position there, Jones wired Hunt not to commit himself until he had received a letter, which would follow. From the letter he learned that Jones had gone to Shinn and had offered to resign in favor of Hunt. In his generosity, Jones had told Shinn that because Hunt had made Bessemer steel he would be a wiser choice to run the Braddock plant. Jones believed that he, himself, could find other employment with John Fritz at Bethlehem. [26]

The exodus from Cambria probably would have been even greater than it was, except that on September 19 the influential Philadelphia banking house of Jay Cooke and Company closed its doors. The Panic of 1873 had begun. Railway construction halted suddenly, and the iron and steel industry was the first to feel the impact.

Difficulties abounded for Holley, financially and personally, with the death in that same year of his daughter Alice, who had been born in 1869. One year later, almost one-half of the once-burgeoning steel industry had shut down. Men walked the streets in despair and confusion. Even Andrew Carnegie was hard-pressed for ready money, and work on the new steel plant had to be stopped for a few months. Carnegie even sold his stock in the profitable Pullman Company in order to raise capital, and many of the workers arriving from Cambria were given only board money until work on the plant could be resumed. [27]

After the works at Braddock had been laid out, there remained the question of what to call it. The canny Carnegie understood the importance of maintaining Edgar Thomson's friendship. He could anticipate the business that would flow to the steel plant from the Pennsylvania Railroad. Moreover, the new plant would need special freight rates for bringing in raw materials and shipping finished products. It was obvious that it would be advisable to name the new plant after Thomson. Thomson, who was not certain that high quality steel rails would issue from Carnegie's mill, at first refused the honor. Later, undoubtedly flattered by Carnegie's gesture, he eventually agreed, and the plant was named the Edgar Thomson Steel Works.

William Shinn proved to be a capable general manager and kept the new enterprise running smoothly while Holley and Jones worked night and day to construct the plant. The two men worked unusually well together in planning the new works, making the best possible use of each others' creative talents. Shinn, too, had many sessions with Holley, and was continually astonished by Holley's skill as an engineer. Moreover, he frequently reminded himself that Holley's expertise reached far beyond steelmaking, and that this was also the Holley of the *Railroad Advocate.* Shinn later recalled in writing about this period at the Edgar Thomson Works, "I never knew which to admire most, the versatility of his genius, the depth of his professional knowledge, or that geniality of disposition and quickness of perception which made personal intercourse with him a pleasure, whether the object to be attained was strictly of a business nature, or merely social pastime. He appropriated and imparted valuable information with equal facility and apparent gratification, and in the amenities of private and social life he was as happy as in his professional knowledge he was thorough."[28]

Although money still was scarce, the Edgar Thomson Works slowly began to take shape. The designs of the cupola house, converting house, blowing engine house, boiler house, gas generator house, and the rail mill, as well as the office and shop buildings, all testified to the increasing importance of steel. The buildings were sturdy. All of them, except the generator house and the rail mill, were brick structures with iron roofs. There were two six-ton converters with all the necessary auxiliary equipment. Railway tracks connected the plant buildings. Of all his plant designs, Holley took greatest pride in the Edgar Thomson Works. Here, he said, he had begun at the beginning. He had taken a clean piece of paper on which he drew the railroad tracks first and then placed the buildings and contents of each building with prime regard to the facile handling of material so that the whole became a body shaped by its bones and muscles, rather than a box into which bones and muscles had to be packed.[29]

While serving the Edgar Thomson Works as consultant, Holley also was working on the design and building of additional Bessemer plants. He had

Alexander Lyman Holley about 1875.

begun plans for the Lackawanna Iron and Steel Works at Scranton, Pennsylvania, in July 1873 but, due to economic conditions, the company's first blow was delayed until 23 October 1875. The other plant, built for the Vulcan Iron Works in St. Louis, Missouri, and later a part of the St. Louis Ore and Steel Company, made its first blow on 1 September 1876. This plant, the last that Holley would build, is of special interest since it embodied the latest improvements in the Bessemer practice and represented the high point in what was called the American Bessemer plant. Hunt described Holley's innovations in the design of the plant.

He did away with the English deep pit and raised the vessels so as to get working space under them on the ground floor; he substituted top-supported hydraulic cranes for the more expensive counter-weighted English ones, and put three ingot cranes around the pit instead of two, and thereby obtained greater area of power. He changed the location of the vessels as related to the pit and melting-house. He modified the ladle crane, and worked all the cranes and the vessels from a single point; he substituted cupolas for reverberatory furnaces, and last, but by no means least, introduced the intermediate or accumulating ladle which is placed on scales, and thus insures accuracy of operation by rendering possible the weighing of each charge of melted iron, before pouring it into the converter. These points cover the radical features of his innovations. After building such a plant, he began to meet the difficulties of details in manufacture, among the most serious of which was the short duration of the vessel bottoms, and the time required to cool off the vessels to a point at which it was possible for workmen to enter and make new bottoms. After many experiments, the result was the Holley Vessel Bottom, which, either in its form as patented, or in a modification of it as now used in all American works, has rendered possible, as much as any other one thing, the present immense production.[30]

Holley now possessed several thousand drawings depicting every part of a Bessemer plant. His collection included drawings of both American and European plants. Holley had found in Gram Curtis a superior draftsman who also managed the office and in Phineas Barnes, a devoted assistant who acted as resident engineer as the new plants were going up. While Holley still followed carefully each construction detail, the difficulties that were intrinsic to the primitive early Bessemer plants no longer existed.

Once the Edgar Thomson plant was well under way, Holley felt that once again he should observe European operations. Ten years had passed since he had done so, and he knew that during that period many changes in steel manufacture had taken place. Another of his objectives in Europe was to raise money for the underfinanced Edgar Thomson Steel Company.[31]

Holley crammed every minute of his few weeks abroad with activity. Before going to Barrow, where he was to present a paper on American rolling mills before the Iron and Steel Institute, he attended meetings of the Society of Mechanical Engineers at Cardiff, where he was entertained handsomely by the important Welsh companies of Dowlais and Landore Siemens.

Barrow had created a gala atmosphere for the institute meeting, and flags were flying everywhere to welcome the 600 guests from all over the world. The Duke of Devonshire and the Barrow Haematite Iron and Steel Company entertained the members at a sumptuous dinner in the company drafting office. Holley's position in the steel world was manifested when he entered the large room ablaze with light from 450 gas jets, and was conducted to a place at the center table. From this time on, whenever he attended meetings such as this, Holley was honored with a seat at the head table. No other American was shown such deep respect. This esteem reached its zenith a few years later at an elaborate banquet that the lord mayor of London gave in honor of the president and members of the Iron and Steel Institute; Holley was seated with Bessemer and Siemens at the speaker's table.[32]

Depression and constant lack of funds notwithstanding, the first Bessemer blow at the Edgar Thomson Works took place on 25 August 1875, and a few days thereafter the rail mill produced the first rails. The formal opening took place on September 4, and the company made it a festive occasion by bringing the guests from Pittsburgh on a special train.

Captain William R. Jones. From Herbert N. Casson, *The Romance of Steel* (New York: A. S. Barnes & Co., 1907), opp. p. 32.

Many of the workmen were seeing the process in operation for the first time, but the precision of the operation gave the visitors the impression that the converter and the men had worked together for years.[33]

In Shinn, whom McCandless had brought into the company as general manager, the partners had found a man of extraordinary ability. He introduced the efficient system of accounting that the railroads had been using for a number of years. By this method, a steel company knew the exact cost of every item of raw material that went into a product and could calculate the cost of an order even before it was entered in the books. This information was used constantly in every phase of the Edgar Thomson activities.[34]

Backed by Shinn as general manager, Captain Jones began to practice the principles in which he believed.[35] When the Edgar Thomson Works first went into operation the large number of accidents there worried and exasperated Jones, since most of them were due to the men's carelessness. Jones encouraged his workers to take out accident insurance, and the company helped some of the men to do so in order to get the program under way. Jones did not favor coddling the American workingman. "Let him learn to be prudent like other men," said Jones. "If we can train the workman to be self-reliant, it will be better for the manufacturer and his people. Let the capitalist deal fairly and squarely with the laboring man and leave him his independence and his responsibility—not first make him helpless and then nurse him because he is helpless."[36]

Jones's beliefs about the proper way of conducting the works were clearly defined not only in his own mind. He also had put them in writing in a letter that he wrote to the son of McCandless before the plant began operations. The first part of the letter discussed some of the cost figures at Cambria and the large profits of the rail producers at Harrisburg and Newburg. He then continued:

> I will give you my views as to the proper way of conducting these works.
>
> 1st. We must be careful what class of men we collect. We must steer clear of the West where men are accustomed to infernal high wages. We must steer clear as far as we can of Englishmen who are great sticklers for high wages, small production and strikes. My experience has shown that Germans and Irish, Swedes and . . . young American country boys, judiciously mixed, make the most effective and tractable force you can find. Scotsmen do very well, are honest and faithful. Welsh can be used in limited numbers. But mark me, Englishmen have been the worst class of men I have had anything to do with; and this is the opinion of Mr. Holley, George and John Fritz.
>
> 2nd. It should be the aim of the firm to keep the works running steadily. This is one of the secrets of Cambria low wages. The workmen . . . do better at Cambria than elsewhere. On steady work you can calculate on low wages.
>
> 3rd. The company should endeavor to make the cost of living as low as possible. This is one bad feature at present but it can be easily remedied.
>
> . . . The men should be made to feel that the company are interested in their welfare. Make the works a pleasant place for them. I have always found it best to treat men well, and . . . my men are anxious to retain my good will by working steadily and honestly, and instead of dodging are anxious to show me what a good day's work they have done. All haughty and disdainful treatment of men has a very decided and bad effect on them.
>
> . . . I am afraid that unless the policy . . . is followed we need not expect the great success that is obtainable. These suggestions are the results of twenty-five years experience obtained in the most successful iron works in this country: —Crane and Thomas Iron Works, Port Richmond Iron Works and the Cambria Works.[37]

A new era of competition was dawning at the time that the Edgar Thomson plant went into production. Holley's old friends, both his students and former associates, were now superintendents of plants that he, himself, had designed. Robert Hunt was running Troy, John Fritz was at Bethlehem, John E. Fry had succeeded Daniel Jones at Cambria, Bill Jones directed the Edgar Thomson Bessemer plant, and Robert Forsyth, the one at North Chicago. Holley made rounds from one to the other, both as consulting engineer and as good friend and advisor.

Henry Howe wrote of Holley's hold on the affections of this group. It was this, he thought, almost more than his uncommon mechanical skill, that was responsible for the extraordinary way in which the Bessemer process developed. "With his leadership, this remarkable group even though competitors vying for records, nevertheless joined enthusiastically to establish traditions, standards and aims and their generous rivalry, *esprit de corps,* free trade in ideas, mutual help, quickness to adopt whatever was good in his neighbor's practice whether cis- or trans-Atlantic and a willingness to give and take was extraordinary in any time or place."[38]

The competition among the plant superintendents was fierce. Forsyth, at the North Chicago works, started it when he sent a telegram to John Fry at Cambria announcing that in the week of October 16 his Bessemer plant had made 196 blows and that in one day of twenty-four hours, they had achieved 52. Fry did not reply immediately, but a month later Forsyth heard from Cambria that they had made 231 blows between Monday morning and Saturday evening. At first the rivalry had been between the "Pittsburgh boys" and the "Chicago boys," but Robert Hunt at Troy, eager to join the contest, announced to the competing Forsyth and Fry that in the first week of 1875 Troy had produced 232 blows in ten turns that yielded 1,140 tons. Mischievously, Hunt added that he and his men were so exhausted that the following week they could make only 220 heats, or around 1,100 tons of ingots from one pair of converters. Hunt was doing remarkably well, since his was the oldest plant. When the newest Bessemer installation, the Edgar Thomson Works with Jones at the helm, passed its breaking in stage, the friends were really in competition. As records fell like clay pigeons, the steel industry entered a new era.[39]

NOTES

1. James Howard Bridge, *The Inside History of the Carnegie Steel Company* (New York: Aldine Book Co., 1903), p. 2. James Howard Bridge was born in Manchester, England, in 1859. He was educated in France. After a period of newspaper work he became the private secretary of Herbert Spencer, the English philosopher. In 1884, when Bridge determined to try his fortune in the United States, Spencer gave him a letter of introduction to Professor Edward L. Youmans in New York City, a close friend and one of the leaders in promoting Spencer's work in America.

   Youmans sent the young man to Andrew Carnegie. It was an apropos time since Carnegie was contemplating the writing of an history of material development of the United States during the preceding years. Bridge was engaged as a literary assistant to Andrew Carnegie and went to live with Carnegie and his mother until 1887 when Carnegie married. Bridge continued as literary assistant until 1889, doing much of the research for *Triumphant Democracy.* Later he was editor first of the California magazine *Overland Monthly,* then of *Commerce and Industry,* published in New York. During his last years he was curator of the Frick Collection.

When *The Inside History of the Carnegie Steel Company,* written by Bridge, appeared in 1903 certain criticisms were leveled against the writing of it. In response a note appeared in the third edition, which read:

To meet certain criticisms which have been made concerning the propriety of his publishing this book, Mr. Bridge wishes to say that he was never Mr. Carnegie's "private secretary" as that term is usually understood. For several years he assisted Mr. Carnegie in literary work, especially in the preparation of *Triumphant Democracy,* and during this time he had neither the opportunity nor the inclination to learn the business secrets of the steel companies. This book does not contain a single fact that was acquired by Mr. Bridge in a confidential capacity; nor has any fact been included that was improperly obtained by anyone else (Bridge, Author's Note, *Carnegie Steel Company,* n.p.).

2. Burton J. Hendrick, *The Life of Andrew Carnegie,* 2 vols. (Garden City, N.Y.: Doubleday, Doran & Co., 1932), 1: 132-33.

3. Ibid., p. 133.

4. Ibid., p. 134.

5. Andrew Carnegie, *Autobiography of Andrew Carnegie* (Boston: Houghton Mifflin Co., 1920), pp. 57, 63.

6. Hendrick, *Andrew Carnegie,* 1: 87, 89. The traits of Carnegie that first interested Scott are described by Hendrick: "Carnegie had a quickness of decision, assertiveness, absolute confidence in himself and his willingness to accept responsibility amounted almost to audacity. His self-reliance was so great that those who didn't like Carnegie, and through the years there were many, regarded it as abnormal egotism."

7. Carnegie, *Autobiography,* pp. 79, 139. According to Carnegie, his mother mortgaged their modest house in Allegheny to secure the necessary funds. He invested next in a sleeping car company that grew into the Pullman Palace Coach Company. Investment income then paid for an interest in the Columbia Oil Company, which had struck oil on Pennsylvania farmland. Forty thousand dollars had been paid for land on the Story farm on which in one year the wells earned $1 million. A $40,000 investment in the company is said eventually to have grown to $5 million. By 1863, at the age of thirty, Carnegie had participated in enough projects to have an income of nearly $48,000 yearly from his salary and investments (Hendrick, *Andrew Carnegie,* 1: 120).

Another of Carnegie's biographers states that there is evidence that Carnegie's mother did not mortgage their house at that time. An IOU note indicates that Thomas A. Scott advanced the money for Carnegie's stock purchase. Later, when he could not meet Scott's note, Carnegie borrowed at a higher rate of interest to repay it. Some time later, even after he had made other investments, instead of paying off that loan, Carnegie turned to his mother for help, and it was then, in 1858, that she mortgaged the house (Joseph Frazier Wall, *Andrew Carnegie,* [New York: Oxford University Press, 1970], p. 134).

8. Hendrick, *Andrew Carnegie,* 1: 128.

9. Carnegie, *Autobiography,* p. 116.

10. Bridge, *Carnegie Steel Company,* pp. 12, 16-17.

11. Ibid., p. 21. Bridge dryly comments that Carnegie's settlement of the quarrel was a little like the old tale of the lawyer and the oyster. The litigants got the shell.

12. Ibid., pp. 23, 31, 49.

13. Ibid., p. 28. An amusing story of the time concerned Phipps and his black mare Gypsy. It became routine for Phipps to visit the banks to borrow money to meet payrolls. During these weekly forays Gypsy became accustomed to following a zig-zag course down Pittsburgh's main street as Phipps went from bank to bank. She would wait patiently before one bank until Phipps appeared, and then cut a diagonal across the street to another. In later more prosperous years, Gypsy still could not be taught to traverse Wood Street in a straight path.

14. Hendrick, *Andrew Carnegie,* 1: 46, 179.

15. Wall, *Andrew Carnegie*, pp. 363–64; Hendrick, *Andrew Carnegie*, 1: 130. The success of Carnegie's method was evidenced by the large amounts of business that he obtained for Keystone from the fast-growing Baltimore and Ohio Railroad of which John W. Garrett was president. The two men shared not only what Carnegie described as "a wee drap of Scotch bluid between us" but also an overwhelming fondness for the poetry of Robert Burns. Carnegie's memory was phenomenal, and he took pride in his ability to quote, when the occasion was opportune, nearly all of Burns's poetry.

16. David B. Steinman and Sara Ruth Watson, *Bridges and Their Builders* (New York: Dover Publications, 1957), p. 192.

17. Bridge, *Carnegie Steel Company*, pp. 54–56.

18. Ibid., pp. 72, 73. Wall, however, believes that Carnegie was one of the original sponsors of the Edgar Thomson Steel Works and that he had to force his reluctant partners into joining the new enterprise (Wall, *Andrew Carnegie*, p. 1063.)

19. Bridge, *Carnegie Steel Company*, pp. 73, 74.

20. Ibid., p. 33.

21. Wall, *Andrew Carnegie*, p. 314.

22. Bridge, *Carnegie Steel Company*, pp. 77, 78; Wall, *Andrew Carnegie*, p. 315.

23. Herbert N. Casson, *The Romance of Steel* (New York: A. S. Barnes & Co., 1907), pp. 23–24.

24. Robert W. Hunt, "Cambria Steel Workers' Reunion," *Iron Age* 88 (5 October 1911): 743. For years friends and mere acquaintances had turned to Holley when they were in need of jobs. Since he traveled widely and met many people, he customarily was consulted by both employers and out-of-work employees. Among his correspondence are many letters recommending men for positions or promising others that he would find them jobs. Not only did Holley serve as an employment intermediary, he also on one occasion searched the steel mills in Cleveland in an effort to locate a man who had deserted his wife in England.

25. Bridge, *Carnegie Steel Company*, pp. 78–79. Over two hundred men left Cambria to join Jones. It was a raid of major proportions and the consequent ill feelings persisted between the two companies for several years.

26. Hunt, "Cambria Steel Workers' Reunion," *Iron Age* 88 (5 October 1911): 742.

27. Bridge, *Carnegie Steel Company*, p. 79.

28. *Memorial of Alexander Lyman Holley* (New York: American Institute of Mining Engineers, 1884), p. 26.

29. Ibid., p. 135.

30. Robert W. Hunt, "A History of the Bessemer Manufacture in America," *Transactions of the American Institute of Mining Engineers* 5 (1876–77): 214–15.

31. ALH to father, 29 August 1874, CHS. Holley writes to his father from London: "The money raising part of my business here was finally abandoned by the Edgar Thomson Steel Co. as they found they could get the money on as good terms at home. I am, however, asked to help to raise money for a lot of other American things and good ones too but American investments are sick."

32. "Meeting at Barrow," *Journal of the Iron and Steel Institute* (London) 1 (1874): i; Scrapbook, Author's Collection.

33. Z. S. Durfee, "Description of the Bessemer Process," *Bulletin of the American Iron and Steel Association* 8 (10 September 1874): 265–67.

34. Hendrick, *Andrew Carnegie*, 2: 45. That the accounting system had its effect is evident from an anecdote told by Bridge: As Shinn passed a workman who was building a furnace, he heard the man swearing under his breath. The worker commented that if he used a dozen bricks more than he had in his last relining job, Shinn would know about it and come around to ask why. A conversation between Carnegie and the publisher, Frank Doubleday, also is of interest. Doubleday had stated that he wouldn't know until the end

of the year whether or not he had made a profit. Carnegie was aghast. "I couldn't stay in business a week under those conditions," he told Doubleday.

35. Casson, *Romance of Steel,* pp. 31–32. Jones was not an easy boss. His temper was short and any man who had been subjected to his blistering tongue never forgot it. Old hands did not take seriously the tongue lashing and the dismissal that occasionally went with it and would show up for work the next day as if nothing had occurred. Jones might eye the culprit speculatively and shift the inevitable Pittsburgh stogie from one side of his mouth to the other, but nothing more was likely to be said. There was no sham, deceit, or meanness in Jones, and many workmen had benefited from his great generosity of mind as well as pocket.
36. Joseph S. Harris, "The Beneficial Fund of the Lehigh Coal and Navigation Company," *Transactions of the American Institute of Mining Engineers* 12 (1883–84): 599, 600.
37. Bridge, *Carnegie Steel Company,* pp. 81–82.
38. Henry M. Howe, "Notes on the Bessemer Process," *Journal of the Iron and Steel Institute* (London) 37, no. 2 (1890): 125.
39. *Engineering* 19 (5 February 1875): 108–9.

XVIII · Holley and Education

nly ten years had elapsed between the building of Holley's first converter plant and his eleventh, the Edgar Thomson Steel Works. Since the early days at Troy, many changes had taken place in the machinery necessary to operate the Bessemer converters efficiently, and no one realized that fact more keenly than Holley. Since his hand and mind had guided many of the improvements, he often was asked to speak concerning them. Although formerly he had been overwhelmingly shy, he now had become one of the most sought after speakers of the time. His ability to describe technical matters with clarity and precision was outstanding, and, from the mid-1870s until his death, he appeared before professional groups in both America and England more often than any other person from the industry. His wit and charm made him a sought-after guest whenever steelmen gathered. His extraordinary ability to find the appropriate phrase always brought his listeners to a shared feeling of deep respect, pride, and even affection for their profession and for each other.[1]

In one of numerous speeches, Holley discussed the changes that had occurred in the Bessemer process since he had known it. He said that he believed increased knowledge was the prime reason for the improvements in practice. Workers now had a better understanding of their work; for instance, they had learned to recognize the warning signs indicating that a lining was about to burn out. Because the men knew its probable behavior, the fluid metal no longer held such terrors. Formerly, after achieving four or five blows, workers had quit rather than risk their luck further. Now they could produce thirty or even forty blows during one twenty-four-hour period.

Refractory materials, Holley next observed, had changed comparatively little since his early intense efforts to discover the best. But even though the composition of refractories had not improved much, skill in the lining of converters and ladles had developed to a higher degree.

Increases in production and operating skills Holley attributed to a multitude of improvements in mechanical devices for which no one person could be given credit. The major improvements were Holley's, but most of the others were the product of the long talks during the interminable hours that the pioneers in the Bessemer installations had spent together. The Fritz brothers, Robert Hunt, Bill Jones, and numbers of unnamed workers all had contributed.

The melting department, Holley reminded his audience, was one of the first in which changes had been made. The air furnace used such large amounts of fuel that it almost immediately had been replaced by a foundry cupola, which required only 25 percent as much.

Holley discussed the need to replace burned out linings and tuyeres, which continued to cause problems in operating the Bessemer process. Since these parts might fail after ten or fewer heats, long delays with consequent production losses were frequent, all of which Holley considered a serious deficiency in the process.[2] Fixed costs, he said, continue whether or not a steel mill is operating. Lost time must be avoided by all possible means.

Holley was not alone in his preoccupation with the problem of time lost because of vessel bottom failure. Bessemer had devised a system of removable and duplicate vessel bottoms but Holley brought certain refinements and additional arrangements to Bessemer's patents with his development of the Holley Vessel Bottom, which was perhaps his single most important contribution to the Bessemer industry. By means of his invention, a replacement vessel bottom always was kept in readiness. The failed bottom was removed, the duplicate inserted, rammed into place, and the joint filled from the outside. Holley's patent made it unnecessary to cool the hot converter, and, since the replacement was prepared before a bottom failed, it was already dry.[3] By 1875, the Holley Vessel Bottom was in use at all American plants with the exception of those at Newburg and Joliet, and at many European plants. Each company paid Holley $5,000 for the right to use the patent for its term.

Holley also patented a method for the bottom casting of ingots. The molten metal was conducted through a runner that connected with a riser into which the steel was teemed and around which ingot molds were arranged in a circle. The royalties for this patent amounted to $3,000. Holley's enthusiasm helped launch this casting method, which was popular at first but later proved expensive and was abandoned.[4]

As the steel industry forged ahead, a group of engineers connected with it recognized that the number of men working in mining and metallurgy had grown along with the industry. For some time, the group had been interested in forming an association whose membership would be composed of men engaged in those fields. There was at that time no American organization devoted to the technology of the iron and steel industry. The American Iron and Steel Association had been organized in 1855 mainly for collecting industry statistics and acting jointly on tariff matters. Its membership was composed only of executives of iron and steel companies.

In England, the Iron and Steel Institute, which had been organized in 1869, had demonstrated the importance of exchanging information and practical experience. The secrecy that had been accepted company procedure for many generations had broken down. Through the influence of the institute, companies frequently permitted inspection of their practices by competitors.

The American group of engineers admired the manner in which the British organization had grown in its two years of existence and decided to form an organization in the United States that would provide a place where men with like interests could meet and where they could discuss their ideas together. Holley was not a member of the organizing committee, but he became one of the first members of the American Institute of Mining Engineers.

The institute began in a modest way in 1871 but, with the wholehearted support of Holley and others who were influenced by his enthusiasm, it steadily gained professional standing.[5] On European trips, Holley had observed benefits to the members of such societies of the opportunity for group visits to plants and he was eager to establish the same practice in America.

By 1875, when Holley was elected president of the American Institute of Mining Engineers, he had become both the leader and the conscience of his profession. He felt not only the honor of the position, but, also an obligation to use it to bring about changes in the engineering profession that he believed had been needed for some time. His presidential address, "Some Pressing Needs of Our Iron and Steel Manufactures," synthesized the results of many years of thinking about the past, the present, and the future of the industry. It was a forthright speech that was intended to dispel complacency and self-satisfaction. He first discussed the need for fuel efficiency. The cost of coal to drive the machinery of a Bessemer plant was averaging $1.50 per ton of ingots, he said. By using better steam engines, economies of as much as $50,000 per year could be achieved there; and in steel rail mills, as much as $175,000. He chided his audience for trying to avoid radical change by patching up old engines. He embarrassed many of the members by saying, "No man ever throws out an engine which he has just rebuilt, however bad it may be."[6]

Holley said that he considered improved heating furnaces as important as better steam engines. He listed the reasons for the enormous heat losses in all plants and discussed remedies. In each instance he bolstered his arguments with accounts of the practices at other plants in the United States and Europe.

Refractory materials remained the problem they had been ever since the building of the first converter. The cost of maintaining refractory linings and fixtures in the Bessemer process amounted to nearly $1 per ton. Holley stressed that, merely by doubling the life of the bottoms, a single works could save $6,000 annually. "Our metal manufacturers," he said, "could copy the steam engine results of others but the refractory material problem is all their own."[7]

Holley then made a radical suggestion. Inasmuch as the solution of refractory problems would require both long and careful chemical studies and long-term systematic experimentation, individual companies should not attempt that work. It should instead be undertaken jointly for the general benefit, even though he conceded that there could be valid objections to a group effort.[8]

Holley next admonished the engineers, as he had the producers, "Enlarge the range of manufacture and stop wasting." In the United States, he told them, 350,000 tons of Bessemer steel were being made each year. Yet all of this amount, except not quite 6,000 tons, was used in the production of rails. When rail orders fell off, the Bessemer plants were affected almost immediately. In Europe, there had been a rapid adaptation to using steel for structural purposes. In 1873 France already had built three successful men-of-war and three more had been ordered in 1874. Of the locomotives recently built for the London and North-western Railway, the entire engines, with the exception of the wheels and some castings, were of Bessemer steel. In England, railways, machine shops, bridge works, and shipyards, had learned how to treat, heat, and shape steel for numerous applications; while in America, few Bessemer plants even possessed suitable machinery to produce a shape other than a rail.

He bluntly stated that on this side of the Atlantic, where the best steel works in the world had been built and where increased production and decreased costs were a source of envy to the Europeans, the producers were concentrating more effort on schemes to prevent overproduction than on finding other uses for the steel pouring from the converters. Holley emphasized the need for companies to conduct experiments in order to discover the behavior of their products in tension, compression, and elasticity. "Steelmakers cannot expect a very brisk demand for materials which they do not know whether they can make or not or what they will cost or be like when they are made," he advised.[9]

Holley's closing words recalled pioneering days as he spoke of the men who had been willing to back the construction at Troy, Harrisburg, and Cleveland of the first plants he had designed, where by trial and error he had found his profession and perfected himself in it: "I cannot close these remarks without bearing testimony to the gallant manner in which the commercial promoters of our early Bessemer works poured out money and encouragement to us, who were so long floundering in the slough of technical uncertainties."[10]

During Holley's term as president of the American Institute of Mining Engineers there was much discussion about the proper nomenclature of steel. In 1875 a series of articles by Henry Howe appeared in the *Engineering and Mining Journal*. A distinguished metallurgist and teacher, Howe had made certain recommendations with which Holley could not agree.[11] At the October meeting of the institute in Cleveland, he presented a paper entitled "What Is Steel?" which served as a rebuttal to the Howe proposals. A lively discussion ensued, and an International Committee on the Nomenclature of Iron and Steel was formed to consider the subject.[12]

A supremely distinguished group of metallurgists constituted the committee: I. Lowthian Bell of England; Hermann Wedding of Germany; Peter Ritter von Tunner, the great teacher from Leoban, Austria; Richard Äkerman, Sweden; L. Gruner, France; and with Thomas Egleston of Columbia University and Holley representing the United States. Although the committee made assiduous efforts to standardize nomenclature, its recommendations were not accepted by the membership of the institute. The committee was asked to try again, but there seemed to be no desire among the membership to change their accustomed terminology, and the committee eventually passed quietly out of existence.[13]

Much of Holley's attention during his presidency was devoted to the subject of technical education. He had observed and admired the methods used by many of the leading technical schools in Europe, and he hoped to see an educational institution of similar quality established in America.[14]

One of Holley's best lectures has become known as a near classic in its field; it was delivered in 1872 at the Stevens Institute of Technology. The subject was Bessemer machinery. The lecture was typical of the extraordinary care he took to prepare a technically competent lecture for what he always presumed to be a mature and receptive audience. His exuberance and interest in his subject then held the attention of an audience of any age or background.

He pointed out to his young listeners the uniqueness of the Bessemer process. While it had been dependent for its improvement on chemistry and metallurgy, he considered the Bessemer process an outstanding example of what the invention and adaptation of mechanical appliances and processes could achieve. He spoke of the relatively short sixteen years during which the process had been transformed from a floundering effort to an enterprise that then produced one million tons of steel per year.

He described the mechanical problems of handling seething, molten metal, the demands on the mechanical engineer, and, finally, explained the working of the process. He described the machinery, pumps, and engines required by the process and illustrated his points with his own detailed drawings.[15]

Holley never lost interest in the problems of students of engineering. He continued throughout his career to meet with both individuals and groups who sought his advice. Nevertheless, he had strong opinions about how students ought to go about getting information. He advised a student that he should attempt to write a thesis on the new open hearth process only if the student were able to spend a week or two studying the process at first hand. The student, he wrote, might best observe the process at the Otis Steel Company in Cleveland, where Holley would help him gain entrance through his good friend Samuel Wellman, who was in charge there. In addition, Holley offered to lend the young man a paper on the open hearth process by William Hackney, the only one in existence at that time. "But," Holley warned, "merely copying this will be of no use to you or anyone."[16]

To another student seeking information about certain details of the manufacture of Bessemer steel he was equally frank.

> I received your request for a report on Bessemer manufacture and for a long time debated what I ought to do in the matter. Having lately had much conversation with instructors and experts on the general subject of technical education I am convinced that your ferreting out these facts for yourself would put you in a much more likely position to build a Bessemer plant than your saving yourself the trouble by asking me. You are wrong in saying that "it is impossible to get running estimates from any mill." Some details of cost you might not get, but it would not be proper for me to give you such and general details you can get as nearly as you can ask me to give them."[17]

Toward the end of his presidency, Holley spoke on a subject that had concerned him since his days at Brown University. In an address in Washington, D.C., in February 1876, he discussed the state of technical education. The teaching profession, the managers of iron and steel plants, unions, and students all were considered in a frank discussion that made many of his audience uncomfortable and others angry.

Holley first dealt with what he considered the underlying weakness of technical education: the teachers were disseminating second-hand material. "The expert who delights to call himself 'practical' is honestly amazed at the attempts of experts by school graduation, who have not been graduated in works, to solve the engineering problems of the day. . . .

The attainment of powerful and sufficiently hot blast by means of waste heat, the adaptation of shape and proportion of stack to different fuels and ores, labor-saving appliances and arrangements—all these have grown out of the constant handling, not of books, but of furnaces."[18]

Holley then pointed out that Bessemer had developed his process with little more knowledge than a schoolboy might possess; that he and his followers had brought the difficult process to a high degree of commercial success with little help from the literature and none from teachers. Similar circumstances, he asserted, surrounded the development of machinery and the rest of the growth in the steel industry.

> When one can feel the completion of a Bessemer "blow" without looking at the flame, or number the remaining minutes of a Martin steel charge from the bubbling of the bath, or foretell the changes in the working of a blast-furnace by watching the colors and structure of the slag, or note the carburization of steel by examining its fracture, or say what ore will yield from its appearance and weight in the hand, or predict the lifetime of a machine by feeling its pulse; when one in any art can make a diagnosis by looking the patient in the face rather than by reading about similar cases in a book, then only may he hope to practically apply such improvements as theory may suggest, or to lead in those original investigations upon which successful theories shall be founded.

Holley conceded that perhaps a neophyte in industry with a scientific education could learn, from what he observed, a little faster than the workman could. But he was convinced that unless the graduate studied in the plant as long as he had in school, he could not in the long run compete with the men who had come up through the ranks.

Holley expressed scorn for the efforts of teachers to further their academic standing through inventions and scholarly papers.

> Another evil growing out of the inadequate regard of mere schoolmen for practice, is the frequent failure of their works or their inability to complete them. Inventions and constructions, designed after a scientific method and under the light of organized facts and detailed history as laid down in books, may fail simply in default of a practical knowledge of how far the capital at hand will reach, or what the means at hand will do, or what the materials at hand will stand, or what the labor and assistance at hand can be relied on to accomplish. A vast number of facts about the operation of forces in materials are so subtle, or so incompletely revealed or disentangled from groups of phenomena, that they cannot be defined in words, nor understood if they could be formulated. But after long familiarity with the general behavior of materials under stress, a practical expert, can by a process more like instinct than reason, judge how far and in what directions he may safely push his new combinations. . . . In this chasm, between science and art, how much effort and treasure, and, even life, are swallowed up year by year.

At the same time that Holley belabored the smugness of the teaching profession, he looked closely at the attitude of the managers of the iron and steel industry. Holley considered their attitude toward practical training even more reprehensible than that of the teachers.

> A greater obstacle is the combined misapprehension and ignorance on the part of a large class of "practical" men, of what they are pleased to call "theory," meaning by theory something which is likely to be discordant with fact — or possibly with the interests of the craft. We can hardly complain that their objection is ill-grounded, as far as it is grounded upon the practice of theoretical men; but the world has a right to complain of their narrowness of observation, of their stolid incomprehension of the results of science, of that pride of ignorance, of that bigotry, of that positive fear of the diffusion of knowledge, which is the normal condition of those who range only within the sphere of their own practice, and to whom analysis and generalization, in their business affairs, as well as in morals and politics, are an unknown thing. It is unfortunately true that a large number of managers of metallurgical enterprises — men who are deemed indispensable, and who probably are indispensable, in the average state of practical science, are thus not incorrectly characterized. Conscious of their power as conservators, ignorant of the elements of improvement, and not unfrequently jealous and blindly fearful for the interests of their craft, they sit triumphant on an eminence (the steady undermining of which they cannot observe), and sneer at the too frequently condescending magniloquence of recent graduates and book men.

Holley described an incident that to him exemplified the attitude of management. Years before, as an outgrowth of a series of elaborate experiments that covered both the civil and mechanical engineering aspects of railroad operation, the Erie Railroad had accumulated thousands of drawings and voluminous records. Yet when James Fisk, Jr. gained control of the Erie, he had ordered all of this mass of information to be sold as waste paper, and had thus destroyed information that was both invaluable and irreplaceable.

Throughout his talk Holley was endeavoring to demonstrate that scientific education existed on many levels, and that the nature of education deserved serious study by everyone concerned with the future of engineering. Holley himself had concluded that the technical school should preferably teach artisans of good general education.

> The art must precede the science. The man must feel the necessity, and know the directions of a larger knowledge, and then he will master it through and through. Mark how rapidly the more capable and ambitious of practical men advance in knowledge derivable from books, as compared with the progress of bookmen, either in books or in practice. Many men have acquired a more useful knowledge of chemistry, in the spare evenings of a year, than the average graduate has compassed during his whole course. These men realized that success was hanging on their better knowledge. Familiar with every changing look of objects and phenomena, they detected the constant

play of the unknown forces which underlie them, and longed for a guide to their operation, as a mariner longs for a beacon light. This practical familiarity and judgment at once revealed the importance of scientific facts and methods, promoted their acquisition, and guided their application. Under what comparative facilities does the mere recitation-room student, or even the mere analyst of the hundred bottles, study applied chemistry? It is to these a matter of routine duty, without a soul; they are neither stimulated nor directed by a previously created want.

Holley spoke in favor of establishing schools in the engineering works themselves, with a small group of students gaining skills and knowledge of materials and machinery under the discipline and leadership of a master. Holley, the master, had a parting remark for the students themselves.

> It seems proper to say a word about the royal road to learning, which a few ill-advised students attempt to pursue. I do not refer to their availing themselves of professional data and drawings on file in engineering offices, but I do refer to their asking engineers and managers to furnish them special reports on subjects regarding which their own observation would be vastly more useful to the applicants, and quite as convenient for the respondents—reports on the number and duties of workmen in each department, and the particulars of operation and of relative cost, which can only be profitably investigated by a student, when not only the facts but the reasons are ferretted out by himself, rather than transmitted to the academic grove through the post office.

The extent of the discussion that Holley's paper provoked indicated that there had existed an unexpectedly great concern for the quality of technical education. Holley moved quickly to capitalize on that interest and was instrumental in bringing about a two-day meeting, which was held at the Franklin Institute in Philadelphia under the joint auspices of the American Society of Civil Engineers and the American Institute of Mining Engineers.[19]

The meeting took place just prior to the opening of the Philadelphia Centennial of 1876, the first large international exhibition to be held in the United States.[20] Leaders in the engineering profession, management, and educational institutions in both Europe and America were brought together for the first time through the efforts of a committee of which Holley had been secretary to discuss how best to educate young men entering the engineering field. At the opening session Rossiter Raymond, the chairman, first acknowledged that Holley had been the moving force behind the discussions.

The debate centered on a question that Holley had raised earlier: when should a student begin scientific studies? There were three points of view. One group believed that practical instruction should precede theory; another group thought theory and practice should be simultaneous; the third group advocated that theory precede practical application. The majority agreed with Holley that a program that combined theory and practice was preferable. Alternating the two also was considered acceptable.

The group generally agreed that liberal education was an essential part of education for engineering. As Holley put it, "... the want, not of high scholarship but of liberal and general education, is today the greatest of all the embarrassments which the majority of engineering experts and managers encounter. ... It seems of first importance to promote, if not almost to create, a public opinion that liberal and general culture is as high an element of success in engineering as it is in any profession or calling."

No definite program came out of the sessions. However, if Rossiter Raymond had not declared such an action out of order, the group would have moved to dissociate the engineering degree from formal education after Ashbel Welch, a leading railroad engineer, asserted that the title of Civil Engineer was meaningless if it placed a boy just out of school on a level with men of many years experience in the profession. He moved that the title be granted only after a requisite period of experience.

The conference, nevertheless, could be considered successful. Holley had wished to arouse in both the teaching and the engineering professions a recognition of their mutual responsibility for the solution of the problems of engineering education, and he more than succeeded in his aim.[21]

Holley also had been named chairman of the committee to arrange the metallurgical engineering section of the centennial. The appointment was a signal honor, but a vast expenditure of both time and labor was required. Remuneration consisted of a mere $600 for personal expenses and the sum of £3 per page for articles he was to write describing the exhibition for *Engineering* in London.

The centennial would attract engineers from all over the world and Holley worked energetically to make the engineering section outstanding. He persuaded companies to submit suitable exhibits; he had a room opened in Philadelphia for the use of visiting engineers; he persuaded railroads to run special trains at reduced rates to the various steel plants; and he went so far as to ask Abram Hewitt to pay for lunches on one of the trips.[22]

For the first time, the American engineers had the privilege of meeting their distinguished European colleagues. The Americans were hosts at numerous receptions and dinners to welcome the famous men whom they previously had known by name only.

The centennial made the rest of the world aware of the enormous advances that had been made by American industry. Dr. Hermann Wedding, the leading German metallurgist, was openly astonished at how far American metallurgy had progressed. He was amazed to discover that, although Great Britain, as expected, was first in steel production, the United States was second, while Germany was third. "In particular," Wedding said, "the achievements of the never-to-be-forgotten Holley in the field of Bessemer practice set us an inspiring example for imitation in our own country."[23]

Holley's term as president of the American Institute of Mining Engineers was to expire before the opening of the exhibition, but there had been a move by certain members to keep him in office. James Drown, the secretary of the organization, wrote Holley about the move, to which Holley replied, "You put it very seductively but as Raymond (with Plymouth culture) has found a way out for me of my difficulty, I must go out. Everybody in Washington understood that Hewitt would go in. I couldn't take time to get up the necessary historical address that's a clencher. . . . Hewitt expects to go in and it would therefore be awfully awkward for him and for me if I should hold over. I was first elected with the understanding that I was a good fellow and wouldn't mind giving up the place to another at Centennial meeting."[24]

NOTES

1. Alexander Lyman Holley, "Lecture on Bessemer Steel," *New York Evening Post,* 1 April 1872.

    The large attendance on Saturday evening at the Cooper Institute to listen to the lecture of A. L. Holley was a pleasing evidence of the interest felt in scientific and mechanical progress. Notwithstanding the very inclement weather over one thousand persons, including many ladies, listened with unvarying attention to the clear and practical explanation of Mr. Holley. By the use of a camera diagram of the various forms of furnaces, machinery, etc., adopted in this country and Europe for the manufacture of Bessemer steel, were clearly set forth, and the advantages gained by American skill and experience were fully appreciated by the audience. During an intermission of a few minutes between the two sections of the lecture Abram S. Hewitt, Esq., addressed the audience stating that Mr. Holley was himself the inventor of the American improvements referred to but that his modesty had prevented his reference to the fact. The entire lecture was listened to with much attention.

2. Alexander Lyman Holley, "Recent Improvements in Bessemer Machinery," *Transactions of the American Institute of Mining Engineers* 2 (1874): 263-64. Earlier methods of replacing the bottom were awkward. Some European plants knocked out the old tuyeres, inserted new ones, and filled any opening with semifluid ganister (lining material), a silica refractory. Other plants permitted the vessel to cool enough to allow a workman to crawl inside to ram the ganister between the brick. The vessel then was fired until the material was set and dry. Either method resulted in loss of much time and tonnage.

3. Ibid., p. 268. Prior to that time, when the bottom still had to be replaced from inside, it was difficult to know when the bottom had dried. If any moisture remained, dreadful explosions were caused by the formation of steam.

4. Ibid., pp. 269, 272. ALH to W. W. Scranton, 4 November 1875, Author's Collection (hereafter cited as AC).

5. Holley also became a member of the Iron and Steel Institute in London in 1873, the American Institute of Civil Engineers in 1876, and the Institution of Civil Engineers in London in 1877.

6. Alexander Lyman Holley, "Some Pressing Needs of Our Iron and Steel Manufactures," *Transactions of the American Institute of Mining Engineers* 4 (1876): 78, 81.

7. Ibid., p. 86.

8. Ibid., pp. 86-87.

9. Ibid., pp. 93-94, 98.

10. Ibid., p. 99.

11. Henry M. Howe, "What Is Steel?" *Engineering and Mining Journal* 20, no. 9 (28 August 1875): 213; 20, no. 10 (4 September 1875): 235-36; 20, no. 11 (11 September 1875): 258-59; 20, no. 12 (18 September 1875): 282-83.

12. Alexander Lyman Holley, "What Is Steel?" *Transactions of the American Institute of Mining Engineers* 4 (1875-76): 139-49.

13. "Committee Report on Nomenclature," *Transactions of the American Institute of Mining Engineers* 5 (1876): 19-20.

14. Samuel Rezneck, "The Engineering Profession Considers Its Educational Problems," *Association of American Colleges Bulletin* 43, no. 9 (October 1957): 410-18. After his election as a trustee of Rensselaer Polytechnic Institute in 1865, Holley had shown a great interest in engineering education. Rensselaer, at Troy, was one of the outstanding American technical schools. Although he was young and relatively inexperienced, in 1870 Holley served as chairman of a committee of the Board of Trustees. He contributed a major portion of the suggestions that the committee offered for changing and improving the course of study at Rensselaer.

The proliferation of technical schools concerned Holley because of what he saw as an accompanying diminution of educational quality. As late as 1878 he suggested to a correspondent at Stevens Institute in New Jersey that, in view of the existence of so many, a good school certainly ought to protect itself by affixing its own identification to the degree (ALH to Robert H. Thurston, 13 February 1878, Holley Family Papers [hereafter cited as HF]).

15. Alexander Lyman Holley, "Bessemer Machinery," *Journal of the Franklin Institute* 44 (1873): 252-65; *Engineering* 14 (12 November 1872): 357; (29 November 1872): 374-75; (10 December 1872): 416, 417, and *Engineering* 15 (14 February 1873): 112-14.

16. ALH to Student, 18 March 1876, HF.

17. ALH to J. Holm Maghie, 29 February 1876.

18. Alexander Lyman Holley, "The Inadequate Union of Engineering Science and Art," *Transactions of the American Institute of Mining Engineers* 4 (1876): 191-207 passim.

19. Ibid.

20. In its way this commemoration of America's centennial was as important for its impact on science and industry here as the Great Exhibition of 1851 held in the Crystal Palace in London had been in Great Britain. But annual fairs had been held in New York City by the American Institute beginning in 1828. These fairs displayed the newest machinery and many important inventions, including the Morse magnetic telegraph, the McCormick reaper, and the Singer sewing machine.

21. Holley, "Inadequate Union," pp. 191-207 passim.

22. ALH to Abram Hewitt and several presidents of railroads, March and April 1876, HF.

23. *Journal of the Iron and Steel Institute* (London) 37, no. 2 (1890): 491.

24. ALH to James Drown, 3 April 1876, HF. Rossiter Raymond was an active member of Plymouth Church in Brooklyn of which Henry Ward Beecher was the minister.

XIX · William Siemens and the Open Hearth

lthough Holley had been closely identified with the Bessemer process and had been its sponsor and advocate from the time he first had seen a converter in Sheffield, over the years his view of the steel industry had broadened. Scarcely five years after Bessemer introduced his process in England, another method of making steel began to attract attention: the open hearth process or, as it sometimes was called, the scrap process. The process later was called Martin-Siemens on the Continent, but in England it was called Siemens-Martin.

Holley first became acquainted with William Siemens and the Siemens process in 1874, and thereafter had watched its development closely. A year later he began advising his clients to install open hearth furnaces. "I have little doubt that your farther examination of the latest developments of the steel manufacture will confirm you in your idea to establish it for your Road," he wrote to one railroad executive in October 1875. "Undoubtedly the Bessemer process will make steel cheaper just as things stand here today, but the Siemens-Martin manufacture is gaining on it fast, and seems capable of the greater improvement. One of my clients has resolved to try it for rails, and I am furnishing him drawings, partly in accordance with the latest French practice, which is rather beating the French Bessemer manufacture. The Martin can be begun on a smaller scale than the Bessemer can be, and the Martin product could be adapted to more kinds of Railroad work."[1]

Holley's activity on behalf of the Siemens-Martin process attracted the attention of the inventor. In November 1875, Holley wrote to Siemens.

> I got the impression from Mr. A. S. Hewitt, that you felt some dissatisfaction about my giving results of Landore practice to my clients. I published nothing. The facts did not get into the hands of competitors of the Landore Company or of the steel works of Scotland. I have, however, made these facts the means of stimulating a strong determination (already put into practice in some cases) to largely increase the open hearth steel manufacture and to compete with the Bessemer, in this country. . . . I think I have worked for, rather than against your interests.
>
> I think I am justified in stating that I have received very little encouragement

274

from anyone connected with the ownership or management, in my endeavors to increase its use in America.

I should be glad to know when you are prepared to make any statement on the subject, what you are doing in the matter of better refractory materials. If we can make these satisfactory here, the Siemens-Martin process will come very largely into use here without delay. It will be the key to the situation; so that I do not hesitate to ask you to help me, and in doing so, to help all concerned, in this direction.[2]

Holley also advised steelmen to prepare themselves to operate open hearth plants. In 1876, he told A. L. Rothman, who had been superintendent of a shut-down mill, that he ought to try to find a place in an open hearth plant. He told Rothman to accept a position on any terms in order that he might become thoroughly acquainted with this new method of making steel.[3]

In the same year, William Siemens came to Philadelphia to serve as a judge at the Centennial Exposition. Holley spent an entire day with the inventor. Siemens confirmed much of what Holley had come to believe regarding the economy and efficiency of the open hearth furnace. Convinced that the steel industry was on the verge of even greater accomplishments than it had achieved yet, Holley determined to visit the leading open hearth installations in Europe as soon as he could.

It is a curious fact that many of the men whose efforts transformed the iron and steel industry were not metallurgists or even, at the time they began to experiment, familiar with the behavior of metals. Bessemer was a novice; Kelly was an amateur; only Mushet could be called a metallurgist.

William Siemens, the next great innovator in the steel industry, was an inventor in the Bessemer mode. It was said of Siemens that, if given a problem, he would suggest a half dozen solutions, two of which would be complicated and impracticable, two difficult, and two perfectly satisfactory.

Siemens was born in 1823 in Hanover, Germany, into a family of eight boys, four of whom, working closely together, were to achieve distinction in special fields of science. Their father was a farmer who is said to have been a man of some culture and intelligence. Werner, the eldest, served in the United Engineers Artillery School in Berlin, where he received the scientific training that laid the foundation for his later brilliant work in electrochemistry.[4]

William, seven years younger, received his early education in Lübeck. Although William previously had shown no interest in scientific studies, Werner was convinced that he should study science and persuaded their parents to send him to the Polytechnic School at Magdeburg. After the father's death, Werner Siemens became responsible for the education and welfare of all the brothers. He arranged for William to study at the

University at Göttingen. Having finished at Göttingen at the age of nineteen, William Siemens entered the engineering works of Count Stolberg at Magdeburg as an apprentice.[5]

Werner Siemens now was stationed at Magdeburg, and the brothers worked together on developing their scientific ideas. In 1842, Werner had patented in Prussia a process that used electric current for the deposition of metals, and he was casting about for a possible user. When he learned that George Elkington, the leader in that field, also recently had developed such a process, the forthright Siemens brothers decided that William should go to Birmingham to see Elkington. A stranger to English customs, William thought that there might be some central bureau there that would inspect his brother's invention, and if it had merit, would represent him in finding a market for it or would arrange an introduction to Elkington. In Birmingham he eventually found his way to the office of Poole and Carpmael, where he was received politely and given a letter to Mr. Elkington.[6]

Elkington was interested in Werner's invention even though it was similar to his own. He had, however, been unable to deposit a certain coating with a noncrystalline appearance upon a dish cover by means of his process. When William Siemens demonstrated that he was able to do so with his brother's process Elkington purchased it for £1,600.[7] William returned to Germany in high spirits. His success gratified them both. The money also was gratifying because Werner needed it to provide for the other brothers.

William wanted to learn everything he could about practical engineering, and he remained at the Stolberg works until the end of 1844, at the same time working on new inventions with Werner. By the end of the year, they had developed two more inventions, a chronometric governor for steam engines and a practical development of the anastatic printing process. Because of his early success in England with the electro-gilding process, William decided to return there to patent the new inventions.[8]

The two inventions brought only small financial return to William Siemens, but they did provide him with entrée to the engineering profession in England at a time when from one end of the country to the other, the natural sciences were being explored by scientists such as the great Michael Faraday, the distinguished physicists John Tyndale and Lord Kelvin, and engineers like William Fairbairn and Joseph Whitworth.

The chronometric governor was so ingenious that it attracted the admiration of many outstanding engineers and later won a prize at the Crystal Palace exhibition, although it never was widely used. Sir George Airy, the astronomer, did adopt it for use in regulating the movement of certain delicate instruments at the Greenwich Observatory, but it brought almost nothing to the inventors. The printing process was hardly more successful, although it won the attention of Faraday, who used it in 1845 as a subject for a lecture before the Royal Institution.[9]

William Siemens spent his first five years in England in scientific studies. He constructed an air pump and began research relating to fuel economy, but became so discouraged over his financial position that he considered joining his brothers Karl and Frederick who, after 1849, for a short time considered emigrating to California.[10]

Siemens moved to Manchester in 1847 and worked for Hoyle and Sons of Mayfield. It was here that his attention first was drawn to steam power, and he became particularly interested in the various means of heat recovery. On the basis of this study, he built a regenerative steam engine in the factory owned by John Hicks at Bolton, near Manchester. Shortly afterwards, he devised a regenerative condenser that was considered sound in its principles but was too complicated to make its way. However, the invention earned him a gold medal for 1850 from the Society of Arts.[11]

At the same time he was developing his idea for the regenerative evaporator that could be used in the process of evaporating liquids in distilling operations and in the production of salt and sugar. However, Siemens's biographer writes, "... the manufacturers were ... shy of adopting the novel process. It required an entire remodelling of their works, and a great outlay of capital, which they did not feel warranted in undertaking, without an unmistakable assurance of the success of the plan."[12]

During 1847, Siemens's brother Werner had joined forces with a young engineer named Halske to organize a telegraphy and electrical company in Berlin. Due to Werner Siemens's talent for invention, the business prospered, and in 1849 he was able to carry out his intention to resign from the army. At the outset of the Crimean War in 1853 the company won the contract to install overland telegraph lines from St. Petersburg to the war area. This was the beginning of the work in communications that, in later years, saw the company of Siemens and Halske responsible for telegraphic installations all over the world.[13]

Just as Bessemer's metal powder invention had provided the money for the experiments with steelmaking, so William Siemens's invention of a water meter provided the royalty income that made it possible for him to give up his position in Manchester in 1852 and return to London.[14]

William Siemens's intention had been to enter practice as an independent civil engineer, but at that time his brother's prosperous new company opened a London office, and he became its manager. As their London agent, he was to receive one-third of the profits from any business he brought the firm as well as a 12.5 percent commission. This was not an ideal arrangement for William Siemens since the business took time that he preferred to direct to his inventions. A modern historian has written, "All his life 'business' as such was something he only turned to when he had to." Yet, in later years, Siemens was known as a "shrewd and clear headed man of business."[15]

The four Siemens brothers, Werner, William, Frederick, and Karl worked together so closely and so generously gave credit to each other that it was not always certain which one of them had originated an idea, and which had carried it to fruition.[16] However, letters written to William, who was equally brilliant in research in electricity and the utilization of heat, indicate that he carried on his electrical experiments with Werner and the development of the furnace with Frederick.[17]

In London, the two continued their work on problems of heat utilization to which William already had devoted about ten years. It was Frederick who developed a simpler way of applying the regenerative principles to the furnace in which the fuel was consumed, and for this he was granted a patent in 1856, the year in which Bessemer read his paper at Cheltenham. It was probably the simplest of the several heat utilization patents that the Siemens brothers had been granted. William Siemens's biographer comments, "It is often found, in tracing the history of an invention, that the simplest form is the latest in presenting itself to the Inventor, and this case furnishes a striking instance of the fact."[18]

In the new application, the heat from the products of combustion did not escape up the chimney. Instead it was directed into a chamber filled with a checkerwork of refractory brick through which, when the brickwork had become sufficiently hot, the air supply to the furnace was led, thus heating the air before it reached the fuel. By using two chambers, the process was made continuous: one chamber was heated while the other was in use. The increase in heat was spectacular. At the same time, 50 percent to 80 percent less fuel was required.[19]

When in 1861 the brothers introduced gaseous fuel as an alternative to solid fuel, the value of the regenerative furnace was expanded further. After five years of work they had realized the extent to which their process could be used. It was used first in a glass factory, where it proved unexpectedly simple and economical to operate. However, William Siemens stated in his application for the furnace patent that it was especially adaptable for melting steel on the open hearth.

In 1861, Bessemer was producing steel in Sheffield, Göransson was turning out steel successfully in Sweden, Mushet was doing his best to revive a failing forge business in the Forest of Dean, while Holley, with no notion that his future would be affected by the inventions of any of them, desperately was searching for work in New York. In the same year, William Siemens suggested to Abraham Darby at the Ebbw Vale plant that Siemens's regenerative furnace might be used in the produc-

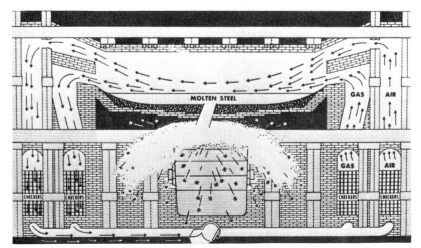

Diagram of open hearth furnace. (Courtesy of the American Iron and Steel Institute.)

tion of steel on an open hearth. Whether the suggestion seemed impractical or whether Darby may have considered it only one more of the ideas for steelmaking processes that were flooding the patent office, he neglected to take any action.[20]

Although William Siemens already had made his mark as an inventor, even greater fame awaited him. The regenerative stove principle attracted the attention of Michael Faraday, now in the twilight of a long life. In his farewell lecture before the Royal Institution in 1862, the scientist chose to speak about the regenerative gas furnace. In his usual vivid manner, Faraday described its simplicity, power, and economy. However, in his confusion at the end of the hour, Faraday accidentally burned his notes. Later he could manage to reconstruct only an abstract of what he had said. Such recognition by the "prince of pure experimentalists" provided the impetus that brought the furnace to general acceptance. From then on, Siemens's ideas met with little opposition, and the furnace came to be used in many industries where intense heat was necessary to melt metals or glass.[21]

One of the first persons to be attracted by the possibilities of the Siemens furnace was M. Le Chatelier, Inspecteur-General des Mines in France. He believed that, by using bauxite for the furnace bed, this new furnace might be ideal for making steel on an open hearth. However, the bauxite shrank when heated and was not sufficiently cohesive when it was used for repairs. Nevertheless, Le Chatelier maintained his interest in the furnace and helped Siemens form an association with Boigues, Rambourg et Cie of Montluçon. A large furnace designed by

Sir C. William Siemens. From *Iron and Steel Magazine,* January 1906.

Siemens was built at Montluçon in 1863. Otto Siemens, another brother, supervised the construction and remained there to direct the production of steel. A small amount of good quality steel was produced, but when the silica roof melted from the intense heat, the owners did not consider that the production results warranted rebuilding the furnace, and they abandoned the project. Two other companies also had taken out licenses and had produced steel, but neither continued long enough to achieve commercial success.[22]

After several other companies had begun and then abandoned the use of his furnaces for steelmaking, Siemens realized that, if his furnace and method of producing steel were to be successful, it would be necessary to direct the experiments himself. As Siemens writes, "I was glad to obtain possession of some works at Birmingham, since known as my Sample Steel Works, which I started with the object of having experiments made under my own supervision. Here I gave my attention to the fusion and production of steel, both in crucibles and upon the open hearth, with the entire or partial employment of ores and scrap metal."

The first furnace at the Sample Steel Works was one in which crucible steel could be melted in closed pots. William Siemens experimented in this way until 1867, when he expanded and built an open hearth furnace capable of melting a charge of twelve tons every six hours. In these trials he used both iron ore and scrap in the numerous charges, and the results were encouraging.[23]

Since this method of producing steel was unique, Siemens, like Bessemer and Holley before him, found it necessary to assemble a working force that could be persuaded, in spite of their doubts, to follow his directions. Unlike those of Bessemer and Holley, his experiments succeeded in a relatively short time. Siemens was able to send samples of his steel to the Paris Exhibition of 1867, where he was awarded a grand prize for both the furnace and the process.

Among the first to secure a license from Siemens for the regenerative furnace had been Emile Martin and his son Pierre, operators of an ironworks at Sireuil, France. In April 1863, with the help of Siemens, they built a regenerative furnace, as well as an open hearth constructed from their own designs. They experimented in producing steel by melting wrought iron and steel scrap in a bath of molten pig metal. By 1867 they were producing steel of such high quality that they also sent several samples to the Paris Exhibition and also were awarded a gold medal.[24]

Neither Siemens nor the Martins can be said to have invented the open hearth process.[25] As Siemens's assistant, William Hackney, wrote, "The melting of steel in large masses on the open hearth of a reverberatory furnace had long been a favorite dream of inventors; but it was not until the regenerative gas furnace gave the ready command of a temperature practically limited only by the fusibility of the materials of which the furnaces could be built, of intense heat without intense draught or cutting flame, that any open hearth steel melting process became a practical success." The idea had been proposed by Reamur in 1722 and even the unfortunate Josiah Marshall Heath took out a patent in 1845 for the production of steel by the fusion of scrap iron, or iron sponge, in a bath of cast iron on the sole of a reverberatory furnace heated by jets of gas, but Hackney doubted whether a practical attempt ever was made to carry the plan out. Inasmuch as the high temperatures produced by the regenerative gas furnace exacerbated the old and constant problem of the fusibility of furnace linings, the search for suitable refractory materials continued.[26]

Siemens described his own search for refractory materials as follows.

Considerable difficulty was experienced to find a material to resist the excessive heats necessary for carrying out this process; ordinary Dinas bricks, which are considered the most refractory material in general use, would be rapidly melted, but a brick, specially prepared by crushing pure quartz rock, and mixing it with no more than two percent of quick lime to give cohesion, answers well. The hearth of the furnace is made of white sand with a small admixture of more fusible fine sand, which mixture sets exceedingly hard at a steel-melting heat, and possesses the advantage of combining into a solid mass with fresh materials introduced between the charges to make up for wear and tear.[27]

The similarity of the open hearth processes that were carried on by Siemens and by the Martins is evident in descriptions of them by a contemporary historian of steelmaking. Of the Siemens version he wrote:

> A charge of about five tons of pig iron of the best quality is put on the bed of the melting furnace and in four or five hours when it is melted and at a good heat, iron ore broken to about the size of road metal is thrown in, making the melted metal boil violently; more ore is added, in successive portions, so as to keep up a good boil, until the metal has become partly soft enough for spiegel, and it is then allowed to stand for a short time, that as much as possible of the iron may work out of the slag. A small quantity of broken limestone is also added at intervals during the process, to aid in throwing down the iron from the slag. When the metal is at the proper pitch of softness, spiegeleisen is added, and as soon as this is melted the charge is tapped out.[28]

The Martin process is described as follows: "A sufficient quantity of pig iron is first melted down on the bed of the furnace, and malleable iron scrap, or steel scrap, is put into the bath of liquid metal, in successive portions, as fast as it dissolves, until a sample taken out in a small ladle, and quenched in water, indicates by its toughness and the appearance of its fracture that the metal has been brought to the required pitch. From 6 to 8 percent of spiegeleisen is then added, and as soon as this is melted the charge is tapped out. The time taken to work a 10 or 12 ton charge is from seven to eight hours."[29] It is interesting to note that Robert Mushet's discovery, the use of spiegeleisen, was necessary to bring both methods to a satisfactory conclusion.

At first, the two open hearth processes were called "the pig and ore," which was used principally in England, and the "pig and scrap," which was developed by the Martins and used chiefly on the Continent, but later in the United States. Eventually, the use of scrap increased, and the term Siemens-Martin came into use.

The Siemens exhibition in Paris attracted the attention of engineers from several of the railroad companies in England. John Ramsbottom, the well known chief engineer of the London and Northwestern Railway Works, was particularly interested in this new manner of producing steel. After spending some time studying the process at the Sample Steel Works, he decided to adopt it and to build a furnace at the railway shops at Crewe. In 1868, using pig metal and scrap for his charge, he began to produce steel. This had been cited as the first use, on a commercial scale, of the process that was to become known in England as the Siemens-Martin process.[30]

Since the advent of the Bessemer process, the problem of what to do with the ends cropped from the rails made from Bessemer steel had become serious. Although many schemes were tried in which this mounting tonnage of scrap might be used, it remained a drug on the market. With the introduction of the open hearth process, it soon became evident that, at last, a means was available for using the large quantities of scrap that the Bessemer process created but was incapable of using.

When the directors of the Great Western Railway discovered that it now was possible to convert the quantities of iron rails piling up in their railroad yards into steel, they asked Siemens to undertake the task. The rails were high in phosphorus, and Siemens encountered some difficulty in handling them. But he eventually was successful, and the ingots were shipped to Sir John Brown and Company in Sheffield to be rolled into rails.

In spite of the problems inherent in the process, the results were such that Siemens decided to enter production on a larger scale than was possible at his Sample Steel Works. In 1870 he formed the Landore-Siemens Steel Company with L. I. Dilwyn, a director of the Great Western Railway, as manager. During its first year, the company produced steel at the rate of 75 tons per week. By the end of four years that figure had risen to 1,000 tons. Landore-Siemens became a leading steel company, but it was not commercially successful and ceased production in 1888.[31]

NOTES

1. ALH to William Kuper, 2 October 1875, Author's Collection (hereafter cited as AC).
2. ALH to C. W. Siemens, 5 November 1875, AC. Siemens had organized the Landore-Siemens Steel Company in 1870. Holley refers to the Steel Company of Scotland located at Newton near Glasgow.
3. ALH to A. L. Rothman, 13 May 1876, Holley Family Papers.
4. William Pole, *The Life of Sir William Siemens* (London: John Murray, 1888), p. 10. William Siemens had been baptized Carl Wilhelm Siemens, but used the Anglicised version of his name when he decided to live in England.
5. English Electric Company, *Collection of Letters to Charles William Siemens* (London: English Electric Company, 1953), p. 10.
6. Pole, *Siemens*, p. 45. As he walked along the streets of the city, Siemens noticed a sign over a doorway giving the name of the proprietors followed by the word *undertaker*. To a young man, uncertain of his English, this seemed just the person to see in regard to marketing his invention. To his embarrassment, however, the word had an entirely different connotation from the one he had expected and he hastened back to the street, "having come decidedly too soon for the kind of enterprise there contemplated."
7. English Electric Co., *Letters to William Siemens*, p. 11.
8. W. T. Jeans, *The Creators of the Age of Steel* (New York: Charles Scribner's Sons, 1884), p. 138. A factor that contributed to William Siemens's decision to work in

England was the English patent system under which it was possible to patent an invention and, under favorable circumstances, to receive remuneration for it. English patent law, in spite of its intricacies, was still much more dependable than was the Prussian system, which offered little protection. Only short term privileges were granted—these only occasionally. Siemens knew of important inventions that had been refused rights and negligible improvements that had been granted protection.

9. Ibid., pp. 138, 139.
10. Pole, *Siemens*, p. 71.
11. W. T. Jeans, *Age of Steel*, p. 140.
12. Pole, *Siemens*, p. 78.
13. J. D. Scott, *Siemens Brothers, 1858-1958* (London: Weidenfeld and Nicholson, 1958), p. 25.
14. For many years the English royalties amounted to £1,000 per year and, in addition, there were substantial royalties from the Continent.
15. Scott, *Siemens Brothers*, p. 28.
16. Ibid., p. 29. The Siemens were devoted to each other, but all were wilful and hot tempered. According to a biographer, they quarreled easily and loudly: "The noise produced by a Siemens in a rage could be terrifying; even their associates and subordinates picked up the habit of sudden outbreaks; objects such as books continued to be hurled across rooms at Woolwich [the English plant] at least until 1914" (Ibid.).
17. English Electric Co., *Letters to William Siemens*, p. 10.
18. Pole, *Siemens*, p. 99.
19. J. S. Jeans, *Steel: Its History, Manufacture, Properties, and Uses* (London: E. and F. N. Spon, 1880), p. 90.
20. W. T. Jeans, *Age of Steel*, p. 150.
21. J. S. Jeans, *Steel*, p. 90. The furnace took time to make its way and Werner Siemens frequently complained that the firm of Siemens and Halske was carrying more than its share of the financial burden.
22. Ibid.
23. W. T. Jeans, *Age of Steel*, pp. 156-57.
24. Ibid., p. 164.
25. Ibid., p. 165. The way in which the name of Martin came to be connected with the process is an interesting part of its history. At the Paris Exhibition of 1878 the Society for the Production of Martin Steel—the successors of Emil and Peter Martin—published a pamphlet in which they claimed that the credit for having invented and perfected the open-hearth process belonged to Martin alone. Sir William Siemens, of course, demanded a retraction and issued a statement of his claims to priority. The company declined to retract, whereupon Sir William published the entire correspondence in a pamphlet illustrated with drawings. In one of these letters he stated that he had worked at the solution of the problem of smelting steel on an open hearth since 1856, and that in 1861 he took out a patent that was put into practical operation by Atwood in Durham, and by Messrs. Boignes, Rambourg, and Company of Montluçon in France.

He called the attention of Messrs. Martin to the fact that on the occasion of his first negotiations with them, in a letter dated 26 May 1863, he informed them of the experiments at Montluçon. The correspondence showed that the question of the application of the Siemens regenerative system by the Martins referred only to the crucible steel furnace. William's idea of applying the open hearth furnace to steel-making was a novelty to the Martins, who, in accordance with his original request, agreed to the condition that the furnace was to be a reheating one, which might at a small expense be altered to an open hearth steel furnace. The first furnace at Sireuil was begun, according to William Siemens's plan, on 17 April 1863, and put into operation by his engineers. He acknowledged that after 1864 the Martins followed up the process with great perseverance, especially the proper mixture of the materials. It was only in 1867, after having concluded their experiments and begun regular work, that the Martins took out a new patent, which included many points of the open hearth steel process and which also were included in the patent granted in the same year to William Siemens.

26. William Hackney, "The Manufacture of Steel," *Minutes of Proceedings of Institution of Civil Engineers* 42, part 4 (1874-75): 21, 27. Hackney reported, "A careless furnace-man, with plenty of heating power at command, may bring the furnace roof down in a day; but in practice and with reasonable care, a roof of good bricks will last, even in working pig iron, from one hundred to a hundred and fifty or two hundred charges."

27. C. William Siemens, "On Smelting Iron and Steel," *Journal of the Chemical Society,* July 1873, p. 2.

Dinas ... was the name given to the sandstone which was found above the lime measures in the Vale of Neath, Wales by Mr. Weston Young, the pioneer silica brickmaker who mixed the sandstone and limestone together and made silica bricks with the natural lime bond, from the beginning to the middle of the nineteenth century. His process was imitated. Quartzite and sandstone were ground and mixed with limestone which did not occur with the silicious rock but was especially quarried for the purpose, and from the mixture the dinas-silicon brick was made.

These dinas bricks achieved a reputation for strength and infusibility which carried them all over the world and resulted in later years, in the establishment of quarries and factories in Germany, America and other countries for the production of the silica brick with lime bond. ... Curiously the name dinas was maintained in Germany, but discarded in America, where the name ganister was given to the lime-bond silica bricks. The clay-bond silica bricks, which were called ganister in England and German dinas in Germany, became quartzite brick in America, and in this class were included all brick having 70 percent and above silica and the rest clay, or clay mineral substance. Now the silicious rock which is used in the preparation of lime-bond silica (the America ganister) brick may be quartzite or sandstone (F. T. Havard, *Refractories and Furnaces* [New York: McGraw-Hill Book Co., 1912], p. 32).

28. J. S. Jeans, *Steel,* pp. 100-101.
29. Ibid.
30. W. T. Jeans, *Age of Steel,* p. 157.
31. Alan Birch, *The Economic History of the British Iron and Steel Industry 1784-1879* (London: Frank Case and Co., 1967), pp. 375-76.

XX · The Open Hearth Process in the United States

iemens's regenerative furnace had become known in the United States several years before the Paris Exhibition of 1867. In 1862, Lewis Powe, manager of the copper works at Park, McCurdy and Company of Pittsburgh, had gone to England to study the manufacture of tinplate there. Even though industry in America consumed considerable amounts of tinplate, it never had been produced competitively here, and America still looked to England for its supply. In Birmingham Powe saw a Siemens furnace. He was impressed by its possibilities and obtained a pamphlet that described the invention. On his return Powe urged Park to use the booklet for guidance in building a furnace for the smelting and refining of copper. Park did so without obtaining a license for the use of the patent.

Two other Siemens furnaces also were built without licenses: one for glassmaking by James B. Lyon and Company and the other for steelmaking by Park, Brother and Company. Both of the furnaces were constructed using only the limited description available in the pamphlet. Neither was successful and both soon were abandoned. The first licensed Siemens gas furnace to be built in America was installed in 1867 at Troy, where it was used as a heating furnace for the rolling mill. Another furnace was installed shortly after at the Nashua Iron and Steel Company in New Hampshire, and a third was built at Anderson and Woods in Pittsburgh for melting steel in crucibles.

William Durfee, who also had pioneered with the pneumatic process at Wyandotte, was first in the United States to puddle iron in the Siemens furnace, an event that took place in 1869 at the rolling mill of the American Silver Steel Company of Bridgeport, Connecticut.[1]

Although Siemens's activity in steelmaking was well known in the United States, little had been done to push the use of his regenerative furnace in conjunction with the open hearth process, since practically all American steel producers were committed to the Bessemer process.[2] Abram Hewitt, at Trenton, was one steel producer who was not. He had attempted to experiment with the Bessemer process, but when it fell into bad repute in England, he quickly abandoned the effort. Since then, the New Jersey Iron and Steel Company had been producing steel using an adaptation of the Mushet-Heath process. According to Hewitt's biographer, the process was "a fusion of wrought iron and special Ringwood blooms rich in carbon, effected in a high-temperature furnace. The details have been lost. We do not know the proportions of metal used or the exact nature of the furnace. We do know that by the beginning of 1866 steel in small quantities but of high quality was being produced, and that Hewitt relied upon it and his gunmetal as the two pillars of the business." The gun metal was used for locomotive tires as well as gun barrels. In 1866, the Trenton company first produced steel-headed rails. These found such favor with the railroads that in a brief time as much as 24,000 tons were being rolled each year. This type of rail was significantly cheaper than imported steel rails and also could be rerolled whenever wear became evident. Most of the companies still were rolling only iron rails, and Hewitt's mill soon gained a solid place in the competition for the booming railroad business.[3]

Hewitt, always a distinguished representative of the United States, was appointed as one of the commissioners to the Paris Exhibition of 1867. In spite of his poor health, Hewitt accepted the honor since he was anxious to see for himself what the European iron and steel industry was doing.

For some time, Hewitt had hoped to relocate the steelmaking portion of his business. As steel production moved westward, the New Jersey Iron and Steel Company was becoming increasingly isolated and freight transportation via the Camden and Amboy Railroad was becoming costly. Hewitt believed that it would be wise to move at least part of the plant further west.[4] However, the problems involved in making such a move proved difficult. Some years before, Hewitt's father-in-law, Peter Cooper, had built Cooper Union, his renowned educational center. Cooper could not bring himself to leave it, and Hewitt felt that he could not ask his father-in-law, at his age, to make the change. He decided instead that some means of making operations at Trenton more efficient would have to be developed. In the hope that he might find some useful ideas that would help to do so, Hewitt was eager to study European operations.

Some years had elapsed since Hewitt's last visit to Europe, and he was amazed by the changes that had taken place in the iron and steel industry. He made a tour of the Paris Exhibition and then visited plants on the Continent and in England and Wales. Hewitt returned to Paris to review the steel section at the exhibition more carefully. All the famous companies had their wares on display, and some of the exhibits were spectacular. Krupp, particularly, made a dramatic showing with a cast steel ingot that weighed forty tons and a steel crankshaft that weighed more than nine tons. With patriotic fervor, all of the French steel companies, which had previously trailed the rest of the world, had submitted entries.[5]

At the exhibition, Hewitt's eye was caught by a singularly bright ingot marked *Sireuil*. He learned that it was made of Martin steel and that it came from the plant of Emile and Pierre Martin at Sireuil in France. When he ferreted out further details of their process and learned the advantages that it might have over the Bessemer process, Hewitt wrote his brother-in-law, Edward Cooper, that he believed that this was the process he had hoped to find. Learning that the Siemens furnace formed an essential part of the process, Hewitt decided to visit the famous Schneider Works at Creusôt where one of the furnaces recently had been installed. Schneider had not yet tried the Martin process, but his enthusiasm was so great that Hewitt became even more convinced.

In Paris, Hewitt immediately entered into negotiations with Pierre Martin for the American rights and wrote to Cooper.

> Today I have made the following arrangement with Martin. We will grant licenses to all who may want them, retaining for ourselves the exclusive use of the patent for gun-barrel iron and any other things for which we will undertake to supply the demand, especially tyres. The royalty is to be the same as the Bessemer royalties, from which is to be deducted the royalty paid for the use of the Siemens furnace. After we are refunded the cost of taking out the patents we are to divide the royalties equally with Martin, we paying our royalty into the common fund, the same as any other parties.[6]

Hewitt believed that he had obtained a major business advantage for the Trenton mill. His company now controlled the only process, other than the Bessemer, by which steel could be produced on a tonnage basis. The installation was cheaper than the converter plant, it could make use of materials rejected by the older process, and, throughout the process, the open hearth furnace was more easily controlled than the better-known Bessemer furnace. Furthermore, the new process could make use of iron and steel scrap.

Sometime before this, Alexander Holley's brother-in-law, Frederick J. Slade, then in his twenties, had begun work for Hewitt as his personal emissary. Slade's keen and brilliant mind had attracted Hewitt, who gave Slade the responsibility for projects that he, himself, was too busy or physically unable to carry out. In 1867, after Hewitt secured the Martin rights for the United States, he dispatched Slade to France to study the process with the view to its early installation at Trenton. But it was not until 10 December 1868 that Hewitt was able to write to the Martins that the first ingot made in America by the new process had been poured.[7] Fifteen months had elapsed between the day that Hewitt obtained the rights and the day that the first steel was poured.

Samuel T. Wellman, who in later years would occupy the same eminent position in open hearth practice that Holley now was occupying in Bessemer practice, recalled his first experience as a young man engaged to help J. T. Potts, the Siemens engineer at Trenton.

I was sent by Mr. Potts, to assist Mr. Slade in the starting of their gas producers, with which they were having some trouble, thus having good opportunity to see the first efforts to make open hearth steel, and I have a very vivid recollection of some of these experiments. Mr. Slade was the only man connected with the works who had ever seen a heat of open hearth steel melted, and as he was not a practical steel man, he probably overlooked some of the important points while watching the manufacture in France. At any rate, everything bad that could happen in the manufacture of open hearth steel seemed to fall to their lot. The best man among the workmen was the melter, who had been an old puddler, or heater, but of course he was unaccustomed to such high temperatures as were necessary to melt mild steel. The first trouble encountered was with the making of good gas. This rose from a too literal interpretation of the printed instructions for running the producers, which had been sent them by Messrs. Siemens' agents.

When I took charge of the producers, I found they were full to the very top, the coal being some six or seven feet deep; so thick a fire that about all the heat in the bottom of the producers could do, was to drive out the moisture from the upper layers. Water and thin tar were running out of the joints through the whole length of the gas flue, or cooling tube, which was of wrought iron plates, placed above ground. We ran the producers for two days without charging any coal. After about twenty-four hours work, the producers began to do good service, and we had no further trouble with them. The only difficulty was to get men to attend faithfully to the making of the gas, a very disagreeable job, and one that demands constant watchfulness in order to get the best results. . . .

The great difficulty in these pioneer days of steelmaking was to teach the melters to maintain the steel at the proper temperature. The margin between the proper casting temperature of the steel and the melting point of the furnace itself was very narrow, much more so than today, as at that time no silica brick was available, and we were obliged to use clay firebrick for the roof and sides of the furnace. Shortly after the heat I have mentioned, I saw an attempt made to cast a charge which had chilled in the nozzle so that it could not be started again. By the time the shoe could be taken off, the steel had chilled solid, clear back into the furnace, which was then shut down half full of steel. The next day the furnace was started up again. A quantity of pig iron was added and the steel finally all melted, but it was so slow in coming to a steel melting heat that the carbon was reduced, and several more additions of pig iron had to be made, until finally the furnace was overfilled, making it all the more difficult to get the proper temperature. It was finally tapped, but it was so cold that it chilled in the nozzle and the forehearth, so that when the latter was taken off the steel ran down over the moulds, cars, and tracks, welding the whole together, and making a mess that I was afterwards told took two weeks to clean up so that the furnace could again be started. These are only examples of the difficulties and rough experiences which were encountered in these initial trials, and it was no wonder that poor Mr. Slade was discouraged.[8]

Legal aspects of the Martin patent also played a large part in the delay in getting the operation under way. Because the open hearth always is spoken of as a unit, it is easy to overlook the fact that there were two distinct patents and two groups involved. The Siemens patent for the regenerative furnace was held at first by Tuttle, Gaffield and Company of Boston and, later, by Richmond and Potts of Philadelphia; Hewitt held the Martin rights. Before he could build his open hearth, it was necessary for Hewitt to secure the furnace rights from the Boston group. After prolonged negotiations with both the Boston company and Siemens, it was agreed that Cooper and Hewitt, acting as agent for the New Jersey Iron and Steel Company, would pay Tuttle, Gaffield and Company a royalty of three dollars per ton on all steel produced. There was further delay while a Hewitt lawyer attempted to make the Martin patent secure against attack in the American courts. In addition to legal difficulties, there also were the inevitable troubles with the refractories. Only after Hewitt was able to import a special type of brick and cement from Wales could production begin.[9]

Following the first satisfactory heat in their relatively small furnace, the Trenton mill for a time conducted its steelmaking activities with modest success. As Wellman commented, "As is very often the case, this experiment was not very profitable to the pioneers, Messrs. Cooper, Hewitt & Co. The furnace was operated at intervals for a year or two, but was finally abandoned, and the manufacture stopped, as it was not a commercial success, though those who saw the experiments were able to profit by them and avoid many of the difficulties encountered in them."[10]

A number of years passed before Hewitt succeeded in granting licensing arrangements to other companies in the United States. Although Hewitt was convinced that the open hearth eventually would surpass the Bessemer converter, there were reasons for the delay. Many brilliant engineers, Wellman in particular, were connected with the open hearth process in later years, but at the beginning there was no one man who knew all the peculiarities, the weaknesses, and the strengths of the process as Holley knew those of the Bessemer process. The open hearth process had none of the brilliance and drama of the Bessemer process. The open hearth operation was much slower than the Bessemer process; the latter averaged twenty minutes for a blow, while the open hearth heat might take hours. Furthermore the most established and successful steelmakers already had committed their capital to the Bessemer converter, and banks had little money to lend to an outsider who intended to risk it building an open hearth plant. The open hearth royalty arrangement was far from pleasing to the companies. And finally, they objected to the large double fee involved: one for the Siemens furnace and another for the Martin process.

Even though Hewitt's company was the first to build an open hearth furnace, the Bay State Iron Works of South Boston, Massachusetts, is credited with being the first commercially successful operation. It produced its first steel in 1870. The company had been producing iron rails for some years, but with the introduction of steel rails the company was attracted by the idea of producing steel-headed rails. Neither puddled steel nor Bessemer steel, brought from Troy, proved satisfactory for this purpose. Ralph Crooker, the superintendent, believed that perhaps the new open hearth steel might be the product he needed. He visited the Cooper & Hewitt works to watch the process and came away convinced that steel made in the open hearth furnace would be excellent. He began negotiations with Siemens's agents for the plans of a furnace to be designed to his liking. When he was unable to come to satisfactory terms with them, he engaged young Wellman, who had left Siemens's employ some time earlier, to construct the furnace.[11]

Wellman had served an apprenticeship both in various machine shops and under his father, who was general superintendent at the Nashua Iron and Steel Company in New Hampshire. He had learned from experience that which could not be gained from books. Like Alexander Holley, he would design plants and invent improvements.[12]

Although Hewitt had no connection with the Bay State Iron Works, evidence suggests that he may have supported the enterprise indirectly. In a letter that he wrote in 1876, Holley recalled that, some years before, he had been in conversation with a Mr. Richmond of the Philadelphia firm that had succeeded Tuttle, Gaffield and Company as Siemens's representative in the United States. The subject of the Martin royalty came up and Richmond intimated that no royalty had been collected up to that time. Holley reminded Richmond that Bay State certainly must have paid it. Richmond rejoined that he knew more about it than anyone else, and he believed that Cooper and Hewitt, as part of their promotion efforts for the open hearth process, had arranged with Bay State to have it said that full payment was made.[13]

NOTES

1. James M. Swank, *History of the Manufacture of Iron in All Ages*, 2d ed. (Philadelphia: American Iron and Steel Association, 1892), pp. 421-23.

2. One of the important reasons for the acceptance of the open hearth process in England and on the Continent was that iron ores satisfactory for Bessemer production could be obtained only in a few countries. In America there were plenty of good ores, primarily from the Lake Superior deposits.

3. Allan Nevins, *Abram S. Hewitt, with Some Account of Peter Cooper* (New York: Harper & Brothers, 1935), pp. 234–35. Ringwood, an estate of 22,000 acres in northeast New Jersey, was a celebrated iron ore property. The first iron ore had been discovered there soon after 1700, and from that time until the property was purchased by Peter Cooper in 1853, between three hundred and five hundred thousand tons of iron ore had been taken from the mines, and large reserves were still in the ground. The rich magnetic ore made iron of a superior quality.

    During 1865–66 the Trenton Iron Company discontinued rolling iron and concentrated on the production of pig iron. At this time Abram Hewitt with the aid of Peter and Edward Cooper bought the rolling mill. The Trenton Iron Company remained in possession of the mines, furnaces, and the wire mill. Hewitt, along with Peter and Edward Cooper, incorporated the new company as the New Jersey Steel and Iron Company. Hewitt was the principal owner (Nevins, *Abram Hewitt,* pp. 254–55).

4. Nevins, *Hewitt,* p. 234.

5. Abram S. Hewitt, *Production of Iron and Steel* (Washington, D.C.: U.S. Government Printing Office, 1868), pp. 10–11.

6. Nevins, *Hewitt,* pp. 243–44, 245.

7. Ibid., p. 246. Hewitt wrote to Martin:

We beg leave to state that we made our first ingot of steel yesterday at Trenton, and that the furnace worked well in all respects, except that the bottom was too porous to hold the melted steel well, and a considerable portion of the charge was lost. The quality of the steel, however, appears to be good, and we are now making a new bottom of more fusible sand than we used at first. The heat in the furnace appears to be all that could be desired. We have a demand for all the steel that the furnace will produce, and as soon as the product is regular, we shall license other parties who are waiting to see the results of our operations.

8. Samuel T. Wellman, "The Early History of Open-Hearth Steel Manufacture in the United States," *Transactions of the American Society of Mechanical Engineers* 23 (1902): 79–80, 82.

9. Nevins, *Hewitt,* p. 247.

10. Wellman, "Open-Hearth Steel Manufacture," p. 82.

11. Ibid., pp. 82–83.

12. Ibid., pp. 92, 94. In 1873, when Wellman was engaged to build the open hearth department at the Otis Steel Works in Cleveland, Ohio, he recommended that Holley be appointed as consulting engineer in the planning of the new works.

    In later years, Wellman took pleasure in reminiscing about his early relationship with Holley.

I had seen Mr. Holley but once before he came to Cleveland on the occasion of his first visit as consulting engineer. That was seven years before, when he was in charge of the works at Troy, and I, wanting to learn the steel business, had asked him to give me a position there. He told me that he had no vacancies at that time. On his visit to Cleveland I recalled this circumstance to him, and laughingly said, "As long as you would not give me a position, I have at last succeeded in getting you one." This was the commencement of one of the most pleasant friendships of my life which lasted until Mr. Holley's death.

13. ALH to Rossiter Raymond, 15 March 1876, ALH to D. E. Garrison, general manager of the Vulcan Iron Works, 7 August 1876, Author's Collection. The payment of

royalties was always a source of controversy. As Holley traveled continuously from plant to plant and had installed most of the machinery in each Bessemer mill, he could observe easily whether or not the companies were meeting their obligations. Frequently he was forced to remind plant managers that they were delinquent in paying the required fee. At one time he wrote the president of the North Chicago mill to remind him that they were using the Worthington cranes without having paid the royalty of $200 per crane for the term of the patent. These patent rights fees were not excessive in relation to the overall cost of a plant. For example, when the Vulcan Iron Works built its Bessemer plant in St. Louis, the fees involved amounted so some $20,000: $7,000 went to Holley for royalty on his vessel bottom, about one hundred and fifty drawings, and his services until the plant began operations; $2,500 was paid to James Moore for the blooming mill train; $200 to Worthington for each hydraulic crane; $5,000 to the estate of George Fritz for blooming tables; $3,000 to Robert Hunt for a bottom casting process; and $500 to Thomas Critchlow of the Pennsylvania Steel Works for his hydraulic valve.

XXI · Holley and the Open Hearth Process

Because business remained slack and the Edgar Thomson Company was proving to be a difficult competitor after it began production in August 1875, tensions developed among the companies producing steel rails. Carnegie's biographer wrote that he may have been a pacifist as far as international relations were concerned, but that in the sale of rails he was a warrior. The officers of many of the steel companies had underestimated Carnegie. They considered him an upstart and, until the building of the Edgar Thomson Works, were not inclined to take him seriously. But the Edgar Thomson plant, the most modern in the country, soon proved to be a dangerous competitor. Carnegie, with his railroad connections, always knew ahead of the others of prospective rail orders and consistently was able to underbid the market. He could do this and still make a profit, since the Edgar Thomson Works had been built at a time when costs were low, and it was efficiently operated under the skilled direction of Captain Jones.

Under the provision of the steel rail pool, which existed at the time, the producers parceled out orders among themselves, based upon an agreed percentage determined on the basis of each company's importance and previous accomplishments. "Although these pools were not illegal, Carnegie disliked the pools," his biographer wrote, "not because they were wicked but because they conflicted with his free ranging spirit. To let an opponent have a hand in regulating one's output or determining the price was intolerable. He did enter now and then but was inclined to break at the first opportunity. Carnegie took a considerable pleasure in making his fellow steel rail producers squirm."[1]

The friendship among the operating men still was solid. However, Holley was beginning to find himself in a situation that, as time went on, made his relations with the individual companies increasingly difficult. All the rights to the pneumatic process rested with the Pneumatic Steel Association, which collected royalties from the companies producing Bessemer steel. At the urging of Daniel Morrell, the amount assessed the licensees had been decreased somewhat, but the Bessemer producers still considered the rates excessive. Holley served as consulting engineer to the Pneumatic Steel Association, which paid him an annual fee of $1,000 and also had given him a certain amount of stock. From the scanty information available, it appears that his position with the Pneumatic Steel Association had been established in recognition of his pioneering efforts in the industry, rather than from any real need for his labors, since he also was employed as a consultant by various members of the association.[2]

Holley, however, was not making enough money to cover his expenses. He was forced to draw constantly on his savings in order to keep his office open and his small staff intact. Since the completion of the Edgar Thomson plant, Holley had had no further connection with the company. In October 1875, he wrote Jones that he had not been able to get his salary. "You are perhaps aware," he said, "that I have no connection with the company since the completion of the works, nor have I been able to get my pay yet. When the company concludes to carry out their contract with me I shall visit the works, and if they want to hire me as Consulting Engineer, of course I shall be happy to cooperate with you." He suggested that he would be willing to join the company in that position for $2,500 a year.[3] His offer was not accepted and two months later he wrote Shinn, the general manager, that, since the company was not pushing its new construction as rapidly as he expected, he would be willing to accept them as a client on the same basis as Cambria, Bethlehem, and North Chicago, each of which paid him $1,000 per year.[4]

Apparently Shinn was able to accept this proposal.[5] Holley wrote to Jones in December, "I am happy to inform you, if you don't already know, that I have arranged matters with the E. T. Co. and am to occupy a consulting capacity—one of the chief functions of which, I apprehend, will be to get your various good ideas—tho I hope and trust I may give you some in return."[6]

In spite of Holley's close relationship with many European steel companies, he strongly favored the use of American steel for American engineering projects. When he learned in January 1876, that Washington Roebling was planning to use crucible steel made in England for the cables for the Brooklyn Bridge, he wrote to several steel companies warning them of the situation. The letters were similar. The one to Otis Iron and Steel Company in Cleveland was typical.

> It now seems probable that the Brooklyn suspension bridge is to be made of crucible steel—at least Washington Roebling wants it. He evidently knows nothing about open hearth steel and thinks the world has stood still since he was in England two years ago, where he got into the hands of some Sheffielders who persuaded him that their pot steel alone would do. You may as well make the wire for the bridge. Roebling is too sick to learn anything now, or even to

read his correspondence carefully. I think you had better see Martin, the next man to Roebling and especially Senator Murphy, one of the Board. ... It is whispered that the Roeblings want to draw the wire and roll it at Trenton in their mill. One of them is appointed inspector. Perhaps they have arranged to get the billets in England. But you can furnish the billets, if they *must* roll the wire. I hope you will make some wire according to their specifications and *then* if they go abroad for wire we will make a devil of a fuss about it, anyhow.[7]

Learning that Bolckow, Vaughan & Company in England was contemplating the building of a new Bessemer plant in the spring of 1876, Holley endeavored to interest the company in using his designs and services. The letter that he wrote seeking the position is interesting since it explains many of the reasons for the high level of production to which the American Bessemer plants had been brought.

As I presume it is ... no longer a matter of doubt that our works do actually and regularly make the large output (and hence at the small cost) with which they are credited; and as I learn that you are going to erect a Bessemer plant in the spring I would like to furnish you complete detailed drawings of our latest and best plant and machinery.

I can give you at very short notice a set of say 150 sheets of general and detailed Bessemer drawings, so complete that no farther [sic] Engineering work will be required except to see that the parts are correspondingly built and erected. These designs would be improvements, in many details, on the plant I am now building for the Vulcan Iron Co. at St. Louis, the latter being a considerable improvement on the Edgar Thomson Works at Pittsburgh, which I recently completed, and which are already producing at the rate of over 40,000 tons of ingot per year, out of two $5\frac{1}{2}$ ton vessels in one pit. I should have no hesitation in guaranteeing 50,000 tons per year from the Vulcan Works, out of a similar pair of vessels, slightly larger, and making less than 6 tons per heat.

I should say, that although we are very glad to show our works to our English friends, and to give them the features of our practice, we can assure them, from costly experience, that any attempt to engraft American improvements upon the English type of works, will be disappointing in its results. The success of a works, as I need hardly say to you, is measured by the capacity of its weakest point. We therefore began anew and radically changed, not only details, but system and arrangement. Now our works, are homogeneous and harmonious, and it is as easy and as common to make 20 to 25 heats in 11 or 12 hours, as it is in the average English Works to make half this number. We have made 62 heats in 24 hours, in one 2-vessel pit; and about 40 is the regular practice.[8]

Evidently the company took advantage of Holley's offer to some degree, since in a report on the direct use of blast furnace metal in the Bessemer process he mentions that the general features of the Bolckow, Vaughan plant were copied from his designs.[9]

During these slack times Holley busied himself with numerous small unprofitable but time-consuming projects, such as checking delivery dates of engines for Edgar Thomson, suggesting changes in a blowing engine for Troy, reviewing patent applications or attending to the vast number of small details that were necessary to bring the Vulcan plant to completion. The successful operation of a furnace or converter rests on a delicate balance made up of countless details. This was especially true during the years when the Edgar Thomson and the Vulcan Iron Works were being built. The care that Holley lavished on the size of a valve or the positioning of a vessel was part of a competence that earned him the respect of plant managers everywhere.

Holley never planned a mill or a piece of machinery without considering both its efficiency and beauty of design. He was deeply interested in both design and architecture. To him a functionally designed machine or steel plant was as beautiful as a fine building. During the time that the Vulcan steel plant was being built to his designs, he became exasperated when he discovered that they were considering changing the method he had designed for transporting hot metal to the converter. He wrote them that they were about to spoil the symmetry of the plant and to make it look like a "patched up affair."[10]

Rossiter Raymond discovered Holley's intense interest in architecture only after the two had been close friends for many years. While waiting for Raymond in his lecture room at Lehigh University, Holley began to outline a Gothic cathedral on the blackboard. By the time Raymond appeared, the building was almost complete down to the last flying buttress and intricate spire. Although Holley often had considered the idea of presenting a course of lectures on architecture as related to mechanics, he never did so. Nevertheless, the theme often was in his mind.

Holley was reminded of the subject on a train trip down the Hudson River Valley, during which he passed numerous examples of Hudson River architecture. Holley remarked that he long had been disgusted by the habit of putting different styles of architecture, Gothic and what not, into steam engines and other machines. "We know perfectly well that what is most fit is the most beautiful," he said.[11]

In 1875, Holley's close friend, Rossiter Raymond, became consulting engineer to Cooper & Hewitt and shortly after Holley also made an arrangement with the same company. Holley was to interest American steel manufacturers in the open hearth process, particularly in a version that utilized the new Pernot furnace, the rights to which the company in Trenton now also owned. The innovative feature of the Pernot furnace, which recently had been developed by Charles Pernot of St. Chamoud, France, was a revolving hearth inclined at an angle of five degrees to the vertical. The pig iron was heated to a red heat, placed on the bed of the furnace, and covered with scrap. As the bed slowly revolved, the pig gradually melted, and the scrap was alternately exposed to the heat of the flame and then dipped in molten pig iron. The mass became fluid in about two hours. The heat then was completed in the routine open hearth method.[12]

Speaking from long experience in introducing new furnaces and processes, Holley cautioned Raymond that, if Cooper & Hewitt were going to build a Pernot furnace, they must build one that licensees could copy—that is, the best it was possible to construct. They might build a cheap version for experimental purposes, but since they had secured the American rights and were recommending the furnace as an assured success, they must construct only the best.

Holley had made a similar suggestion to Slade, but was told that Cooper & Hewitt considered themselves perfectly competent to set up a Pernot furnace. Holley felt that he could talk more bluntly to Raymond than to his brother-in-law, and wrote him as follows.

> Perhaps I am wrong but it seems to me that your people hardly realize the necessity of taking the lead in this new process of steel manufacture. In the ordinary open hearth manufacture you have remained in the rear and have developed no improvements, so that it was once remarked to me that the process had developed in spite of you. Of course you are not bound to ... take the lead but as the process is steadily gaining on the Bessemer and in my opinion bound to beat it at last, it would seem good policy to make a good showing here and get people to build furnaces while the patents last. Every day you delay reduces the value of the patents. I have spent many weeks and traveled thousands of miles since our arrangement began by trying to get people to start the open hearth business. The general question especially re: the Pernot is, "Why don't Cooper & Hewitt put one in—Do they expect us to develop the patent.[13]

In the same year, Holley was engaged as consulting engineer by the Springfield Iron Company. The letter that he wrote to Charles Ridgley, president of the company, exemplifies the scope of Holley's work as an engineering consultant.

> My engagement with you as Consulting Engineer (including furnishing you open hearth drawings etc.) at the rate of $2000 per year, should perhaps date from my last visit to your works or, to make even months, say September 1. Is this satisfactory?
>
> Glad to hear you did so well on roofs. I think I told you that I have now received full detail drawings of Terre Noir new furnaces and plant. I am also copying Pernot drawings that Hewitt brought over and Slade's and Maynard's reports to him. I shall before long put all this new information into a private printed pamphlet.

Hewitt says there will be no more Bessemer works built abroad, and that Open Hearth—either with stationary or revolving furnaces is going to beat the Bessemer here. Several of my Bessemer clients are going into Open Hearth, they say.

I shall certainly work out the details of [the] plant with great care, and shall make them satisfactory to Wellman and other experts. The great thing is *good brick*, to be made where used, as they can't well be transported, being too tender—pure silica just stuck together with the least bit of lime.

It seems to me not to be very risky to go into open hearth because there is a good business outside of rails, if you can't make rails cheap enough, and also because the estimates I sent you *do not contemplate any of the following economies*:

1st The Pernot Furnace which at least doubles the product mentioned in the estimates;

2nd Using cast iron, direct from Cupolas;

3rd Using cheaper Fero Brick;

4th Working in old iron rails as at Terre Noire, by means of that antidote to Phosphorus, Ferro Manganese, which is now quite cheap—5 cents gold, per lb for 40 percent Manganese.

5th The various advantages of an improved general arrangement.

If the original estimates are right for a five-ton stationary furnace plant, and Wellman says they are—Your plant ought to make steel *below* the cost of Bessemer.[14]

The comparative status of the rival processes in early 1876 is analyzed in a letter that Holley wrote to a prospective client.

In response to your request for a general statement of cost of Bessemer and Open Hearth Plants at Youngstown, Ohio, I would state:

1st. Cost of Bessemer plant. Experience fully proves that it is much cheaper to run a pair of 6-ton vessels of the American type, than to run a single vessel or smaller vessels. A pair of 6-ton vessels will turn out 60,000 tons of ingots per year, and the cost of such a plant in complete working order, including buildings, engines, boilers, tools and appurtenances would be, at present prices of materials, about $280,000. The plant would be fitted to remelt all the metal or to take it all or in part from the blast-furnaces. These costs are pretty accurately got at from those of the Edgar Thomson Works which I have recently completed. All the works of this type are now turning out from 1200 to 1300 tons per week.

2nd. The cost of a pair of 6-ton Open Hearth furnaces of the ordinary type, and buildings and appurtenances, to turn out say 5000 tons of ingots per year, would be about $68,000. The cost of a 4-ton furnace plant would be about $125,000. The latter could be run with great success and economy, on a product of say 10,000 tons per year of fine steel, while no Bessemer plant could be run successfully on so small a product of any grade or kind of steel. More open hearth furnaces could be added as required. Thus an Open Hearth works of 60,000 tons annual capacity would cost over twice as much as a Bessemer plant of the same capacity.

3rd. But if the cheaper as well as the finer grades of Open-Hearth steel were required, and especially if the production were to be large, the common open-hearth furnace making but 2 heats in 24 hours would probably not be used. It seems probable that the Pernot revolving-hearth furnace now successfully making, in France, 4 heats per 24 hours will come into general use. I am now getting out improved drawings of a Pernot plant with 2 furnaces, for one of my clients, but I cannot yet estimate the cost very accurately. I should say that it would be about $90,000, to produce say 10,000 tons per year, and that a 4-ton furnace plant of twice the capacity would cost $170,000. Even then the Bessemer plant would be the cheaper for a 60,000 ton annual product.

4th. But while the Bessemer plant and process are very fully developed and highly perfected, the Open-hearth is to be the subject of considerable improvements, chiefly in the increase of product due to better refractory materials and to the use of blast furnace metal direct. So that probably the Open-hearth process will be in every respect a rival of the Bessemer, in a few years, it already has the advantage of the Bessemer in providing more varieties of steel. I would therefore advise your going into the Open Hearth Manufacture rather than the Bessemer, although exact costs of plant and product, and the direction of improvement cannot be so closely estimated. The manufacture of *good* open hearth steel of all varieties is no longer experimental. The manufacture, of *cheap* open-hearth steel is somewhat experimental in this country.

5th. Bessemer ingots now cost in the U.S. from $5 to $10 less than the open-hearth ingots which are made in the old stationary furnaces, all the material being melted in the furnace. But I have no doubt that by using good refractory materials, you could after a while make just as good steel just as cheap, by the pig-and-ore open-hearth process at Youngstown.

6th. I consider Youngstown a first-rate place to establish the steel manufacture, because good and suitable ores are quite accessible, and excellent fuel is cheap and abundant. The utilization of waste furnace gases, from your black coal, would also save notably in the cost of producing and farther manufacturing steel.[15]

At this time the Bessemer steel industry was entering a new phase characterized by both new competition and new faces. Holley had been devoted to the Bessemer process since the day he procured the rights for Griswold and Winslow. Now he wondered if members of the industry still would value his good faith, or even remember his pioneering efforts.[16]

From the production of less than 8,000 long tons of Bessemer steel in 1868, the industry had grown so that in 1876 it produced almost 470,000 long tons. This figure represented about two-thirds of the converter capacity of the eleven Bessemer steel producing companies. The price of Bessemer steel rails was now $68.75 per long ton, a great difference from the 1868 price of $158.50.[17]

The Pneumatic Steel Association and its supporters believed that the country now possessed sufficient facilities for the manufacture of Bessemer steel. Others believed that the reason no more Bessemer steel plants would be built was that the association would not permit their construction. Whatever the reason, the Vulcan works in St. Louis, which made its first blow on 1 September 1876, was the final Bessemer plant to be constructed. A letter Holley wrote to Daniel Morrell in October of 1876 expressed his worry over the effect the association's plans might have on his own future.

> I should judge from looking over the proposed $6.75 pneumatic royalty scheme that the stock is to be watered and that I not being a licensee shall get no equivalent value—also that there is no provision for my $1000 per year salary as engineer for the company. They do not need the engineer, but the salary was, you know, part of the settlement with me. As you saw my first claims carried out before, I naturally rely on you to see that in this new deal I am not squeezed. I am very anxious to have the new scheme carried out—I think it will be the salvation of all the Bessemer works and I am willing to make some sacrifice in order to have it carried out. I do not think that any of the licensees would intentionally wrong me. I simply am anxious that they should not forget my claims. When they consider that the Bessemer patents will soon run out and that my patent of January 26, 1869 is in use and cannot be got rid of by all the works: that it has more than nine years to run and that it will soon be all that secures to them this valuable monopoly, I do not think they will do anything to reduce the value of my already small interest. Would it not be best for the Pneumatic Company to buy my stock at its present unwatered value and so avoid further embarrassment about it?[18]

Other patents owned by the association were, however, due to expire soon. Bessemer's first patent—which covered the rotating vessel combined with the hydraulic piston, the tuyere box, and the ladle crane—was due to expire 1 March 1877. An additional patent covering all existing arrangements of converters operating in pairs and the ladle cranes would not expire until 1879, and one covering the system of removable and duplicate vessel bottoms would remain in force until 13 January 1880. The Kelly patent, which had run out in 1871, had been extended for seven years and now was due to expire on 23 June 1878. These vital rights also covered the pneumatic process up to a certain point and were thus an essential part of the whole patent fabric. The Holley patent, which covered the entire system of raised pit, raised vessels, arrangement of cranes, the handling of vessel bottoms, and the rest of the many improvements known as the American system, would not run out until 1886.

The Pneumatic Steel Association, in one of its circulars issued on 7 November 1876, acknowledged that the only way any American works could avoid the use of Holley's inventions would be to reconstruct its converting department. Chester Griswold, in his capacity as secretary of the organization, said he had issued the document because a number

Bessemer converters in 1876. From *Harper's Weekly,* 25 March 1876.

of licensees under these Pneumatic Steel Association patents had requested that he state in detail which claims were still in force and on what grounds their validity rested.[19]

There were serious rumblings in the Pneumatic Steel Association membership that were to lead to its eventual reorganization. Certain letters written by Holley indicate that with the first patent on the verge of running out, the group seemed to be making an effort to consolidate its position by purchasing patents that would terminate much later. In November, he advised the widow of George Fritz that the Pneumatic Steel Association was trying to make a "new deal about patents." They apparently did not approach Mrs. Fritz directly, but were using Holley to secure her husband's patent for the blooming mill table. In his letter to Mrs. Fritz, Holley said that he had offered his vessel bottom patent for $5,000, and he understood that the group was willing to pay $10,000 for her husband's patent. Holley advised her to take it in cash or equivalent. No more Bessemer works were likely to be built, he said, and it would be a long time before she would get that sum from the use of the patent.[20]

Financial pressures slowly but steadily increased. But Holley frantically tried to keep his organization together in the hope of better times. He had been forced to employ the trustworthy Phineas Barnes only part time. Gram Curtis, his valued office assistant, wanted higher pay. He could only tell Curtis exactly why he could not pay him more. He reminded Curtis that drafting expenses had been heavy, that there was not enough demand for Holley's services as consultant, and that proceeds from the sale of drawings did not even pay the office rent. If he would stay and accept the best Holley could offer, he promised Curtis he would be suitably rewarded when things improved.[21]

Holley's old worries over his status with the Pneumatic Steel Association persisted. A letter from Morrell questioning Holley's relationship with other steelmakers seemed to indicate that the Bessemer producers might hope to possess him completely. Because of their longstanding association, Morrell should have been the last person to question Holley's motives. But from Holley's reply he must have done so.

> Dear Mr. Morrell. In reply to letter. I can only say that I have not given any information directly to consumers but only to my few clients who are producers. I have refrained from reading any papers or furnishing any articles for technical newspapers regarding what I have seen abroad. I quite agree with you that all this information furnished to one works alone would be much more valuable to that works than if the same were furnished also to rival works but please to remember that no one works pays me expenses of collecting this information not a tenth part of it and in order to get it at all I must distribute the results of my investigations among my clients at large. I should be very glad to devote my time to one works exclusively if I could get the pay that I now do from my several clients.
>
> I should be glad to receive any suggestions you may make in this matter.[22]

There is no indication that any action resulted from the letter.

After his meeting with Siemens in Philadelphia, Holley determined to go to Europe to study the operation of the open hearth process. He made plans to be away three months and left the United States in January of 1877. Holley ranged the Continent in his usual way, using London as his base. He was now as familiar with the plants abroad, and with the men who operated them, as he was with those at home. Everywhere he went, he was welcomed heartily, and information that might have been refused others was given him gladly. He succeeded in breaching even the Krupp fortress. The Vickers plant in England refused him entrée the following year.[23]

When he returned to New York in April of 1877, Holley was brimming over with information about the improvements in steel operations that he had seen abroad. He barely had landed before he began writing detailed reports for his clients. Unfortunately, he returned to an industry that was sinking day by day into inactivity. While the demand for some iron and steel products was higher than it had been in 1876, prices had fallen so much that little or no profit was possible. An industry publication reported that "More than half the furnaces and many of the rolling mills were idle the whole year. Failures and suspensions were frequent and more were to come." As the price of steel rails dropped from $49.00 per ton in January to $40.50 in December, the year turned out to be more disastrous to the steel industry than any since the panic of 1873.[24]

In spite of plant shutdowns, one record was established. For the first time, the production of Bessemer steel rails not only overtook that of iron rails but also exceeded it by close to 100,000 tons. In addition to the eleven companies producing Bessemer steel, nine companies now were turning out open hearth steel, but only to the extent of some 25,000 tons in all. The Bessemer process still reigned.[25]

Holley spent the first months after his return in catching up with office affairs and endeavoring to find additional business. In spite of the many demands on his small resources, he performed an act of generosity and human kindness for an old acquaintance that would have been unusual under any circumstances. John Thompson had been the superintendent at Troy, at the Cleveland Rolling Mill Company at Newburg, and again at Troy, but had been replaced by Hunt in October 1873. Thompson and Holley had worked together through many trying times and Holley respected his ability. He knew, however, that a drinking habit was responsible for Thompson's inability to keep a job. In June of 1877, Holley wrote Thompson an extraordinary letter in which he offered not only to take him into his office, but also to supervise him both in his professional development and in an attempt to break his habit.[26]

Thompson accepted the offer and entered Holley's office to try to come back as an engineer. He apparently kept his part of the bargain for some time, perhaps even for the stipulated period, since the following year Holley wrote several letters of introduction to friends in the Chicago area where Thompson had gone to work as inspector for the Chicago, Milwaukee and St. Paul Railroad. But the reform probably was only temporary, since Thompson dropped out of sight for several months. He reappeared in 1879, sick and unable to work. Again he was helped by Holley, who collected money from friends and obtained passes on the railroads so that Thompson could be sent to Minnesota to start anew. It was too late, however, and Thompson died a few months later at the age of thirty-six.[27]

Despite financial difficulties, Holley continued his longstanding practice of refusing for ethical reasons to accept commissions on equipment ordered by clients. At one time when Creusôt had asked Holley to become the American agent for one of their patents, Holley replied, "In view of the fact that my position as a consulting engineer is so largely based upon my entire independence in a commercial view, of any patent, I do not think that I could afford to be commercially identified with any invention which my clients are using. You will please understand that whatever I may do in this matter for the present is in the service of my clients and is not subject to any commission from you."[28]

The reorganization of the Pneumatic Steel Association had been under way for almost one year. In June of 1877, the final plans were completed, and a new association called the Bessemer Steel Company was organized. Evidently the eleven licensees joined together and for $825,000 purchased the patents that still were valid. A letter to his financial advisor John Slade provides the only clue as to how Holley fared as a consequence of the now obscure transactions that resulted in the new organization. Holley's comment was terse: "I think we are real fortunate to have got out of the thing so well."[29]

Although Holley always had been regarded as the apostle of the Bessemer process, he had become increasingly convinced that the open hearth process had many advantages.[30] Sixteen companies had adopted the open hearth process by July 1877, and were using it with fair success. Holley believed that Cambria was the logical company to install the process next, and he had spent months to no avail attempting to convince Morrell of the advantages of the open hearth process using the Pernot furnace.

During 1877, the financial condition of the entire country grew more insecure. Holley experienced considerable difficulty in collecting the fee that the shut down Joliet Iron and Steel Company had owed him for some time. This was the first of Holley's Bessemer clients to go. Late in August, Holley started out on one of his periodic trips in which he traveled from plant to plant. Since the fall of 1876, Holley's friends had noticed a gradual deterioration in his health. In recent months he had suffered from neuralgia, but with his professional accounts in a precarious state, he did not dare to relax or even to curtail his activities. By the time he reached Pittsburgh, he was in great pain, and he knew that, even with all the fortitude he could muster, he could not continue the trip. He returned to New York. Holley's physician insisted that he drop everything and get away to the woods for at least a month, away from mail and the problems of steel.

At that time, a small group composed of Frederick Church, S. R. Gifford, H. W. Robbins, and Lockwood De Forest, all fellow members of the Century Club in New York, was about to set off on a sketching trip to Mount Katahdin in Maine, and they persuaded Holley to join them.[31] With the addition of L. G. Laureau, an open hearth expert who recently had joined the Holley organization, the group set out. The trip was a welcome change for Holley although he was so exhausted he found it difficult to keep up with his companions. The few vacations he had taken in his lifetime usually had been combined with work.

Now, with nothing to do, night after night, he lay awake for hours looking at the night sky, too tired to relax. Yet, despite its strenuous nature, the trip was a great pleasure for him. He treasured the opportunity of being with the artists and sharing temporarily in their free way of life. Holley's own art was turning everything he saw and experienced into prose. His essay "Camps and Tramps about Ktaadn," which described the adventure, was published the following May in *Scribner's Monthly*.

Holley returned to New York refreshed, but worries and uncertainties returned almost at once. With the slackening in his business, he had resumed writing, and he now continued the series on American iron and steel works, which began in the London magazine *Engineering* in March 1877, and ran for forty-one issues. Each article described in detail, with illustrations, the leading American Bessemer works. Lennox Smith, also an engineer, helped Holley assemble the material for some of the articles that document the iron and steel industry of that period.

Captain Jones was anxious about his friend's flagging morale, and when he saw Holley in December broached to him the idea of a testimonial dinner. Totally unprepared, Holley was shaken by the suggestion. He wrote to Jones a few day later.

> Upon further reflection concerning the testimonial you were so kind as to mention the other evening I hardly know how to exactly express my feelings. I will only say that while I fear that the occasion and general feeling are unfavorable to such an expression I should be all the more gratified at the friendliness of the few who would do me such a courtesy in the absence of any special act of mine or of any occasion to call out any testimonial.
>
> In the excitement of the occasion the other evening, I mentioned asking a lot of people to dinner. That, of course, upon a moment's reflection, I observe to be an improper thing for me to do—an immodest thing. Nothing would give me more pleasure than to entertain a few friends, such as a committee who might do me the honor to present the testimonial, perfectly privately at dinner. For me, however, to make any other than the most private demonstration to a few friends would, I now see, be an advertisement of myself not in good faith.
>
> I am feeling a little blue about my connection with the Bessemer works. Three of my clients—North Chicago last—have dispensed with my services. They may not have got much good out of them but that was their fault I worked hard enough and gave them facts enough. But I suppose prosperity will return again.[32]

The loss of these clients seemed to presage Holley that Bethlehem was about to dispense with his services as well. Holley was distressed and he wrote to Alfred Hunt, the president, in an endeavor to ward off this possibility. The letter was to no avail. Hunt replied that the board had decided not to keep him on after the end of the year.[33] Yet despite the hardship that resulted from the loss of these important clients, his deep sense of honesty forced him to write to Erastus Corning at the Troy works that, since the drawings that they had needed were finished and since the depression was so bad, he would suggest that beginning in January 1878, his salary from Troy be reduced to $1,500 from $2,500.[34]

With the number of Bessemer clients steadily dwindling, Holley began to concentrate on the possibility of developing more interest in the open hearth process using the Pernot furnace. Morrell at Cambria was still the most likely candidate and Holley centered his efforts there. The high royalties were the stumbling block; the companies all objected to having to pay them. Although the duration of the patents was lessening steadily, Richmond and Potts, now the licensing agents, were still asking the full price for the Siemens's furnace rights. On the other hand, Cooper & Hewitt did make a reduction in the Martin rights, lowering the price for the first furnace from $10,000 to $8,000. Holley asked Hewitt to forego the royalty on the Pernot furnace since none had been built in America, and if Cambria put in a pioneer demonstration installation, it would be of immense value to Cooper & Hewitt. Although Abram Hewitt certainly was interested in furthering the use of the patents that his company controlled, he did not relish the idea of giving Cambria the rights without payment.[35]

Exasperated, Holley tried to show Raymond, their consultant, the advantages of encouraging Hewitt to be more positive.

> I wish I could make the Cooper & Hewitt people see as well as I do that the Cambria people are in just the condition to be turned to or from the open hearth by the breadth of a hair. I have worked up Morrell and the folks at Johnstown to quite an enthusiasm. They and I together have for the first time got Townsend and the Philadelphia people to admit the possibility of going into open hearth. Rather than lose all the results of this hard work I would start a subscription with the amount of my year's salary at Cambria to pay the royalty on a second furnace in case it is not given free. If Cambria goes in at all it will be in gorgeous style and an assured success whatever it costs.[36]

Holley finally convinced Hewitt of the possible advantages for his company if he would cooperate with Cambria regarding the Pernot furnace royalties. Although Hewitt considered Cambria's request unreasonable, the only requirement he set was that the two ten-ton furnaces would be built within twelve months and be the first in the United States. The Siemens's royalty on the regenerator furnace, which amounted to $1.75 a ton for rails and $2.50 for other steel products, still had to be paid.

Now that Cambria had taken action after many months of hesitation, Holley directed his attention toward some of the other plants. Using Cambria as a lever and emphasizing the immense value of the open hearth for using bad pig iron and scrap as raw materials, he persuaded Springfield Iron Company in Springfield, Illinois, to follow Cambria. Well situated near a large bed of bituminous coal and reasonably close to iron ore deposits, the company had committed itself to the open hearth.[37]

Holley was less successful in his attempts to influence the managers at the Edgar Thomson and Troy plants to install the new furnace. Although they recognized its importance, neither was willing to adopt it.[38]

For many years one of Holley's best sources of information had been the Compagnie des Founderies et Forges de Terre Noire, the second largest steel company in France. Many metallurgical discoveries and plant improvements had emanated from Terre Noire. It had pioneered in both the Bessemer and open hearth processes, and, in recent years, the company had developed a steel casting process that had attracted Holley's attention. In a letter to his friend, George Maynard, who now had settled in London, Holley said that he was getting so very little consultation work that he hoped to move more into commerce.[39] He had been dropped as a consultant by many companies and he probably felt that he could no longer depend for a livelihood on anything so arbitrary as the vote of a board of directors. Holley and William Sellers each secured a one-quarter share of the right to the Terre Noire casting process. Holley did his best on its behalf, but the process, although considered a good one, proved too involved to be really profitable. His account books indicate that Holley received very little in royalties from this source.[40]

While Holley's relations with his clients were traveling a rough road, his professional prestige continued to increase. In 1878, on the twenty-fifth anniversary of his graduation, Brown University awarded him an honorary Doctor of Laws. That Holley had received his first degree from Brown University by a narrow margin could have been known to no one else at the ceremony. This honor, with its recognition of his worth, brought extreme pleasure to Holley.

Holley's letters indicate that during the first six months of 1878 he continued in his role as consultant, mediator, and intermediary to the industry on a variety of problems. Just as some of the steel producers were looking for ways to circumvent the open hearth patents, others would have liked to circumvent the Bessemer patents, which now were controlled tightly by the Bessemer Steel Company.

In response to an inquiry about the relative qualities of steel produced by the two processes, Holley was quite specific. The Bessemer had progressed so that a number of grades of steel, including that suitable for nails, could be made. Troy was supplying the Springfield Armory with Bessemer steel for gun barrels, he told his client. "However," he warned, "the Bessemer patents are owned by the eleven Bessemer works and I do not think that they will grant any licenses. They might license for some speciality other than rails but a general consent must be obtained. I think you may conclude that you could not obtain a license."

"The cost of a Bessemer plant to turn out two hundred and fifty tons of ingots would cost about $250,000 exclusive of land," Holley estimated. "A smaller Bessemer plant would be a losing proposition. On the other hand although an open hearth plant of the same capacity would cost about the same if built all at once, it would be possible to build a smaller one that could produce sixty tons per day for about $90,000 which could be a paying proposition."[41]

When the various superintendents of the Bessemer plants started their rivalry in quantity production, it was done in a friendly way. Little did they foresee the ultimate results of their friendly needling of each other. Of all the plant superintendents, Captain Jones of Edgar Thomson was particularly troubled. Andrew Carnegie, more than any other company head, was demanding and aggressive; and it was in his nature to be first, whether in stock held or in production. He received daily reports from every department of Edgar Thomson no matter whether he was in New York, Scotland, or Singapore. If there was slackening of output or it fell below that of another plant, he required a reason. Carnegie never was satisfied. A report of record production could expect to receive the response, "Congratulations! Why not do it every week?"[42]

Holley on several occasions had served as an intermediary between Jones and the Edgar Thomson management. Even though Carnegie understood Jones's importance to the operation of the Edgar Thomson Works, Carnegie was unable to resist constant meddling. Jones had submitted his resignation several times only to be enticed back by promises of more money and less interference. Holley had served as an emissary between Jones and Shinn and McCandless when the question of Jones's salary first came up in January 1876. After that interview, Holley wrote Jones, "I spoke to Shinn and D. McCandless about you and Walton, pretty fully. They appreciate that the present pay is low and fully agree that your good is to be properly compensated, but they want you to grow with the Company and not expect your full reward when they are yet struggling. They think the desire for more pay is a little premature, but the matter stands in good shape and will be pushed along as fast as possible."[43]

Jones's dissatisfaction must have reached a peak in June of 1878 when

Holley wrote him: "I had a long talk with Shinn in Boston about salaries and the interference of a certain party and other matters. I feel more certain than I ever did before that Shinn intends to be Boss of the works after this and that in his quiet and judicious way things will be satisfactorily arranged."[44] Evidently Holley's response failed to satisfy Jones. Holley wrote him again a few days later that he fully appreciated the difficulties and could only repeat that he thought they would be remedied.[45]

In July of 1878, Holley decided to go to Europe to see Siemens once again and to discover whatever was new and valuable in open hearth practice. Mary Holley had been in poor health for months, and it was decided that their older daughter, Gertrude, by then a beautiful and vivacious girl of sixteen, would accompany him. Scraping together the ready funds he would need for a few months of business and pleasure travel in Europe, the Holleys departed in midsummer for what was to be a most enjoyable trip for both.[46]

Holley had been asked to work on the American exhibit for the Paris Exhibition of 1878 but had refused, for reasons of health as well as lack of time. He also may have recalled the onerous tasks and lack of remuneration for large expenses in connection with this work for the centennial at Philadelphia.

NOTES

1. Burton J. Hendrick, *The Life of Andrew Carnegie*, 2 vols. (Garden City, N.Y: Doubleday, Doran & Co., 1932) 1: 210–11.
2. ALH to Daniel Morrell, 18 October 1875, Author's Collection (hereafter cited as AC). At times, Carnegie's efforts to achieve economies seemed penny-wise but pound-foolish. During the first year of E.T.'s operations he had heard that another steel manufacturer had dispensed with some of Holley's improvements on the Bessemer process. He suggested to Shinn that they do the same. Then they would not have to pay the yearly patent royalties to Holley. Shinn balked at this suggestion. He told Carnegie bluntly that on this issue, "You are entirely mistaken. We could not abandon the use of Holley's patents for a cost of $100,000, as it would involve a reconstruction of our converting works and then it is doubtful if we could manage economically to carry on the process," In this issue Shinn stood his ground, and he won. (Joseph Frazier Wall, *Andrew Carnegie* [New York: Oxford University Press, 1970], p. 343.)
3. ALH to Captain Jones, 16 October 1875, AC.
4. ALH to W. P. Shinn, 6 December 1875, AC.
5. ALH to W. P. Shinn, 18 September 1876, AC. Holley's letter gives a hint of the financial stability of Carnegie's company. Holley had a note for $5,000 from the Edgar Thomson Company, and this may have represented his royalty fee. When it became payable, Holley suggested to Shinn he was willing to take bonds at their market price instead of money if Shinn believed that the investment was a perfectly safe one. He suggested that he would leave the decision to Shinn and if the check were sent he would know that Shinn considered it advisable to put the money into

better established securities. Holley assured Shinn that if the bonds were sent, he would, of course, not hold his friend responsible for their future value. Four days later Shinn sent Holley five bonds and a check for $512.

6. ALH to Captain Jones, 11 December 1875, AC.
7. ALH to Otis Iron and Steel Company, 26 January 1876, AC.
8. ALH to Bolckow, Vaughan & Company, 27 December 1875, AC.
9. Alexander Lyman Holley, "The Direct Use of Blast-Furnace Metal in the Bessemer Process," Private report, no. 2, 2d. ser., New York, June 1877, p. 17.
10. ALH to Vulcan Iron Company, 19 February 1876, AC.
11. *Memorial of Alexander Lyman Holley* (New York: American Institute of Mining Engineers, 1884), pp. 20, 89. His friend Coleman Sellers remembered that this conversation took place not long after the Chicago fire because Holley asked him if he had been to Chicago lately. Sellers had not. Holley then said:

Chicago has risen like the Phoenix from her ashes and yet you will find there the most striking instances of the barbarity of the American people and their low condition in regard to fitness and propriety. In Chicago they have erected numbers of handsome buildings and some of the noblest specimens of architecture have been put into their stores and warehouses, but what else have they done? All these beautiful pieces of architectural ingenuity are plastered from top to bottom with signs advertising cigars, snuffs, tobacco, dry goods—covering the whole city of Chicago with black and gold after putting millions of dollars into those buildings in which the highest skill in construction has been displayed. ... It certainly shows the very low condition of the aesthetic element in this country that they are capable of defacing these beauties with this mere advertisement of their goods in such a glaring way.

12. J. S. Jeans, *Steel: Its History, Manufacture, Properties, and Uses* (London: E. & F. N. Spon, 1880), p. 465.
13. ALH to Rossiter Raymond, 1 July 1876, AC.
14. ALH to Charles Ridgley, 29 September 1875, AC.
15. ALH to W. C. Andrews, 8 January 1876, AC.
16. Inference that Holley's position was becoming uncomfortable can be drawn from his letter to the president of Bethlehem.

I did not mean to manifest any special championship of the Edgar Thomson works last evening in my talk about the price of rails as I have no interest in any works but Bethlehem. I only wanted to shield the company and Mr. Shinn against idle talk which is often exaggerated. For instance, I was positively assured today that you had sold twelve thousand tons for $50 and that your agent acknowledged it. It is none of my business what anybody sells for. I simply express a belief now and then that certain parties who have been accused of going away below the market unnecessarily, have not done so (ALH to Alfred Hunt, 16 September 1876, AC.).

17. United States Bureau of the Census, *Historical Statistics of the United States, 1789-1945* (Washington, D.C.: U.S. Government Printing Office, 1949), pp. 187; *Annual Report of the American Iron and Steel Association* (Philadelphia: AISA, 1877), p. 44.
18. ALH to Daniel Morrell, 21 October 1876, AC.
19. Chester Griswold, secretary, Pneumatic Steel Association, *Patents Owned by the Pneumatic Steel Association and Still in Force* (New York: Pneumatic Steel Association, 1876), 3 pp., AC.
20. ALH to Mrs. George Fritz, 14 November 1876, AC.
21. ALH to Gram Curtis, 24 November 1876, AC.
22. ALH to Daniel Morrell, 21 October 1876, AC.
23. ALH to Joseph Hartshorne, 16 May 1878, AC. Holley never abused a trust and always gave as much or more than he received. A steel plant could only profit from a visit

from him. It, therefore, is not surprising that, during his visit to Germany, he was made an honorary member of the *Verein zur Beförderung des Gewebefleisses*. This organization, the most prestigious society of applied science in Germany, was concerned with the mechanical part of metallurgical works, especially those that used the Bessemer process.

24. *Annual Report of the American Iron and Steel Association*, 1878, p. 9. Erastus Corning, whose father was one of the original owners of the Troy works, had for some time favored selling the plant, which had been shut down for some time. He had suggested to Holley that he feel out the European market while he was abroad. Holley reported that European capitalists might be interested, but they wanted to know whether foreign directors would be permitted in New York State. "Capitalists in Europe," Holley said, "have a keen dread of Yankee management of their affairs. It is almost impossible to get a penny invested in anything American" (ALH to [name illegible], 8 May 1877).

25. *Annual Report of the American Iron and Steel Association*, 1878, pp. 9, 19, 23.

26. ALH to John Thompson, 4 June 1877, AC.

Dear John: Upon consultation with Madam [Holley] I have concluded to make you the following proposition. If you have a mind to come to Brooklyn to live where we can see your outgoings and your incomings so that we can have not only a knowledge but a sort of control of the critical hours of your life, that is your evenings. And if you will go to work in my office or in the field . . . and do the best you can. And if you will make it a part of your contract that the very first or any departure from your determination to drink no spirits, wine or ale, shall be without further notice a severance of our professional connection. Then I will pay you enough to keep you comfortably for one year and if you have not by that time shown any departure from your good resolutions and if your professional work is satisfactory . . . I will give you a percentage on my whole business which can be fairly arranged.

I deem it only fair to us both, to state my fears at the outset. I don't object on general principles to the use of wine (spirits I have long since discarded as a person) in a moderate way at dinners etc. but I am of the opinion . . . that in your particular case and with your particular temperament no alcoholic stimulant whatever, however disguised and diluted is safe. I have heard of your excesses on several occasions and I fear that your sins in this direction have been more habitual than occasional. . . . It is either to be one thing or the other and there can be no middle course.

But there is another thing which gives me doubt and uneasiness. You don't seem to have that professional workfulness and enthusiasm without which none of us, in our profession, can ever get anywhere. I have urged you to keep posted in the proceedings of learned societies, to write papers and to take that active professional position which you could easily enough take if you would spend more of your leisure in a professional way and less of it in the mere amusements of the town.

Now John the whole thing lies in a nutshell. If you are ripe for what I must call a revolution in the two matters . . . I can give you a living and if you are willing at any cost to develop and magnify the power which a fair talent, a good culture and a liberal experience have given you, I can make you prosperous. . . .

Don't seize on this offer, in your distress as an easy way out of trouble. You will have to use great self-denial and to have never done working within the limits of health — just like myself. You will want to dig your way slowly and painfully, just as I have done but you will have a sort of professional help which I never had and you will never have it extended to you again in the way of business if you should accept this proposition and then fail to carry out your part.

27. ALH to A. G. Cassatt, President, Pennsylvania Railroad Company, 5 December 1879,

Holley Family Papers (hereafter cited as HF); ALH to Purdy & Nicholas, 9 December 1879, HF. Whether Holley actually had given up hard liquor or whether his statement in the letter to Thompson was made for emphasis is difficult to say. He obviously had no scruples about keeping liquor in his house, since there is at least one letter in which he ordered five gallons of whiskey from the same dealer who supplied the wants of John Fritz. Holley emphasized in his order that he wanted the same high quality that his friend recently had bought. Evidently Fritz was considered an expert in this field as well as in the construction and operation of steel plants.

28. ALH to A. G. Gustin, 5 January 1878, AC. At another time, he wrote to the St. Albans Iron and Steel Company, "Can you give me a commission for getting your hot-straightening machine adopted — of course by companies that are not my clients?"

29. ALH to John Slade, 29 August 1877. In 1898 a suit was brought by Orville Owen and others against Orrin W. Potter and others to compel an accounting of certain assets from the estate of Eber Ward. An excerpt from the decision in the case concerning the Pneumatic Steel Association and the formation of the Bessemer Steel Company is of interest.

We have shown how the North Chicago Rolling-Mill Company became possessed of the property of the Wisconsin Iron Company and the Milwaukee Iron Company. It is claimed that it absorbed the estate's interest in the Pneumatic Steel Company also. The Pneumatic Steel Company was formed for the purpose of buying the Bessemer steel patents, and profiting from royalties paid by the various manufacturers of Bessemer steel. We do not discover readily the amount of its capital stock, or the amount that had been paid in, but Ward was owner of $75,000 (par value) of the stock. There is evidence tending to show that it was a profitable venture, and paid large dividends after the death of Ward, and nearly up to the time it was sold. On April 3, 1877, Potter filed a petition, asking leave to sell this stock at 80 cents. This petition, and so far as we can learn, the profits in this case, show that some of the patents had expired, others were about to expire, and that some manufacturers were questioning the validity of other patents, and refused to pay royalties. A company had been formed comprising all of the Bessemer steel companies in this country; and this company had offered to pay 80 cents upon the dollar for the stock of the Pneumatic Steel Company. The North Chicago Rolling-Mill Company was one of the eleven companies which were stockholders in this new concern, and apparently the object was to reduce the cost of manufacture by cutting off the royalties theretofore paid. The Pneumatic Steel Company had no tangible property, and the value of its patents depend [sic] entirely on the use that should be made of them. We have no means of knowing whether or not the payment of royalties could have been enforced. We may suspect that this new company was the result of concerted action, to profit at the expense of the Pneumatic Steel Company; . . . there is nothing to indicate that the executor could have obtained more for the stock. We can understand how it was possible that he might have been compelled to do worse. (Owen et al. v. Potter et al. [Supreme Court of Michigan, 25 January 1898] in *Northwestern Reporter* 73 [19 February 1898]: 984).

30. Robert W. Hunt, *"A History of the Bessemer Manufacture in America" and "The Evolution of American Rolling Mills"* (New York: American Society of Mechanical Engineers, 1918), foreword. Reprint of papers presented before the American Institute of Mining Engineers 5 (1876–77) and the American Society of Mechanical Engineers 13 (1891). On one occasion when Holley and Robert Hunt were attending a technical meeting, Holley had spoken eloquently and emphatically in praise of the open hearth. When one of the men present questioned Holley's statement, he had jumped to his feet and, pacing the room with his old vigor, had replied, "I expect to live to see the

American open hearth process attend the funeral of the Bessemer process." The remark was received indulgently by the group, amused at the speaker's earnestness.

31. Cleveland Amory, "The Great Club Revolution," *American Heritage*, December 1954, p. 30. Although he had pleaded that he was too often out of town at the time he refused an invitation to join the Union Club, in 1868, Holley later became a member of the Century Club. The Century Club was organized in 1847 by a group that believed the Union Club was slighting intellectual eminence. "There's a club down on 43rd street," said one member of the Union Club, "that chooses its members mentally. Now isn't that a hell of a way to run a club?" The Century Club membership was composed of a cross section of New York life but the artists and writers formed a nucleus. Holley, because of his interest in painting, particularly injoyed the artists whenever he was able to attend the famous Saturday nights at the club quarters. "Camps and Tramps about Ktaadn," *Scribner's Monthly*, May 1878, p. 33. Although unsigned the article is known to have been written by Holley. He explains the unusual spelling of the mountain's name: "The orthography—Ktaadn—is not that of the maps; the Maine State College people, who ought to be allowed to name their own mountain, insist upon 'Ktahdin.' But those eminent authorities, Thoreau and J. Hammond Trumbull,—the latter our best expert in Indian nomenclature,—prescribe the spelling here adopted."

The artists contributed small sketches for use as illustrations, and this delighted Holley beyond measure. His painter friends were amused at the pseudonyms he gave them in the article. Gifford became Don Gifaro, Robbins was Herr Rubens, Church was appropriately Don Cathedro, De Forest was Monsieur de Woods, Laureau became Monsieur La Rose. Holley signed the article Arbor Ilex. Church, one of the most successful contemporary painters whose landscapes sold for as much as $10,000, painted a view of Mount Ktahdin for Holley as a souvenir of the trip.

32. ALH to Captain Jones, 24 December 1877, HF.
33. ALH to Alfred Hunt, 3 January 1878, HF.
34. ALH to Erastus Corning, 10 December 1877, HF.
35. ALH to Daniel Morrell, 27 December 1877, HF.
36. ALH to Rossiter Raymond, 5 February 1878, HF.
37. Although Springfield was the second plant to install the Pernot revolving and movable hearth, it was the first to adopt the Krupp system of washing phosphorus out of pig iron.
38. ALH to Richmond and Potts, 11 April 1878, HF.
39. ALH to George Maynard, 17 May 1878, HF.
40. Even though Holley was aware of Bessemer's sometime lack of faith in other men's inventions, he was deeply provoked when Bessemer announced that he had practiced the Terre Noire method of casting since 1862. "He couldn't have," Holley complained to Sellers, "I was studying for two months or more in 1864–1865 and he did not make castings nor could he cast sound tyre rings. Everything he made had blow-holes unless it was very hard and brittle. It is very easy for Bessemer to simply assert that his old use of silicon and manganese is the Terre Noire process. He don't [sic] give any proof; on the contrary his practice shows that he never knew how to use silicon and manganese by the Terre Noire method" (ALH to William Sellers, 23 March 1878, HF).
41. ALH to F. J. Hearne, 16 May 1878, HF.
42. Hendrick, *Andrew Carnegie*, 1: 210.
43. ALH to Captain Jones, 3 January 1876, AC.
44. ALH to Captain Jones, 24 June 1878, HF. Jones was not alone in his dissatisfaction. Bridge portrays Carnegie as playing an active role in fostering internal rivalry. "Some of these managers and partners did not speak to each other for years, so skilfully were

their jealousies and rivalries played upon; and there was hardly a man at the head of any department of the Carnegie concerns whose flanks were not ripped open in the fierce race for supremacy."

On the eve of one of his annual trips to Scotland, Carnegie made one of his infrequent visits to the plant where he found Jones superintending a group of mechanics who were repairing a break in the rolls. Covered with grease, with the perspiration running down his face in rivulets, Jones did not present the ideal picture of an executive to Carnegie's eyes. "You have skilled men to do this, why don't you take a vacation?" said Carnegie. "You cannot imagine the abounding sense of freedom and relief I experience as soon as I get on board a steamer and sail past Sandy Hook." To this the blunt and exasperated Jones replied, "My God, think of the relief to us." It was also Jones who made the famous remark, half in disgust and half in admiration, that "Andy was born with two sets of teeth and holes bored for more." James Howard Bridge, *The Inside History of the Carnegie Steel Company* (New York: Aldine Book Co., 1903), pp. 113, 114.

45. ALH to Captain Jones, 1 July 1878, HF.

46. ALH to Captain Jones, 1 July 1878, HF. Evidently Holley was short of ready funds for such an expedition since he wrote many of his clients asking them if they would mind paying his semiannual fee early. In addition, Corning lent him several thousand dollars on Edgar Thomson and Crown Point Iron Company bonds that Holley left with him as security.

XXII · Sidney Gilchrist Thomas

Drawn by the exhibition of 1878, the British Iron and Steel Institute held its September meeting that year in Paris. Sterry Hunt, former president of the American Institute of Mining Engineers, John Fritz, and Alexander Holley were the only Americans present at the banquet that brought the meetings to a close. Holley was asked to speak representing both the United States and the American Institute of Mining Engineers. Recalling the event, Hunt wrote, "I have often listened to him, and we all remember his wit, his wisdom and his eloquence, but those who were not there present have never, it seems to me, heard him at his greatest and his best, as inspired by the occasion, he addressed that brilliant assembly."[1]

The institute meetings always interested Holley both for the information he could garner from the technical sessions and for the opportunities to renew old friendships and to make new ones. One of the papers that was scheduled to be read attracted Holley's attention. It was called "Elimination of Phosphorus in the Bessemer Process," and initially had been listed first on the printed program. However, preference was given to papers by foreign authors, and, as the meetings wore on, the paper gradually was dropped down the list until finally it was eliminated.

The name of the paper's author, Sidney Gilchrist Thomas, probably was new to Holley, despite his wide acquaintance with English steelmakers. Thomas's subject was a vital one. The large deposits of iron ore in Wales and in the Cleveland district of England contained excessive amounts of phosphorus and thus could not be used in the Bessemer converter. For years iron ores low in phosphorus had been in increasingly short supply. It was estimated at that time that only one-tenth of the known iron ore deposits in Europe could be used in the Bessemer process.[2] An outstanding metallurgist wrote in 1872, "The limit to the production of Bessemer pig is the want of ores free from phosphorus. The hematites (non-phosphoric ores) of this country under the sudden demand have doubled in price, and speculators of all kinds are rushing off to Spain, where tracts of land, conceded without any payment a few months ago by the government of that country, are said now to be worth large premiums."[3]

Bessemer and many others, including England's great metallurgist

Lowthian Bell, and George Snelus, manager of the Dowlais Works in South Wales, had spent years trying to find ways to eliminate the phosphorus from the converter. Yet the council of the British Iron and Steel Institute found it possible to drop the reading of a paper on the subject, because several of its members believed that, if none of the leading metallurgists had accomplished the feat, it was unlikely that an unknown person could have done so.

Holley entertained no such prejudices, and when he learned that the Thomas paper would not be presented, he went in search of the author. When the two met, a friendship immediately ensued. They discovered a mutual interest in music and art as well as in the Bessemer converter. The sense of humor and gaiety that both men possessed belied the extraordinary amount of work of which each was capable. In appearance they were quite different. The tall, blond Holley, eighteen years the senior, stood in direct contrast to the delicate, small-framed Thomas, with his unusually large dark eyes and high forehead. His head seemed large for his body, and he had the sensitive face of a poet. The presence of the young man in Paris was amazing, because he never had worked in a steel plant. Moreover, he was attending the meetings while on leave from his position as junior clerk in the Thames Police Court in London.

Sidney Gilchrist Thomas, who was born 16 April 1850, was only six years old when Bessemer read his famous paper at Cheltenham. His Welsh father was a conservative, but his Scottish mother was a keen liberal. As Thomas often said, this enabled the children to have minds of their own. He had two brothers, one older and one younger, as well as a younger sister Lilian. The family shared an interest in books and music, which may have stemmed from their Welsh background. Young Thomas was undecided about a career and wavered between engineering and analytical chemistry. After years of indecision, during which he pursued a classical education, he determined to become a doctor.[4]

In 1867, his father suddenly died leaving a comparatively small estate. Llewellyn, the oldest brother, recently had returned from Heidelberg and was at a London hospital completing his medical studies. Sidney, who was two years younger, decided to seek a position so that his brother might continue with his medical career. He decided that a civil service appointment was safest, and found a place as junior clerk, first in the Marlborough Street Police Court and, the following year, in the Thames Police Court.

The police court was located in the East End of London where Thomas witnessed poverty and human degradation of a kind he never had imagined possible. His beginning salary was only ninety pounds a year but Thomas considered the position temporary. Although his sister insisted that it had not been necessary, Thomas began a regimen of thrift bordering on miserliness. He practiced economy only upon himself, and the pitiful family cases that thronged the court sometimes experienced his generosity firsthand. Thomas himself ate no lunch and walked the distance between his house in Camberwell and the police court in Stepney to save omnibus fare. When he was ready to strike out on his chosen path, he did not want to be handicapped by lack of funds.[5]

Thomas had studied some law, which he considered essential to his position. But science, his early interest, still attracted him more, and, in spite of the grueling days at the police court, he decided to study science independently. Thomas fitted out a small laboratory on the top floor of his house, and by 1870 he felt that he had advanced enough in his work to attend a series of lectures at Birkbeck Institute given by the chemist George Chaloner. During one lecture, Chaloner remarked that the man who one day eliminated phosphorus in the Bessemer converter would make his fortune. That pronouncement fired Thomas's imagination and it became his consuming desire to solve the problem. He set up an improvised converter in his fire grate, but soon found it impossible to produce enough heat to obtain a Bessemer blow, although he had no difficulty in setting the room on fire repeatedly. As Thomas continued experimenting and needed more equipment, Chaloner permitted him to use the college laboratory.[6]

Thomas's search for the solution to the problem of phosphorus continued for a number of years. His daytime position was onerous; frequently the court handled as many as 1,000 cases a month. Thomas was obliged to work at night and during weekends and had little time for his experiments.

In addition to his teaching duties, Professor Chaloner was editor of the magazine *Iron,* and Thomas contributed regularly to its columns.[7] He wrote about any phase of the subject of metals that interested him. Unlike Holley, who enjoyed writing, Thomas avowed that he had no "taste for the pen." In contrast also to Holley, he detested drawing, but commented that "it was probably the best of educational processes to work at a subject that one disliked." As he grew older, his interest in books increased. His reading followed his interests in art, travel, and literature, as well as in science. He expressed a lack of interest in poetry, but his sister noted that he had read more than he would admit.[8]

With so little free time, Thomas planned his holidays carefully. In 1874 he arranged his vacation for August both to attend meetings of the British Association in Belfast and to visit steel plants. His cousin, Percy Gilchrist, was chemist at the Cwm Avon Works in Glamorganshire, Wales, and Thomas spent part of the vacation with him. At that time he visited the Siemens's plant for the first time and spent more than five hours watching the operation of an open hearth furnace.[9]

During the entire winter of 1875, he continued work on dephosphorization, and, although nothing tangible was accomplished, a solution was not far away. For five years every free moment had been devoted to the problem. During these years he had amassed everything that then was known about the behavior of phosphorus in the converter. The question remained of why the phosphorus could not be removed from the Bessemer converter separately from the metal. Other elements could be burned out of the charge or eliminated with the slag, but not the phosphorus. In fact, the percentage of the unwanted element even seemed to increase in the molten metal. This last thought may have given him a clue to the answer. Thomas turned his attention from the contents of the converter to its lining. The lining always had been one of the most frustrating parts of the Bessemer process, principally because it burned out so quickly from the intense heat. Consequently, the operators of the converter were more concerned about the heat resistance of the lining material than about its chemical structure.

Thomas now began to analyze refractory linings, and, although he found the chemistry varied considerably, one fact became increasingly evident. The lining was acid in content. When the phosphorus in the Bessemer charge was oxidized rapidly during a blow, it formed phosphoric acid. It was known that lime had an affinity for phosphoric acid. However, it could not be added to the bath of molten metal in a converter lined with silicious brick because the lime would attack the lining and disintegrate the brick. Since the lining was essentially acid and had no chemical affinity for the phosphoric acid created in the charge, the two could not combine to be driven off in the slag. Thomas was certain that if the lining could be changed to a basic substance that would combine with the acid charge, the difficulty would be resolved. He had worked on the problem for months, but his thinking had not yet passed the theoretical stage. More than anything, at this point he needed access to a steel works where actual experimentation could be conducted.

He turned for assistance to his cousin, Percy Gilchrist, the only person he could trust who might be able to carry on experiments in a steel plant. Thomas was trapped in London and could communicate his ideas and enthusiasm to Gilchrist only by mail, and since Gilchrist was only mildly interested, little was accomplished.[10]

Thomas spent holidays of the summer of 1876 in Germany where he traveled from one steel plant to another and compared the practice in Germany with that in Britain. He took long walks through the mining areas in and around Freiburg. Like Holley, Thomas turned all of his experiences into articles, which later appeared in *Iron*. His interest in anything unusual took him into the silver and zinc mines nearby. He wandered into Bohemia and then into central Germany to see the mining and metalworking operations there, and he visited and compared the various schools of metallurgical education.[11]

During the fall of 1876, he managed to attend the meetings of the Iron and Steel Institute at Leeds. At about that time, Gilchrist left the Cwm Avon Works to take a position as analytical chemist at the Blaenavon Iron Company. Up to that time Gilchrist had been haphazard in his experiments but Thomas persisted in his efforts to persuade his cousin of their importance. The next summer, Gilchrist began to experiment in earnest. A small smithy shed lay idle on a hillside near the Blaenavon plant, and Gilchrist conducted secret tests there after hours and on weekends. In London, Thomas continued in his police court position, but chafed at the bond that prevented him from joining his cousin. He wrote Gilchrist constantly as ideas grew and multiplied, and almost every weekend the two men sweated through innumerable blows in the little shed. As the experiments progressed, Gilchrist depended more and more on his cousin, and seemed to feel most hopeful if Thomas were standing by. Although they occasionally obtained promising results, the scale on which they were forced to operate was much too limited.

Because it was difficult to maintain a constant, high temperature in the small converter, which held only six pounds of metal, the results were erratic. Thomas had never paid much attention to eating or sleeping, and now, he thought even less about his health as he struggled to do his job, write articles for *Iron*, attend scientific meetings, and at the same time continue to think about the progress of the experiments. Often he took the midnight train to Blaenavon on Thursday, spent the weekend working in the shed and returned barely in time to be present when the court opened on Monday morning.[12]

Thomas searched the literature of all the technical societies, half fearful that someone else might have arrived earlier at a satisfactory solution. Tediously he scanned every patent issued prior to 1856 that might in any way touch on the problem he and Gilchrist were attempting to solve. Then, with the help of his sister, they reviewed them a second time. He became a veritable mine of information about the work that had been done on the purification of iron and steel. Thomas was aware of the importance of taking out a patent that could not be circumvented. "I am afraid to claim all basic linings," he wrote Gilchrist, "for the simple reason that if you claim anything that is not both new *and* successful, you lose all. Patent law, the more you study it, gets more full of pit falls. Provisional protection doesn't prevent another fellow stepping in and securing a patent before you, though you may have been protected for months before his application — whence the necessity of excessive caution now."[13]

Ten years of frugality had enabled Thomas to save approximately £800. While the sum seemed large and the experiments up to now had been conducted in a modest way, expenses were mounting steadily. It was becoming evident that there would not be enough money to patent the process in as many foreign countries as would be necessary.

In spite of the efforts of Thomas and Gilchrist to keep their research a secret, even a casual observer could have sensed that something was afoot. After the two had been experimenting for nine months, the manager of the Blaenavon Iron Company, Edward P. Martin, could restrain his curiosity no longer and said to Gilchrist, "I know you young men have some secret work on hand, I think it would be well if you put confidence in me." Gilchrist told Martin about the project and went so far as to give him an analysis of their work, which interested the manager immediately. Martin had a small kiln built to make the special brick, and a converter holding 300 to 400 pounds was placed at their disposal. Martin also bought a share in the proposed patent, which to some extent solved the problem of money.[14]

Thomas now wrote numerous letters to Gilchrist and to Martin describing the progress being made by his patent attorney, William Carpmael,

Sidney Gilchrist Thomas at thirty. From R. W. Burnie, ed., *Memoir and Letters of Sidney Gilchrist Thomas, Inventor* (London: John Murray, 1891), frontispiece.

in London. His sister was pleased to become her brother's secretary. For six pence an hour, she kept his letter books, wrote letters, and copied specifications and plans. Thomas wanted to secure patents in all of the steel producing countries at the same time that he obtained his English patent. He filed the English patent request on 23 November 1877. The patent was granted much more quickly than he had anticipated. He wrote to Gilchrist about this exciting event in an offhand manner. "You may be interested to know that I am in physical possession of No. 1 with its 1 lb. of wax seal." The foreign patents were obtained as he had planned and as he also had anticipated, they cost a great deal of money. Only the German and American patents were to cause trouble.[15]

Although Thomas had attended meetings of the Iron and Steel Institute on several occasions, he was not a member. Now that he was in possession of his first patent, he began to think of attending the March 1878 meeting as a member. The distinguished William Menelaus, head of the Dowlais plant, signed the young man's application for membership. Thomas considered presenting a paper at the meeting, but on reflection decided that he would be wise to wait until the patents were secure before making any public announcement.

At the spring meeting, a paper dealing with the separation of phosphorus from pig iron in a furnace lined with oxide of iron was read by Lowthian Bell. To Thomas's surprise and dismay, during the discussion of dephosphorization that ensued, George Snelus, an outstanding chemist and metallurgist who was manager of the West Cumberland Steel Works, commented that about six years before he had taken out a patent that covered the use of lime in the furnace lining. He had said nothing about it at meetings since he still was working on the method, and with his other activities had not had the time to go into the matter on a commercial scale.[16]

In a confidential report to the Bessemer Steel Company, Ltd., Holley wrote, "Had not Snelus been in the employ of a company specially interested in keeping the Bessemer trade on the west coast of England he would now occupy the place of Thomas. He [Snelus] worked out the problem as a chemist and knew perfectly well the full value of his invention, but being afraid of antagonizing his employers, the West Cumberland Company, he dared not fully explain the purport of his discovery."[17]

Although Thomas had intended not to talk about his work at the March meeting, in the light of the Bell paper and the Snelus remarks, he could not refrain from asking to be recognized.

It may be of interest to members to know that I have been enabled, by the assistance of Mr. Martin at Blaenavon, to remove phosphorus from the

Bessemer converter. Of course this statement will be met with a smile of incredulity, and gentlemen will scarcely believe it; but I have the results in my pocket of some hundred and odd analyses by Mr. Gilchrist, who has had almost the entire conduct of the experiments, varying from the very small quantity of six lbs. up to ten cwt., and the results all carry out the theory with which I originally started and show that in the worse cases twenty percent of phosphorus was removed and in the best I must say 99.9 was removed; and we hope that we have overcome the practical difficulties that have hitherto stood in the way.[18]

Whatever reaction Thomas expected to his statement, he must have least expected that there would be no reaction at all, inasmuch as in 1871, Lowthian Bell himself had said that the limit to the production of Bessemer steel was the want of phosphorus-free ores.[19] A later biographer of Thomas was present at the meeting and described the incident.

I doubt if anyone in the hall knew either by sight or name the young man who had dared to rise in the august body and claim that he had accomplished that after which the great metallurgists of Europe had been unsuccessfully striving ever since Mr. Bessemer's great discovery. I remember distinctly the pitying smile of derision and the stoney stare which pervaded the countenance of the distinguished assemblage. No one thought it worthwhile to refer to Mr. Thomas' claim or ask him how he did it except that Mr. Bell in replying to criticisms on his paper remarked that with regard to what Mr. Sidney Thomas hoped to do with the Bessemer converter, he was so much interested in freeing iron from phosphorus that he would hail as a public benefactor, any gentleman who would come forward and do the work more perfectly or more economically than *he* had been able to effect this subject himself.[20]

Another writer also remembered the occasion. "The meeting did not laugh at the youthful Eureka, nor did it congratulate the young man on his achievement, much less did it inquire about his method of elimination. It simply took no notice of his undemonstrative announcement."[21]

Thomas himself was not particularly disturbed by the lack of reaction, since he was eager to have the patent situation solidly fixed before the process attracted much attention. The money he had saved so carefully now was being used primarily in securing additional foreign patents; he already had spent £300 for that purpose.

During the summer the cousins worked on the paper that Thomas intended to read at the September meeting.[22] Even though the paper was not read, Thomas experienced a stroke of luck. As always, during the course of these meetings, trips were made to outstanding steel works. With a large group from the Paris conference, he visited the great

Schneider plant at Creusôt, about 250 miles southeast of the capital. On the trip, Thomas met E. Windsor Richards, general manager of Bolckow, Vaughan & Company, one of the largest English steel companies, located at Middlesbrough in the famous Cleveland district. With his infectious charm and enthusiasm, Thomas attracted the attention of Richards, and the conversation soon turned to Thomas's work. As a consequence, the two met again in Paris and discussed the possibility of installing the process in Middlesbrough.

Upon his return to England, Richards gained permission from his board of directors to explore the process further, and, along with Dr. Jeremiah Stead, the plant's consulting metallurgist, went down to Blaenavon to see the converter in action. After watching three casts in a small converter, Richards was convinced of the worth of the process, and upon his return to Middlesbrough, immediately made arrangements to conduct a series of experiments at Bolckow, Vaughan. Percy Gilchrist resigned his position at Blaenavon and joined Richards's company to supervise the trials.

For many months Richards struggled to operate the process. The principal difficulties were with the brick lining, which shrank in the drying and burned out under the high temperatures. After patient experimentation, a proper basic brick finally was developed, and the first official blow took place on 4 April 1879. The success of the new process excited great curiosity. The pair of two-ton converters ran steadily for an entire month without requiring a change of lining. Pig iron containing an average of 1.40 percent phosphorus was used as a charge, and the resulting product contained only 0.25 percent phosphorus. There was good reason for observers to be impressed.[23]

Thomas meanwhile still was systematically filing applications for patents. After the first in 1877, there had been more each year. Each patent increased the scope of coverage and lessened the possibility of piracy abroad.[24] The news of the impressive results of the process as demonstrated at the Bolckow, Vaughan plant spread, and soon the company began to receive inquiries from leading steel companies all over the world. Interested foreign companies discovered that it was necessary to deal with Thomas himself.[25]

The development of the Thomas process was responsible for a complete turnabout in the European supply of ore suitable for use in steelmaking. At that time it was estimated that only 10 percent of the known ore deposits in Europe could be utilized. With the coming of the Thomas process, the high phosphorus iron ore deposits of Lorraine and Luxembourg became the most important in Western Europe. Similar deposits in Normandy, north Germany, and the English Midlands also came into their own.[26]

NOTES

1. *Memorial of Alexander Lyman Holley* (New York: American Institute of Mining Engineers, 1884), p. 22.
2. F. W. Harbord, "The Thomas-Gilchrist Basic Process, 1879-1937," *Journal of the Iron and Steel Institute* (London) 136, no. 2 (1937): 78P.
3. Sir I. Lowthian Bell, "Chemical Phenomena of Iron Smelting," *Journal of the Iron and Steel Institute* (London) 1 (1871): 85, 277.
4. Lilian Gilchrist Thompson, *Sidney Gilchrist Thomas* (London: Faber and Faber, 1940), pp. 29, 31.
5. Ibid., pp. 34, 41.
6. Ibid., p. 47.
7. Ibid., p. 56.
8. Ibid., pp. 56, 58.
9. Ibid., pp. 60-61.
10. Ibid., pp. 67-68, 69.
11. Ibid., pp. 74-92 passim.
12. Ibid., p. 103.
13. Ibid., pp. 107-8.
14. Harbord, "The Thomas-Gilchrist Basic Process," pp. 79-80P.
15. Thompson, *Sidney Gilchrist Thomas*, pp. 112, 115.
16. W. T. Jeans, *The Creators of the Age of Steel* (New York: Charles Scribner's Sons, 1884), p. 303.
17. George W. Maynard, "Introduction of the Thomas Basic Steel Process in the United States," *Transactions of the American Institute of Mining Engineers* 13 (1885): 283.
18. W. T. Jeans, *Age of Steel*, pp. 303-4.
19. Ibid.
20. George W. Maynard, "Biographical Notice of Sidney Gilchrist Thomas," *Transactions of the American Institute of Mining Engineers* 13 (1885): 784-86.
21. W. T. Jeans, *Age of Steel*, p. 304.
22. Maynard, "Thomas Basic Steel Process," p. 285. Although Thomas's paper was not read, advance copies had been distributed, and a résumé had appeared in *Engineering*. Obviously Holley sensed the possibilities of the invention since he took the time to write a note to George Maynard in London and send him a copy of the Thomas paper. Holley suggested to Maynard if, upon examination, the substance of the paper looked good to him he might get control of the process for the United States.
23. Thompson, *Sidney Gilchrist Thomas*, pp. 133, 137.
24. When Thomas obtained his foreign patents at the outset, he was profiting from the mistakes of his predecessors. He had observed what had happened to both Bessemer and Siemens on the Continent. The Siemens experience in Prussia was typical of what might befall an inventor. J. S. Jeans described the Siemens incident as follows.

Curiously, however, Dr. Siemens was refused a patent in Prussia for his regenerative furnace on the ground of its resemblance to a particular form of medieval warming apparatus. This refusal is all the more curious from the fact that only one building is known in which this form of warming apparatus has been employed. It occurs at the palace abbey, or preceptory at Marienburg, in Prussia, which formed the headquarters of the Teutonic Knights, and is supposed to belong to the latter half of the fourteenth century. This apparatus was applied to the warming of two rooms in the building in question. A fire was made in the lower part of the furnace, and the products of combustion passing through the stones placed in the upper division, escaped into the flue. When the stones had become thoroughly heated the fire was extinguished and the flue closed by a damper. The apertures in the floors of the apartments to be warmed being now opened, cold air was allowed to pass through the heated stones, and becoming warmed entered the floor of the rooms through the apertures before mentioned (J. S. Jeans, *Steel: Its History, Manufacture, Properties, and Uses* [London: E. and F. N. Spon, 1880], p. 104).

25. An amusing story, often repeated, was an outgrowth of that circumstance. Shortly after word had spread about the successful demonstration of the process at Middlesbrough, two Belgian manufacturers set out for England to secure the rights. They met on the channel steamer, but neither divulged to the other the reason for his trip. Arriving at a very early hour in the morning, they both went to the Royal Hotel on the Embankment. One Belgian decided to take a short nap and have breakfast before approaching Thomas. The other, as soon as he had left his traveling companion, engaged a hansom and by seven thirty was seeking admittance to the Thomas residence in Battersea. At the end of three hours, the men were in the final stages of the agreement when a telegram arrived from the other Belgian announcing that he would arrive by noon. He arrived only to see his competitor leaving with the monopoly for the process for his district secure in his pocket.

26. Thompson, *Sidney Gilchrist Thomas*, pp. 272-73.

XXIII · Bessemer Process versus the Open Hearth

olley carried the news of the Thomas-Gilchrist invention back to America when he returned in October of 1878. Very soon afterwards Andrew Carnegie became keenly interested in it. The following March, a month before the official demonstration at Bolckow, Vaughan, Thomas received a letter from Carnegie asking permission to visit the plant. Thomas forwarded the request to Windsor Richards, who replied that he was willing to permit a visit by Carnegie, but cautiously asked Thomas how much information it was proper to give Carnegie.

Richards thought that the visitor should be shown the metal as it was tapped from the blast furnace and be given a piece of it. Then he could observe the blow and be given a sample of the steel. Any detailed information should be given by Thomas only after Carnegie had agreed to take the license.[1] Thomas agreed and was equally cautious in regard to other visitors. He warned Richards to keep everything close, particularly "the overblow, cement and bottom" until the time set for the demonstration in April. He suggested that Richards might give as the reason for withholding certain information the danger of prejudicing the foreign patents.[2]

In May 1879 in London, Thomas delivered the paper that had been scheduled for the Paris meeting of the British Iron and Steel Institute. It was the largest meeting in the institute's history. Many leading metallurgists from the Continent swelled the ranks of the English who attended. Thomas received a reception far different from that on the earlier occasion when he nervously had made his modest statement.

Now Thomas was confident that he could rely for a livelihood entirely on the income from his invention, and on 10 May 1879, he resigned his post at the Thames Police Court. His time thereafter was taken up entirely by the difficult patent negotiations, which had to be carried on both at home and abroad and for which he was solely responsible, since it had been agreed that Gilchrist's only function as a partner would be developing the process. Thomas worked out a number of flexible licensing arrangements. His biographer commented that, "In some

327

countries and districts he sold his rights, in others he conceded licenses to individual ironmasters: in others again, he appointed agents to receive royalties."[3]

Securing the patents in Germany proved to be both troublesome and time consuming for Thomas. The north German ironmasters, especially the Krupp firm, would benefit greatly from the new process. When he learned that they were determined to contest the German patents for the process, Thomas set about to prepare himself for the struggle. With his customary care, he familiarized himself with both German law and the German language, so that he could work closely with his lawyers there. During the course of the controversy, he made frequent trips to the Continent. Relying on his great influence and financial power, Krupp in turn carried the fight to England.

But Krupp did not prevail. On 22 November 1879, after a hard legal battle, the German patent court held that the validity of the Thomas patents had been established completely. Thomas was gratified that his German patents had not shared the fate of those of Bessemer and Siemens. It was, however, a cruel blow to the German producers, inasmuch as they needed the process in order to utilize their high phosphorus ores and had hoped to use it without payment.[4]

Thomas also encountered difficulties when he filed for his patent in the United States, mainly because a process that was said to be similar already had been patented there. In a statement at the March 1878 meeting of the British Iron and Steel Institute, Snelus had claimed results similar to those of the Thomas process. He had obtained an English patent for the use of a lime lining in the converter. But Snelus, who was then an employee of the West Cumberland Steel Company, had not gone on to perfect his method of eliminating phosphorus from the converter because, as he said later, "I had just taken management of a concern, the interest of which was opposed to the solution of the problem."[5]

A later writer explained Snelus's statement in this way.

> The works in the Cumberland district had their rich 50% hematite iron ore deposits at their door, and at that time could make acid Bessemer steel from this local ore cheaper than by transporting ore or pig iron from Northhamptonshire, Lincolnshire, or Cleveland. The other large works in South Wales and elsewhere, were, it is true, by this time largely dependent upon imported ore, but most of them had acquired an interest in high-grade ore deposits in Spain and under the cheap conditions of mining and transport then prevailing, could produce acid Bessemer steel cheaper than they could basic Bessemer from ore or pig from the Midlands, Lincolnshire, or Middlesbrough.[6]

Snelus had realized in 1872 that, in principle, he had solved the problem of eliminating phosphorus from the converter and had described the procedure to a friend, Edward Cooper of Cooper & Hewitt. Cooper had urged him to take out an American patent, but Snelus's desire to perfect all details nearly lost him his rights. He disregarded his friend's advice until Thomas announced his discovery in 1878. Snelus then joined Cooper & Hewitt in taking out an American patent, which secured him a share in the American royalties.[7]

In connection with the American patent, Holley wrote Snelus in December of 1878.

> Since I sent you patent papers from Cooper & Hewitt to be executed, our attention has been called to a scheme by Thomas and Gilchrist to patent and work their process here. We think it very important for you to claim the use of loose lime in the converter or furnace if you can fairly do so. Your English patent refers only to lime linings, and your American specifications based on this had better stand as it [sic] is. If you invented the use of lime in other forms than the lining of the vessel, we think it would be best for you to make a new specification for such use and to send it to us for examination. We will add amendments if any occur to us and send to you for execution.[8]

Thomas's application for an American patent conflicted not only with the Snelus patent, but also with a patent for a process that used petroleum as a binder for the bricks of converter linings. Fortunately, the three men agreed that the situation would be served best by arbitration. Sir William Thomson, afterwards Lord Kelvin, already one of the world's great scientists, was asked to settle the question of how the profits from the British and American patents should be divided. His decision was accepted without question and the result was an amicable settlement among friends. It was decided that Thomas was entitled to the sole rights on the Continent, but that the rights should be shared in the United Kingdom and the United States.[9]

Even though he had become deeply interested in the Thomas process, Holley had returned from Europe more enthusiastic than ever about the future of the open hearth process. John Fritz, who had been traveling with Holley in Europe, was about to begin the construction of an open hearth installation at Bethlehem. Whether or not he would choose the Pernot revolving hearth feature was still being discussed. Holley described the merits of the process in a letter to Lowthian Bell.

> As to the comparative merits of the Bessemer and open hearth as they now exist, I can only say that if the open hearth has a fair chance, for instance if a plant of Pernot twenty ton furnaces with a premelting and prerefining plant were set up alongside a first rate Bessemer plant in a region of ores of average impurity that the open hearth would best the Bessemer. And certainly when all the possibilities of the open hearth are as worked out as

those of the Bessemer have been, the open hearth seems to have every advantage
if that of cheap ores are [sic] also taken into consideration.[10]

Holley's faith in the future of the open hearth process was the faith
of the pioneer. At this time, the relative tonnage produced by the two
processes certainly gave no hint of what the future would bring. In 1878,
Bessemer production in the United States was 732,226 net tons. Open
hearth steel production was 36,126 tons.[11] The price of Bessemer rails
had fallen as production and competition had increased. From an average
of $158.50 a gross ton in 1868, the price of Bessemer rails had declined
steadily until, in 1878, it reached an all-time low of $42.25. Had it not
been for the tariff of $28.00 a ton, the Europeans would have been
able to ship rails to the United States to sell for less than $42.25.[12]

The high tariff had served its purpose in limiting the import of steel
rails. In spite of this protective tariff, only twenty converters in the
ten U.S. Bessemer plants were being used. The new Vulcan works,
which had been opened with such great expectations, was idle. The
plant, one of the most efficient ever built, had been an example of
Holley's highest skill in plant design, and it was natural that questions
were raised about the circumstances of its being shut down.

Hearings were conducted by the Ways and Means Committee in
Washington in February of 1880 on the subject of reducing the high
tariff on steel rails. Henry V. Poor, editor of *Poor's Railway Manual
of the United States,* appeared on behalf of the Illinois Central and a
large number of other railroads. Poor asserted that in order to avert
competition, the other Bessemer steel manufacturers had promised the
Vulcan management that if they would shut down the steel producers
would pay them a larger sum than they could make by operating the
plant. Samuel Felton of the Bessemer Steel Company insisted that anyone
was free to build a Bessemer plant, but the senators questioning him
had some difficulty in getting definite answers to their inquiries. Senator
Frye asked Felton what would happen if he (the senator) approached the
Bessemer group about building a plant. Felton replied, "I presume we
would say, 'If you want to come in, you must come in and pay your
portion of what we paid and be subject to the same royalty.' I presume
we would say that, but I cannot say."[13]

This statement does not coincide with a remark that Holley made
late in 1879 when he wrote to his good friend, Arthur Cooper, manager
of the Bessemer department at Brown, Bayley and Dixon's Works in
England, concerning a hot blast patent that Cooper was trying to sell to
the Bessemer Steel Company. The company had offered Cooper half
the fee he had asked. Holley suggested that it might be wise for Cooper
to take it, since the chances were slim that any more Bessemer plants

would be built. Holley then added that the Bessemer group had just refused a £20,000 bonus for a license.[14]

Later in the hearings, Felton denied Poor's accusation and claimed that the Vulcan organization simply had gone bankrupt. He asserted that the other steel manufacturers had agreed to pay a certain sum to Vulcan in order to tide the company over. He stated that the company had a mortgage of $1,000,000, on which the interest was $70,000 per year, and that the remaining Bessemer steel companies simply came to the rescue of the beleaguered company by paying Vulcan the necessary $70,000. It was understood that Vulcan would resume operations when-ever the demand for rails equaled the supply, including projected Vulcan tonnage. Until such time, Vulcan would receive the money. An interesting consequence of this arrangement was that the decision as to the proper time for Vulcan to resume operations thus rested with the companies contributing the money and not with the management of the closed plant.[15]

In 1879, after a year's separation, Bethlehem decided to reengage Holley as a consultant. This was encouraging to Holley and aroused the hope that business was going to improve. Grumbling continued among the numerous companies who were obliged to make royalty payments on the open hearth process. Morrell of the Cambria Steel Company and Sellers of the Midvale Steel Company were the most restive. In answer to the latest of Morrell's complaints Holley tried again to explain the situation.

I am informed by Cooper [&] Hewitt, that objection to paying the Martin royalty on the part of the Pennsylvania Steel and others have been so fully answered by their counsel that there is a disposition to pay without further contest. When this matter came up before I told you that my opinion had been that the Martin patent could not be sustained in view of the Mushet patent; but since that time the Martin patents had been sustained in the courts of France against Creusot and that the Otis, Bay State, Norway and several other works had after investigation abandoned their proposed legal contest and paid the royalty. I am aware that the many royalties are onerous and vexatious and I was in hope that we might get rid of some of them but there seems no chance of avoiding any of them. Mr. John Fritz makes great objection to the Siemens royalty but thinks the Martin all right and very reasonable seeing that it has been reduced from time to time as the term of the patent gets shorter. My impression is that you will have to pay the Martin royalty at least and that you would have to pay more after a suit than before it.

But your position is peculiar. You have, according to the correspondence in my letter book agreed to pay the Martin royalty and to make no contest in consideration of a free license to use the Post and Pernot patents for two furnaces.

I would gladly do whatever was practicable to save you from these heavy royalties but my belief is that you will have to pay more in the end if you make a contest. You will remember that two or three years ago I had a very careful legal investigation made of the status of the Siemens patents in the belief that they would expire when the British patents expire but by an ill considered Act of Congress that law was changed in such a way as to allow the patents of that year to run seventeen years from their date in this country.[16]

The managements of member companies of the Bessemer Steel Company were suspicious of each other and now also of Holley. They were particularly sensitive to what they characterized as Holley's excessive fraternization with outside companies. In a letter written in the spring of 1879, Holley said, "Some of my Bessemer clients are going for me because I am making the Martin [open hearth process] so dangerously attractive to their rivals." Holley did not respond to the charge in that letter.[17]

Since 1874 Holley had issued a series of reports on European steel-making practices. The reports were prepared solely for the use of his clients, and they were both detailed and thorough. They demonstrated his ability to secure operating figures and other types of information that usually are held as company secrets. For example, "The Direct Use of Blast-Furnace Metal in the Bessemer Process" began with a description of the practice at the West Cumberland plant in the Middles-brought district, covered the plants at Barrow, Dowlais, Ebbw Vale; and described the plants at Creusôt, Terre Noire, and Saint Chamond in France as well as one at Seraing in Belgium. Frequently, as for the West Cumberland company, he was able to include production and cost figures, items that companies are loath to disclose widely even within their own organizations. Many reports included detailed drawings of new and different apparatus.[18]

In another report, Holley wrote, "Wilson, Cammell and Company had made themselves unpleasantly notorious by their unusual measures for keeping their personal friends out of their works and for preventing the peculiar details of their plant from becoming known. I take especial pleasure in presenting my clients as complete working drawings of more important features of this plant as could be obtained if works and drawing office were thrown open to their inspection."[19]

The editor of *Engineering* later wrote of Holley's European relationships.

> The gates of every steel-works in Europe were open to him, and from him no manufacturer's secrets were withheld. It was quite a unique position which he then occupied; one for which his long training, high intelligence, and perfect integrity absolutely fitted him. Manufacturers knew well that their secrets were safe in his hands; that if they gave him much, he brought still

more to them; that his visits were always useful and suggestive, and that the exchange of American experiences for their own never left Holley their debtor. There was also another aspect to these annual visits. Time, hard work, and the cares of a busy life had never withered the great charm of Holley's char-acter; but on the contrary, his personal influence and indefinable power of attraction had increased with the years. He was loved as much as he was admired; he was as popular in France, in Germany and in Belgium as he was in England or the United States.[20]

The early months of 1879 were troubled ones for Holley. Mary Holley became dangerously ill with pneumonia. Holley feared she might not recover but after many weeks she began to mend. He lost an old friend when the mild and genial David McCandless died. As chairman of the J. Edgar Thomson Steel Company, McCandless had brought dignity and standing to the organization. With his passing, the original group that formed the company had almost disappeared from the steel plant's rolls. First to depart had been Coleman; Kloman had followed; and shortly after him, Colonel Scott. The subsequent death of Edgar Thomson virtually had wiped out the pioneering group.[21]

At the time of McCandless's death, Shinn, who had been the active manager of the company and was responsible for a good share of the organization's success, fully expected to take his place. Carnegie had written Shinn in 1876 that there were few nights in which, before going to sleep, he did not congratulate himself on his good fortune in having Shinn. But Carnegie evidently had changed his mind, and now insisted that the office should go to his younger brother Tom. Infuriated, Shinn withdrew from the company in September of 1879. A long legal duel ensued between Shinn and Carnegie over the value of Shinn's stock. The matter finally was settled in 1881 when a board of arbitration ruled unanimously in favor of Shinn. Holley regretted the departure of his close friend and business associate from Edgar Thomson.[22]

Shinn went to Vulcan as general manager and did his best to take Jones with him. Of this episode Carnegie's biographer writes, "As an added bonus for engaging him Shinn assured Vulcan that he would bring with him Captain Bill Jones, the genius of the E. T. Steel Works. There was a certain ironic humor in the fact that Vulcan, which had been paid a bonus to stop rail production for a year, and that largely at Carnegie's urging, should now plan to go back into production by hiring his general manager and his plant superintendent. Carnegie failed to see the humor in the situation."[23]

Taking advantage of Shinn's offer Jones, who for some time had been dissatisfied with his salary, again wrote Carnegie that he would prefer to stay with Edgar Thomson, but that he needed more money. The threat of losing Jones was frightening to Carnegie and he offered Jones a partnership. Jones was not interested; he was quite content with his position as plant superintendent, but "a hell of a big salary" was his aim. After some discussion, Carnegie decided to pay him $25,000 a year, which was at that time also the salary of the president of the United States. After years of niggling over a pay increase, Carnegie, with no replacement in mind for his superintendent, finally had to give in to Jones's salary demands.[24]

NOTES

1. Carnegie took an option on the purchase of the American rights to the Thomas process before he left England late in the summer of 1879. The majority of the Bessemer patents were due to run out shortly, and Carnegie may have believed that the purchase of the Thomas patents would both bolster the dwindling patent holdings of the Bessemer Steel Company and prevent the Thomas rights from falling into other hands.
2. Lilian Gilchrist Thompson, *Sidney Gilchrist Thomas* (London: Faber and Faber, 1940), pp. 140-41.
3. Ibid., p. 143.
4. R. W. Burnie, Ed., *Memoir and Letters of Sidney Gilchrist Thomas, Inventor* (London: John Murray, 1891), pp. 115, 133.
5. W. T. Jeans, *The Creators of the Age of Steel* (New York: Charles Scribner's Sons, 1884), p. 328.
6. F. W. Harbord, "The Thomas-Gilchrist Basic Process, 1879-1937," *Journal of the Iron and Steel Institute* (London) 136, no. 2 (1937):85P.
7. W. T. Jeans, *Age of Steel*, p. 329.
8. ALH to George James Snelus, 3 December 1878, Holley Family Papers (hereafter cited as HF).
9. Thompson, *Sidney Gilchrist Thomas*, pp. 147-48.
10. ALH to Lowthian Bell, 18 October 1878, HF.
11. *Annual Report of the American Iron and Steel Association*, (Philadelphia: AISA, 1879), pp. 23, 25.
12. *Annual Report of the American Iron and Steel Association*, (Philadelphia: AISA, 1878), pp. 67-68.
13. *Hearings by Ways and Means Committee on Covert Bill, February, 1880* (Washington, D.C.: U.S. Government Printing Office, 1880), p. 13.
14. ALH to Arthur Cooper, 9 September 1879, HF.
15. *Hearings, Covert Bill*, p. 60. The hearings may have had some effect, since Vulcan resumed operations on 10 March 1880, but more than a year elapsed before they achieved complete use of all of their facilities.
16. ALH to Daniel Morrell, 13 December 1878, HF.
17. ALH to Rossiter Raymond, 17 April 1879, HF.
18. Alexander Lyman Holley, "The Direct Use of Blast-Furnace Metal in the Bessemer Process," Private report, no. 2, 2d. ser., New York, June 1877.
19. Alexander Lyman Holley, "The Rail Mill and General Plant and Practice of Wilson, Cammell & Co.," Report to the Bessemer Steel Company, Ltd., no. 3, 1880.
20. James Dredge, "The Holley Memorial," *Journal of the Iron and Steel Institute* (London) 37 (1891): 381.

21. James Howard Bridge, *The Inside History of the Carnegie Steel Company* (New York: Aldine Book Co., 1903), p. 117. The percentage of Edgar Thomson Works stock owned by Andrew Carnegie increased with the departure of each partner. As Shinn once remarked, Carnegie had developed very early "a sentimental desire to have an even half."

22. Ibid., pp. 121, 123-24.

23. Joseph Frazier Wall, *Andrew Carnegie* (New York: Oxford University Press, 1970), pp. 358-59.

24. Ibid., p. 359.

XXIV · The Clouds Begin
to Gather

lthough Holley was far from well, he persevered with all of his usual energy in his work for his clients investigating the efficiency and possible future of various new processes. He also continued his activities for the American Institute of Mining Engineers.

For years Holley had been calling attention to the impossibility of making steel and expecting consumers to use it without knowledge of its physical properties. Some years earlier, the American Institute of Mining Engineers had presented to Congress a memorial written by Holley, which requested that a United States Board for Testing Structural Materials be set up. The request included a petition for a sum adequate for the construction of a testing machine. In 1875 a board was created to which Holley was named, along with Lieutenant Colonel T.T.S. Laidley, U.S. Army, president; Commander L. A. Beardslee, U.S. Navy, chief engineer; W. Sooy Smith, civil engineer; and R. H. Thurston of Stevens Institute, secretary.

The new board's primary problem after it had been granted the authorization, was to find means of conducting tests. Various kinds of machines for testing structural materials had been built and used before this time. Riehl Brothers of Philadelphia, long a manufacturer of scales, had made a testing machine as early as 1866. These machines were of importance in the early 1870s for testing the materials for the renowned Eads Bridge at St. Louis.

The rigidity of the standards for the specifications that the board formulated for the new machine were previously unheard of. About $43,000 was allotted for its development, and the board engaged Albert H. Emery to design it and to supervise its construction. The design for a 400-ton machine was ordered in June 1875. However, there were numerous delays before one was built that was satisfactory to both the board and the designer. On 8 February 1879, it finally was installed at Watertown Arsenal, New York.

While they awaited completion of the machine, committees of the board worked on special problems that could be carried out on other machines that already were available. A complete chemical laboratory was set up at Watertown Arsenal, and Andrew Blair was appointed chemist. He was responsible for 213 complete analyses of irons and steels and 249 analyses of alloys. Commander Beardslee, with other members of his committee, completed and published the results of what Holley termed "the most exhaustive and important series of experiments ever made on chain cable and on wrought iron generally."[1] Using the testing machine of the Navy Department, the Beardslee group conducted more than 2,000 tests. In Holley's estimation, Professor Thurston made the most complete series of experiments on record concerning bronzes. Other members helped to conduct experiments on tool steels, beams, and structural steels.

When the long-awaited Emery testing machine was placed in use, its results were a revelation. For the first time, a testing device was capable of subjecting iron and steel to the cruel punishment that these metals undergo in service. The committee stood fascinated as it watched the machine go through its paces. A forged link of hard wrought iron five inches in diameter slowly was stressed, and, when the tension had reached 722,000 pounds, the metal broke with a loud report. In order to determine whether the weighing parts of the machine had been affected by the recoil in the prior demonstration, the next test involved instead of a large piece of metal, a horse hair 7/1000 of an inch in diameter. Stretched 30 percent of its length, it broke at one pound tension. The machine was tried out for some time by alternating tests of heavy pieces that were subjected to as much as 1,000,000 pounds of compression, with tests of delicate objects such as eggs and nuts.

The machine was built by the Ames Manufacturing Company at Chicopee, Massachusetts. The Nashua Iron and Steel Company supplied the forgings, and the South Boston Iron Works poured the principal castings of gun iron. Holley was very proud of the extraordinary delicacy with which the machine could be operated.

The information gained from it was constantly surprising. At last the strength of large bars could be measured, and construction engineers no longer found it necessary to predicate the physical qualities of large bars from those of small ones, a method which, Holley mused, "was a little like exhibiting a brick as a measure of the strength of a wall."[2]

While Congress had provided for building the machine, it had lacked the foresight to earmark money for carrying on experiments. Having contributed uncounted hours without remuneration and with no prospect of such funds in sight, the United States Board for Testing Structural Materials went out of existence on 30 June 1879. Holley's anger and frustration knew no bounds. He considered the failure to understand the future value of the machine as errant stupidity beyond comprehension. Members of the congressional committeee involved with the project did say that anyone might send material to Watertown Arsenal and have it tested at cost—provided, Holley observed sourly, there was someone there to test it.

The development of transportation facilities had brought with it the need for long span bridges of steel but the state of knowledge regarding steel qualities was such that no one really knew what grades of steel were required to withstand the various stresses. Despite the apparent opinion of the committee members that, if the physical properties of a bar were determined, all bars of that size could be said to have the same characteristics, Holley and the men working in the field knew that such thinking was ridiculous, and that the testing of large numbers of bars was required in order to determine the acceptable grades and forms for bridges and other structures.

Engineers were using the new steel rather sparingly, and Holley was certain that he knew the reason why. Until the builders could know the capabilities of the material, they were not going to risk their professional prestige in using it. Holley understood that no individual company or engineer could bear the cost of such experiments. An engineer engaged on a project might order very expensive tests made of a particular steel and the report on them would become his company's property. This procedure could take place repeatedly in numerous companies.

Instead of spending great amounts of money in this fashion and ending up with no permanent answers, why not have the government provide a sum of money—even fifty thousand dollars would be a step in the right direction—to buy material, make structures, and test them under the direction of a board of engineers representing the various fields of construction? The results of such tests would benefit everyone. "The intrinsically ridiculous factor of safety of six to one, half of which, at least, might be called the factor of ignorance—this enormous excess of material which loads down bridges with their own weight and often exceeds the elastic limit of corporate finances—this dreadful incubus could be so largely removed that the same money would span twice the space," Holley told his fellow engineers.[3]

As the months of 1879 passed, the continuing decline in the price of foreign rails offered serious competition to American producers. Samuel Felton, as president of the Bessemer patent holding group, thought it might be wise for Holley to go to Europe to investigate the reasons for the decline. The two discussed at length the cost of making such a trip. Holley estimated that his expenses might run as high as $3,000 and he asked $5,000 as his fee. This sum caused Felton to hesitate, but it was important to have the information. Finally it was agreed that Holley would leave for Europe early in the summer.[4]

Meanwhile, Holley had been appointed to deliver a series of lectures at $100 per lecture at Columbia College, where his good friend Thomas Egleston was head of the School of Mines. Although he had neither the time nor the energy to deliver the thirteen lectures, Holley, with a characteristic sense of duty to a younger generation, spent a great amount of time preparing the series. He arranged for printing the complete outline of his lectures and also made large wall illustrations for them. In order to be sure that the class would be composed of serious and interested students, he took the time to quiz each one who wished to enroll.

The idea of asking Holley to give these lectures had come from the trustees. As a close friend, Egleston knew that Holley would spend much more time and effort on them than his health would permit, and he suggested that Holley refuse. Holley disagreed, "It is my duty, if I am called upon to lecture upon iron and steel to do so." Egleston did not persist. However, he attended the lectures and later said, "There are some who have heard Holley lecture to young men. How earnestly he talked to young men about how scientific and engineering work should be done. They were full of the practice and experience of a great engineer. He criticized freely the work of others and did not forget to blame himself when any of his own practice had been wrong." As he listened to the lectures, Egleston could not refrain from thinking that the young men sitting with him could not possibly appreciate their quality. When the lectures had first been discussed between them, Egleston told Holley that he thought it would be sufficient to present the lectures in an extemporary manner. Holley could not agree and replied, "I am enunciating principles to be remembered. It is my business, my duty, my pleasure to put them down, with my best thoughts, in the best language, and in the fewest words that I possibly can."[5]

In May of 1879, the American Institute of Mining Engineers held its meeting in Pittsburgh. Although he was tired almost to exhaustion, Holley was urged on by his friends and decided to attend. Shinn had invited a large group to his country home outside Pittsburgh for an evening reception, an occasion that was to include a tribute to Holley that had been planned by his closest associates. As evening came on and it began to rain, Holley decided that because of the bad weather and his extreme weariness, it would be wiser for him to go to bed. His friends confided their secret to Mary Holley who told her husband that his friends were disturbed and disappointed that he was not to be one of them. While she knew how ill he felt, she thought that they should go, if only for a short time. Since his wife rarely made unreasonable demands, Holley decided that perhaps he had been lacking in courtesy and, not wishing Shinn's party to be a failure, he said, "Help me to dress and send down word that we are coming."

When their carriage arrived at Shinn's home, it was greeted by cheers.

The Holleys were ushered into the parlor, where he saw displayed on the table an elaborately chased silver pitcher resting on a matching silver tray. Shinn, as master of ceremonies, made a long and ornate speech into which he wove many allusions to Holley's connection with the Bessemer converter and discussed the importance of the part that he had played in the establishment of the Bessemer steel industry.

Holley, taken completely by surprise, found himself so overcome by emotion that for once he had difficulty in saying a word. On an engraved plate attached to the tray appeared the names of the friends who had contributed approximately $1,300 for purchase of the impressive gift. It was a distinguished list. All of the leaders of the steel industry with one or two exceptions were represented.[6]

After a few moments of silence, Holley mustered the words to thank his friends. It would not have been Holley had he not immediately given credit to Bill Jones and all the others for having contributed as much as he to the successful launching of the Bessemer process. He continued in the gracious style that all knew well, reminiscing and joking, but ending on a sad note.

> Among us all who are working in our noble profession and are keeping the fires of metallurgy aglow, such occasions as this should also kindle a flame of good-fellowship and affection which will burn to the end.
>
> Burn to the end! Perhaps some of us should think of that, who are burning the candle at both ends. Ah! well, may it so happen to us that when at last this vital spark is oxidized, when this combustible has put on incombustion, when this living fire flutters thin and pale at the lips, some kindly hand may turn us down, not underblown—by all means not overblown—some loving hand may turn us down, that we may perhaps be cast in a better mold.[7]

When he finished the room was hushed. Both the men and women present were deeply moved. Many had felt that Holley was chanting his own requiem.

Holley now was caught in the net of his own weaving, which made it impossible for him to stop or even to slow the pace of his activities. On May 28, he sailed on the *Gallia* for the long-planned trip to Europe. For three weeks he concentrated his efforts in Sheffield, studying with great care all the improvements that the English had installed in their rolling mills. He found that in many ways the methods used were superior to the American practice but that in the production of Bessemer steel the United States was by far the leader. Holley and Thomas spent long periods of time together watching the latter's process in operation, and Holley became even more convinced that it would revolutionize steel production of the future. He sent his assistant, Laureau, to visit the French plants, while he himself visited Belgium and Prussia on a rapid inspection tour.

Alexander Lyman Holley at the time of the testimonial, about 1879.

Silver pitcher and matching tray presented to Holley by associates in the steel industry. From "Testimonial to Alexander Lyman Holley, May 16, 1879," privately printed.

Holley returned home feeling in better health, primarily because he had been forced to rest during the ocean crossing. He came back to find that the steel industry had recovered and was working at capacity. Seeing prosperity all about him in the fall of 1879, Holley remarked wistfully in a letter that he hoped business would improve for him as well.[8]

Because of the revival of demand and stabilization of prices for steel rails, the decision had been made to reactivate the idle Vulcan works in St. Louis. Holley was called in on his return from Europe to supervise the overhauling of the plant, and Shinn left the Edgar Thomson Works to manage Vulcan. With Shinn at the helm, great things were expected, and Holley tried to induce Robert Forsyth to leave the North Chicago Rolling Mill Company and join Vulcan as superintendent. He told Forsyth that there was no question that Shinn would make a first class go of Vulcan. He would have the money he wanted, and what was even more desirable, no interference in his management. Although Forsyth decided not to leave North Chicago, John Fry from Cambria took the position.[9]

In the same year, Bessemer received his long-awaited knighthood. The year also would see the end of his American patents. The last was due to expire on 13 January 1880, after which time the Bessemer Steel Company would be left with only the Holley and Fritz patents still current. Although Carnegie had taken an option on the Thomas dephosphorization patent, the actual purchase was several years in the future. One of the reasons for the delay was the perpetual spectre, patent priority. Thomas had obtained his American patent on 9 July 1879, but Jacob Reese, a well known Pittsburgh iron manufacturer, claimed priority under certain patents granted him in 1866 covering the same field. Carnegie bought Reese's patents for $15,000 but the Thomas rights still were not taken up. Although Carnegie might have bought the rights himself in order to give the Edgar Thomson Works a monopoly, Thomas wanted payment in cash. Carnegie felt that the amount was more than he could manage at the time, and instead he turned his attention to persuading the Bessemer group to buy the rights. But Carnegie had not yet succeeded in convincing the members of their worth.[10]

Several factors may explain the attitude of the Bessemer group toward Carnegie. The members had little affection for him. After only six years in steelmaking, he was the leader in the business. Many of the rail producers still considered Carnegie a threatening interloper who outbid and undercut them and was not amenable to group action. Whereas the other steelmen had spent long hours in their plants, Carnegie had never found it necessary to soil his hands. Instead, he had taken his well-known coaching trips through England, journeyed around the world, and spent the hot summer months in Scotland. One historian wrote, "Seldom has international prominence been attained with so little exertion."[11]

Carnegie had secured the option on the Thomas American patent in the expectation of a commission from the Bessemer group. Holley also must have expected that outcome when he advised Thomas at the time, "With the converters working overtime, they will pay the price asked in the end and I advise you to humor them."[12]

The Bessemer group also was considering the purchase of the Pernot open hearth patent from Cooper & Hewitt. They were interested only in the rights which would cover rail production, leaving Cooper & Hewitt free to grant licenses for the production of other steel products. The group had taken an option to purchase the patent for $100,000 but had permitted it to expire. Many of the group believed that, rather than purchase the Thomas patent, it was more important to buy the open hearth patent in order to control the growth of competition from that direction. When it was learned that the Bessemer steel group was interested, outside companies began taking out rights in order to protect themselves. Conceivably, Cooper & Hewitt distributed the rights too freely to suit the Bessemer group. In 1880, they would permit the option to lapse for a second time.[13]

Dissension over the Thomas patent continued among the members of the Bessemer Steel Company. John Fritz and some of the more influential men in the industry were in favor of taking on the process, but the company still held back. The Carnegie connection still may have been influencing the group. However, on September 26, Holley was able to write to Thomas.

> I have not heard yet what you propose to do about the proposition of the Bessemer Steel Company. I want to suggest to you confidentially that when Mr. Carnegie's arrangement with you expires—if it is a limited one—you would get on faster by renewing the offer of the Bessemer Steel direct. The Association would then save the commission but you would get as much and get it with less delay. I do not mean to say anything to impair Mr. Carnegie's position. If he can make his commission he is entitled to it. But if he and the Bessemer Company and you can't make business I am quite certain that you and the Bessemer Company can. I work in the interests of my clients and I ask no commission.[14]

Acting on Holley's advice Thomas made a definite arrangement in October whereby the Bessemer Steel Company agreed to pay $50,000 for an option to buy his patent in a year. They then were to pay an additional $250,000.

Thomas remained uncertain of the arrangements where Carnegie and the company were concerned and expressed to Holley his puzzlement regarding the nature of the patent holding group. Holley tried to explain the situation to Thomas.

> The Bessemer Company is a peculiar organization, but as I understand it, a majority may buy a patent, giving the others a chance to come in. I do not think a friend can stop the purchase even if he does not get his commission. However, I hope he will get it and I think he will. Much will depend on the developments of the next few months especially abroad, because I do not think that any of our people will make a demonstration before next summer. They are too full of orders to afford to stop regular work. I shall probably go abroad in March or April to look into all that is doing in lime linings and additions and on my report suppose much will depend. In fact, I do not believe our people will know of their own knowledge when the year of option has expired much more than they know now. They must depend on foreign results. At De Wendel's I hope to find a good practice and I shall then have no difficulty in seeing it. Do you think I can get at the inside of Bolckow Vaughan?[15]

The fall meeting of the American Institute of Mining Engineers was held in Montreal in September 1879. Holley attended and presented a paper on the method of washing pig iron that he had observed at the Krupp works in Essen. By this process, the excessive phosphorus was removed from the pig iron in a Pernot puddling furnace before adding the pig iron to the open hearth charge. Within limits, the process had possibilities and attracted considerable attention for a time.[16]

Although Holley had devoted much time and effort to furthering the American Institute of Mining Engineers, he and other engineers had decided that there was a need for an organization for the mechanical engineer. A preliminary meeting was called for 16 February 1880 by an organizing committee composed of Holley, John Sweet of Syracuse, and Robert Thurston of Stevens Institute. Holley was named as chairman, and in a short address he set forth the proposed scope and character of the society. He pointed out the place of mechanical engineering in the fields of civil engineering, mining, and metallurgy, not only in railway construction but in industry as well. He believed that the application of engineering was changing so rapidly that the papers and discussions presented before these technical groups now formed the principal sources of information. Holley said that if he had been asked to name a book covering modern methods of manufacturing iron and steel, he could think of none. In his estimation, it was the monograph read before a technical society that formed the current literature of the fast changing industrial art.

He spoke about the importance of a central location where engineers could meet together. "It is not only a reasonable supposition, but it is a matter of history, that men in the same business, thrown together in technical and social meetings, and in excursions among engineering works, gradually and often quickly, exchange friendship for jealousy and helpfulness for rivalry. The grandest work of the British Iron and Steel Institute has been to throw open the works and processes not only of England but of France, Belgium and Germany to the observation of all who are interested."[17]

During the ensuing discussion about a name for the society, Holley was reminded of a similar discussion at the Sheffield Scientific School at Yale College some ten years earlier, when the term *Dynamical Engineering* had been applied to a chair of mechanical engineering. He therefore suggested the name American Society of Dynamical Engineers, but the group decided to call itself the American Society of Mechanical Engineers. Another organization meeting was held on April 17 at Stevens Institute at which, because of illness, Holley was not present. The first annual meeting of the society took place in November. Robert Thurston was elected president. Despite high professional eligibility requirements, the society began with 163 regular members, 17 associates, and 9 junior members.[18]

NOTES

1. *Report of the Committees on Chain-Cables, Malleable Iron, Reseating and Rerolling Wrought Iron* (Washington, D.C.: U.S. Government Printing Office, 1878).
2. Alexander Lyman Holley, "The United States Testing Machine at Watertown Arsenal," *Transactions of the American Institute of Mining Engineers* 7 (1879): 256–66 passim.
3. Ibid.
4. ALH to Samuel Felton, 16 April 1879, Holley Family Papers (hereafter cited as HF).
5. *Memorial of Alexander Lyman Holley* (New York: American Institute of Mining Engineers, 1884), p. 34.
6. The list was headed by Captain Jones, followed by Erastus Corning, Chester Griswold, Robert W. Hunt, Charles Kennedy, John W. Hartman, Thomas H. Lapsley, Henry R. Worthington, William A. Perry, Daniel J. Morrell, William P. Shinn, William Metcalf, S. T. Wellman, Daniel N. Jones, Alexander Hamilton, James Hemphill, O. W. Potter, John Fritz, James Park, Jr., Charles C. Teeter, D. S. Hines, John E. Fry, Selden E. Marvin, John H. Ricketson, Robert Forsyth, E. V. McCandless, and John Rinard.
7. "Testimonial to Alexander Lyman Holley, May, 16, 1879," privately printed.
8. ALH to [name illegible], 6 September 1879, HF. In September 1879 Holley had only ten clients, each of whom paid him $1,000 a year. For this fee he provided any required technical aid and gave them all the information he gathered on his European trips. There was some additional income from the royalties from his patented improvements, but, inasmuch as no plant had been built since 1876, this income could not be counted on.
9. ALH to Robert Forsyth, 23 September 1879, HF.

10. Joseph Frazier Wall, *Andrew Carnegie* (New York: Oxford University Press, 1970), p. 502.
11. Herbert N. Casson, *The Romance of Steel* (New York: A. S. Barnes & Co., 1907), p. 93.
12. ALH to Sidney Gilchrist Thomas, 6 September 1879, HF.
13. ALH to J. M. Kennedy, 11 February and 14 February 1880, HF.
14. ALH to Sidney Gilchrist Thomas, 26 September 1879, HF.
15. ALH to Sidney Gilchrist Thomas, 24 October 1879, HF.
16. Alexander Lyman Holley, "Washing Phosphoric Pig-Iron for the Open-Hearth and Puddling Processes," *Transactions of the American Institute of Mining Engineers* 8 (1880): 156-64. As was usual, Holley was joined at the meetings by many close friends. Eckley Cox, the president, and Sterry Hunt had planned every detail of the program for the meetings. The dinner, which always formed an important part of the session, was an occasion for which the food and wines were as carefully chosen as the speakers. Holley reserved for himself the presentation of the toast to "The Engineering Race."

    Holley had definite ideas as to how these dinners should be conducted. When his friend, James Forrest, was given the responsibility of conducting a similar affair for the Institution of Civil Engineers in London, Holley wrote to him, "Take a leaf out of our book and begin the professional and humerous [sic] after dinner talks at once and do not postpone them until you have been talked asleep by solemn swells about the Queen and the Royal Family and all the branches of the Government and Civil Service. Engineering owes them little enough and they are quite likely to be perpetuated without this sacrifice on the professional altar" (ALH to James Forrest, secretary, Institution of Civil Engineers, London, 23 April 1881, HF).
17. *American Society of Mechanical Engineers* 1 (1880): 3-4.
18. Ibid., pp. 274, 276.

XXV · The Bessemer Steel Company

The extent of coverage through the Resse patents, which the Bessemer Steel Company had obtained in 1879, was a continuing source of concern to the Bessemer producers, who hoped that the patents would cover the use of lime broadly enough so that obtaining the expensive Thomas rights would not be necessary. To Holley's disgust and Thomas's frustration, the Bessemer Steel Company's lawyer encouraged the group in this notion.[1] It now appeared that a satisfactory conclusion to the purchase of the Thomas American rights might never be achieved.[2] In addition, the Bessemer group was negotiating with Krupp's agent, Ignatius Hahn, for the pig iron washing process. All this activity in securing rights probably was related to the impending lapse of the most important Bessemer patents on 13 January 1880, an occurrence that would leave the group weakened in its ability to control Bessemer steel production.

Meanwhile several companies that were not members of the Bessemer Steel Company, among them the newly organized Pittsburgh Bessemer Steel Company, were planning to circumvent the existing rights. The Pittsburgh Bessemer Steel Company had been formed in 1879 after the Edgar Thomson management, which had been supplying various firms in the Pittsburgh area with steel for fabrication suddenly cut off the supply, saying that all of their steel was needed for rail production. The customers who had lost their source of supply immediately banded together to form a company of their own. Among the subscribers to the new endeavor were executives from Park, Brother and Company; Hussey, Wells and Company; Singer, Nimick and Company; Crescent Steel Works; Solar Iron and Steel Works; and the Superior Mill, which Andrew Kloman had leased. All were substantial companies that manufactured tool steel, axles, tires, plates for locomotives, boilers, and fireboxes. Kloman's plant rolled bars, structural shapes, and light rails.[3]

Late in 1879 Holley wrote to J. M. Kennedy, treasurer of the Bessemer group, bluntly setting forth the patent situation as he saw it.

> The Pittsburgh Bessemer Steel Co. Ltd. has upon my advice asked you for terms of license. I infer from some conversation I have had with managers of this company that they will not pay you the royalty you charge one another, perhaps not more than half of it. But every cent you get is so much money

saved. After a careful review of the patents you own, I am satisfied that this company or any other company can build a good steel works without infringing any of your patents. Your last Bessemer patent of any value will expire on January 13, 1880. The only other patent which covers existing plants is my patent of January 26, 1869 which you own. This patent cannot, of course, cover any essential feature of the Bessemer steel manufacture seeing that it was developed after that manufacture had been made more or less successful in Sheffield. But my patent covers so many useful and convenient features that the Pittsburgh Bessemer Steel Company express their willingness to pay a certain moderate royalty rather than to go without these conveniences. The exact value of my various claims is quite too long a matter to discuss in a letter. I will, however, as you suggest meet with your committee or your company at any time you may appoint and go over the matter in detail. In case this new company or either of the other companies proposing to build Bessemer works should infringe upon any of my claims which are pretty broad, I will, of course, inform you, but I fear that in that case the lawyers would get all the money damages, hence I have advised both you and the outside parties to try and agree upon a license.

Although it is a little outside of the present question, I think I ought to remind you, in this connection, that these patents have been worth to you all or more than you paid for them. The Bessemer patents of January 13, 1863 cover the essential features of a successful Bessemer plant. Your ownership of it or license under it was absolutely essential to your making steel at all. My patent is also indispensable not because you could make steel without it, but because you would have to rebuild every existing works on some other plan, unless you worked under the plan.[4]

During 1879 a new contract was being negotiated between Holley and the Bessemer Steel Company. Although it had been submitted in late December, in the middle of the following January Holley wrote to Kennedy about the reasons for the delay. At the end of the month, Carnegie entered the picture and questioned certain of Holley's activities. Holley responded immediately.

My proposition to the Bessemer Steel Company and their offer to me and the final form of contract agreed on, all specified that I was to be at liberty to complete engagements previously undertaken. I had furnished a part of the required drawings and information to Spang Chalfant and Company before I went abroad for the Bessemer Steel Company and I was under contract to complete this work. I have, however, given them notice that my services must be terminated July 1. I have taken no new work since my first proposition to the Bessemer Steel Company but have on the contrary dropped two clients and refused three others. I am acting strictly to the letter and spirit of my agreement.[5]

It was not until 11 February 1880, that the agreement finally was signed.[6]

Holley spent most of his time during the next several months attempting to work out solutions for the problems of all the patents that the Bessemer Steel Company had under consideration. Foremost was the Reese and Thomas conflict. Kennedy insisted that it was necessary to conduct a long series of experiments. Holley showed little sympathy for such a project. He pointed out that trials would be very expensive and also would be needless and wasteful, since Thomas already had covered the ground. Holley suggested, however, that if the members thought the Reese patents would cover the process of dephosphorization and would eliminate the need for the Thomas patents, it might be a good policy to undertake the experiments. However, the various companies were all too busy to give him and Reese the necessary facilities for conducting trials, and the experiments were postponed indefinitely.[7]

The Pernot patents, which the Cooper & Hewitt company owned, again were being considered by the Bessemer group. Large amounts of money already had been spent on them for options that had not been taken up; and again the Bessemer Steel Company members were finding it difficult to arrive at a final decision. In the meantime, while they delayed, numerous other companies were securing licenses from Cooper & Hewitt.[8]

As Holley was arranging his affairs preparatory to a European trip in the spring of 1880, Captain Jones wrote him that Carnegie had decided to send him abroad with Holley. Annoyed, Holley wrote Carnegie.

> I am informed that Captain Jones is ordered to join me in my proposed journey to Europe to investigate the Thomas process. Am I to open the facilities in my reach to Captain Jones as representing the Edgar Thomson Company and if so, will you oblige me by procuring the order of the President of the Bessemer Steel Company to that effect?
>
> There is no expert or friend I would prefer as a companion to Captain Jones . . . but he must be sent by the Bessemer Steel Company or else I am put in a very embarrassing position. You know the feeling between certain companies.
>
> I am also informed that you sent Mr. Lenox Smith abroad to investigate the Thomas process among other things.
>
> In all these investigations from different points of view, the facts ought to be got at pretty fully. It is a relief to me, that the action of the Bessemer Steel Company will not depend solely on my report as I have no doubt we three and any others you may send will agree.
>
> I am going in the Gallia April 21. She is pretty full already and you had better get a place for Jones right off.[9]

Holley's letter infuriated Carnegie but Holley explained to Jones that he was obliged by the peculiarity of his position to ask Andrew Carnegie to obtain formally the sanction of the Bessemer producers. Carnegie was so angry at having his motives questioned, that he refused to permit Jones even to sail on the same ship with Holley. However, the two made plans to meet in London.[10]

As the time for Holley's departure neared, he was in a whirl of activity as he tried to tie up the loose ends of his work. He spent many nights in uncomfortable sleeping cars as he journeyed to St. Louis, Chicago, Pittsburgh, Cleveland, and Troy. Added to his exhaustion was the strain of his new relationship with the Bessemer Steel Company. Early in April he finally could no longer ignore the persistent chills and fever and he was obliged to go to bed for a period of rest.[11]

Several important activities outside his practice had taken both time and energy that he could ill spare. For example, Holley was one of a committee of four that arranged an elaborate dinner at Delmonico's in New York City in honor of Ferdinand de Lesseps, who had scored such a triumph in building the Suez Canal and now was involved with a company organized to construct a canal across the Isthmus of Panama. It was Holley who spoke after the dinner as the representative of the engineers of the United States, and it also was Holley, along with Hewitt, who paid most of the deficit of several hundred dollars that the occasion entailed.[12]

Holley sailed as planned on the *Gallia* for Liverpool on 21 April 1880. Laureau, his assistant, and Lenox Smith from Cambria accompanied him. After a stay in London of a little more than two weeks, he set out for the Continent with an itinerary that would have exhausted a man in the best of health. He went first to Dortmund for a week. Then followed successive stays of one to six days in Essen, Rhurort, Liège, Paris, and Creusôt. Traveling on to Switzerland and down to Italy, he worked his way back through Austria, where he spent time at Witkowitz and Vienna. The third week in June he was back in Germany visiting steel plants at Kattowitz, Koenigshutte, Berlin, and Cologne and making plans to visit one or two plants in Russia.

As the trip progressed, Holley in his eagerness to study thoroughly all that was new, tried to ignore the state of his health. But by the time he reached Cologne, he realized that he was on the verge of collapse. In the hope that after rest in bed he would feel restored, with great effort he managed to reach Morley's Hotel in London. After four days, when he felt a trifle better, he decided to try the sea air at Brighton. However, the move had little effect, and in a few days he was back in his room in London. His many friends were worried, but they could do nothing since Holley, wishing not to be a burden to anyone, graciously but firmly refused all offers of aid.[13]

At last, his good friend James Dredge, who was then editor of *Engineering,* called at the hotel and suggested to Holley that a drive in the country on such a fine day would do him a world of good. Holley accepted, only to find that Dredge was taking him to his home in Clapham Common, where Holley could receive the proper medical and nursing atten-

tion. Too sick by this time to resist, Holley accepted the situation gracefully.[14]

Abram Hewitt, arriving in London and hearing that Holley was
dying, hastened to Clapham. Before permitting him to see his desperately
ill friend, the physician warned Hewitt that it was extremely doubtful
that Holley could live much longer, that it might be a matter of hours,
and at the most three or four days. When Hewitt, completely crushed
by the news, was admitted to Holley's bedroom, he saw that indeed
Holley's condition was serious. "I found him the very picture of dissolu-
tion," wrote Hewitt to mutual friends in New York.

"He smiled his old smile, and, after a few words had passed and, I suppose,
seeing upon my face the expression of profound sympathy, he said, "Well, I
suppose they have told you that I am going to die, and I suppose I am, although I mean to get home first, if I can, but I have been trying to amuse
myself while I am dying by inventing something that will enable me to die with
credit. I want to make the basic process practicable, and I have just perfected
here upon my bed the details of the movable shell; and if I don't live long
enough myself to introduce it, I have explained it in full to my assistant so
that this invention shall be known to my countrymen."[15]

Hewitt at once wrote George Worthington, a family friend as well as
business associate in New York suggesting that Holley's family be prepared
for the bad news that inevitably must follow. But miraculously and with
the determination that was characteristic, Holley began to improve.
He stayed with Dredge for the next two months, of which he spent one
month in bed. Dredge, who sat for many hours at Holley's bedside, said
that, ill as he was, Holley "had from the first a fixed resolution to live. . . .
'So far as I am concerned,' he would say, 'it matters nothing; I have
done my best with the powers that I possessed, not for money or ambi-
tion, but because I just had to do my best. And it is impossible for me
to die for another eighteen months. Those I leave behind me must be
provided for, and I can do that easily in a year and a half. Just now
I could not afford it.'"[16]

Clapham Lodge was about six miles from London. Dredge had transformed the grounds into a luxuriant garden. An archway of ancient trees
led to the house. The greenhouses and conservatory attached to the
lodge overflowed with orchids and rare ferns. The nearby lake with its
golden carp presented a magnificent view from the dining room windows,
which opened on a velvet green lawn. The setting provided the peace
and seclusion that did much to coax Holley back to health. His condition, which the doctors diagnosed as a liver disorder, continued to
improve. By September, he was well enough to go to Tunbridge Wells
for a period of convalescence. Holley returned to London for a week

and then managed a trip that took him to Glasgow, Edinburgh, Saltburn, Sheffield, and finally to Liverpool, from where at last he sailed on October 9 for New York.[17]

Holley wasted little time in making public the details of the removable shell. At the November 1880 meeting of the American Society of Mechanical Engineers in New York, he read a paper that described his invention. No company was as yet using the Thomas process in the United States and very few in the audience could have had the opportunity of seeing it in operation in Europe. Holley thus was describing not only a new process but a means of making it practicable whenever the Thomas process might make its appearance in America.

He reminded his listeners that ever since the first Bessemer conversion in 1865, the maintenance of the refractory linings had been the basis of more experimentation than any other feature of the Bessemer process. He pointed out that, in spite of the fact that the operation of the Bessemer converter had reached a remarkable degree of efficiency, the refractories were still inadequate. The linings were not only eroded by the mechanical action of the charge, but also were chemically decomposed by the various slags. Practice had improved to such a degree that it was now possible to obtain as many as sixty charges in twenty-four hours from a pair of converters. With the use of the interchangeable converter bottoms, which also contained the tuyeres, operations hardly were delayed at all. The lining above the tuyeres, however, remained a source of trouble, and the only way to repair this portion of the lining was to take the converter out of operation altogether.

Holley warned that if the difficulties had seemed great in coping with the acid lining, the new basic process would increase the problem threefold. As with the acid process, the bottoms and tuyeres could be replaced readily but the fixed lining above the tuyeres, where the abrasion was most severe, could be repaired only by cooling the converter and either inserting new bricks or patching. The converter, therefore, would be out of operation for at least a day, at a high loss in tonnage. It was conceivable that this limited endurance of the basic linings could force an American plant to be idle as much as half the time.

Holley believed that the removable shell used with the basic process was an answer to these anticipated difficulties. He described how it could be lowered out of the trunnion rings with ease and with little loss of time. A dozen cotter pins would be knocked out, the shell lowered onto a car and run to the repair shop. The car then would return with a new shell, which would be lifted into place, the cotter pins secured, and that would be all. In certain European plants where the basic process already was in use, the entire converter was lifted from its place, a slow and tedious operation, but Holley believed that with his method of replacing only the shell, the change could be made in a matter of minutes.[18]

The patent for the removable shell was to be Holley's last. During his lifetime he had obtained sixteen patents. Of two granted in 1859, one was for a variable cut-off valve for steam engines, and the other was for railway chairs. Ten later patents were for improvements in the Bessemer plant and process. Two were for roll trains and their feed tables, and two for a water-cooled furnace roof and a steam boiler furnace with gaseous fuel. The two of most importance were, of course, the final one for the detachable converter shell used at Seraing, Witkowitz and other European works and that for the converter bottom, which he had patented in 1870 and which was in general use in both Europe and the United States.[19]

In addition to presenting his paper on the removable shell, Holley also took part in a lengthy discussion among the members of the American Institute of Mining Engineers concerning the production of rails. The question first had been brought up in 1877 when Dr. Charles B. Dudley, chemist for the Pennsylvania Railroad Company, described the results of his investigations into the chemical and physical properties of rails. In his discussion the chemist had outlined what he considered the requisites for the ideal rail. Hunt, Jones, and superintendents of other mills that produced Bessemer rails had disagreed strongly with Dudley and insisted that conclusions could not be drawn from the study of only twenty-five rails, as Dudley had done. The controversy, which simmered for several years, boiled up again in 1880 when Dudley presented a second paper on the subject. This time the Bessemer managers had prepared themselves to refute his claims and the result was the spirited discussion in which Holley participated.[20]

Although still weak from his illness, Holley took a stand on the foolishness of producing a multiplicity of rail patterns, leaving the discussion of the chemistry and physical properties to Jones and the others. Ever since 1876, when he had designed a rail with Octave Chanute, he had been concerned with the problem of the extraordinary number of rail patterns that the Bessemer mills were obliged to produce.[21] Chanute, who later became famous for his work in gliding experiments and for his association with the Wright brothers, was then a noted railroad engineer.

Holley again spoke on the subject of rail patterns before the American Institute of Mining Engineers in February of 1881. At that time the eleven mills that rolled Bessemer rails regularly were manufacturing 119 patterns of 27 different weights. If one took into consideration patterns that were considered obsolete, but still were in use, the number may have been approximately 300.

With his accustomed practical approach, Holley endeavored to point out to both the railroads and the rail producers the seriousness of their course. Each rail pattern required its own set of rolls with guides and guards that cost about $1,500. In order to produce the 119 patterns that the railroads demanded, the mills were forced to keep on hand 188 sets of rolls. Some mills kept duplicates of the patterns, which explained the difference between 119 and 188. These sets alone required an investment of $282,000 by the producers.

Holley then described the problems involved in changing a set of rolls. Transferring from one pattern to another required about two and one-half hours, during which time thirty-five tons of rails could have been rolled. When a mill found it necessary to change its rolls four times in one week over a period of some months, which was not unusual, the lost tonnage could make the difference between profitable and nonprofitable operations. Even if one mill changed its rolls only once per week, allowing $5 per ton profit and time for repairs, he reminded the eleven companies that they could lose $90,000 each year.

Holley was convinced that producing a maximum of ten patterns of each weight, if adopted by all the railroads, would bring about enormous savings. He pointed out, "Mistakes and misunderstandings occur when each customer wants a different rail, different variations in standard length, different percentages of short rails, different sized fish holes, different pitch of fish holes, different patterns of fish plates, different lengths of fish plates, different slotting, different bolts, different everything." He also pointed out that the present practice prevented rail makers from keeping rails in stock. "If railmakers could roll and stock standard patterns, when special orders were slack, they could prevent the excessive rise in prices, which the scarcity of rails from time to time creates." In twenty-one rail patterns of one mill, Holley found six different hole diameters, ten patterns that had varying distances between holes, and fourteen variations in the distance of the first hole from the end of the rail. As he commented, "For one pattern the distance is 2-13/64th inches. To think that the mind of man can hit perfection in spacing fish plate holes within the sixty-fourth of an inch, is almost appalling."[22]

No doubt many in his audience shifted uneasily in their seats when he went on to give his interpretation of the rail situation.

There is another cause of the multiplication of patterns, more potent and more difficult to remedy than any intrinsic cause. It is the egotism of certain engineers and officers of railways. I do not refer to those honestly differing opinions which are founded on observation, even if that observation is incomplete and one-sided. I refer to the determination of certain persons who dictate patterns to railways—and these persons are not a few—to use no pattern that any other manager has invented, but to vary from all standard patterns, for no reason whatever, except to inflict their own individuality upon some feature of the interest confided to their care. I cannot, of course, give names and particulars in a public paper; but there are plenty of names and instances known to rail-makers and to dealers, and to the railway fraternity. There are instances of men signalizing their accession to power by the imposition of a rail pattern which is not only new, but inferior in every way to patterns in current use. Some of these tinkers think they win the admiration of boards of directors by thus showing up the general ignorance and their own technical genius. They fondly believe that having "Stiggin's" rail pattern talked over in the mills and railroad offices gives them a certain immortality. It does.[23]

When Holley concluded his talk, Chanute rose to verify his remarks and to relate how the Erie Railroad had profited from Holley's advice. Yet, despite his efforts to standardize rail patterns, it was near the close of the century before the situation was recognized by the railroads as serious. In 1898, eighteen years after Holley first had suggested standardization, a committeee composed of seventeen engineers and maintenance men from various railroads met in Chicago to discuss the formation of an association that would investigate all matters dealing with rails and their accessories. From this meeting evolved the American Railway Engineering Association, which later became the research branch of the Association of American Railroads. One of the first problems undertaken by the new group was the standardization of rails.

NOTES

1. ALH to J. M. Kennedy, 20 November 1879 and 11 February 1880, Holley Family Papers (hereafter cited as HF).
2. While the purchase of the Thomas patent still was pending in America, Carnegie went to great lengths in his efforts to entertain the young English inventor. When he invited Thomas and his sister to be guests on one of the famous coaching trips that Carnegie and his mother took from England to Scotland, Thomas politely refused. His sister was bitterly disappointed, but her brother told her that "in view of pending negotiations, he was anxious not to feel that his independence of action was tied by obligations, however delightful." Thomas did accept an invitation to visit the group when it arrived at Windsor and joined them for dinner (Lilian Gilchrist Thompson, *Sidney Gilchrist Thomas* [London: Faber and Faber, 1940], p. 169).
3. James Howard Bridge, *The Inside History of the Carnegie Steel Company* (New York: Aldine Book Co., 1903), pp. 150-52, 159. The plant of the Pittsburgh Bessemer Steel Company was built in the record time of fifteen months and rolled its first steel rail on 9 August 1881. In later years the owners experienced so much labor strife that the owners sold out to the Edgar Thomson company. As the Homestead plant, it became notorious during the events of a tragic strike in 1892.
4. ALH to J. M. Kennedy, 20 November 1879, HF.

5. ALH to J. M. Kennedy, 10 January 1880, HF; ALH to Andrew Carnegie, 30 January 1880, HF.

6. ALH to J. M. Kennedy, 11 February 1880, HF.

7. ALH to J. M. Kennedy, 11 February, 25 February, and 16 March 1880, HF.

8. ALH to J. M. Kennedy, 14 February 1880, HF.

9. ALH to Andrew Carnegie, 25 February 1880, HF.

10. ALH to Captain Jones, 21 and 25 February 1880, HF. In his continuing search for social and cultural standing, Carnegie already had begun to establish the relationships for which he became so well known. He became interested in English politics and made many political friends including Prime Minister William Gladstone, to whose party he gladly contributed when it was in need of funds (Joseph Frazier Wall, *Andrew Carnegie* [New York: Oxford University Press, 1970], p. 429). Although Carnegie impressed many people with his charm, camaraderie, enthusiasm, and wealth, those men of the steel industry who knew him best were the least impressed and disliked him heartily. The buoyancy, effervescence, and wit that were useful to Carnegie socially did not impress his colleagues. They were aware of his insincerity and of the sharp treatment that many of his friends had received. When these men died, Carnegie, who was given to tears in moments of emotion, commemorated them with phrases suitable for the occasion.

11. ALH to Captain Jones, 17 April 1880, HF.

12. ALH to Abram Hewitt, 1 and 4 March 1880, HF.

13. James Dredge, "Holley Memorial Address," *Transactions of the American Institute of Mining Engineers* 20 (1891): xliv.

14. William H. Wiley, "James Dredge: The Man and His Work," *Cassiers Magazine*, February 1892, p. 291.

15. *Memorial of Alexander Lyman Holley* (New York: American Institute of Mining Engineers, 1884), p. 25.

16. Dredge, "Holley Memorial Address," p. xlv.

17. Wiley, "James Dredge," p. 287.

18. Alexander Lyman Holley, "An Adaptation of Bessemer Plant to the Basic Process," *Transactions of the American Society of Mechanical Engineers* 1 (1880): 130-32.

19. *Memorial of Alexander Lyman Holley*, p. 137.

20. C. N. Dudley, "The Chemical Composition and Physical Properties of Steel Rails," *Transactions of the American Institute of Mining Engineers* 7 (1878-79): 172-201, 357-413.

21. Alexander Lyman Holley, "On Rail Patterns," *Transactions of the American Institute of Mining Engineers* 9 (1881): 360.

22. Ibid., pp. 360-73 passim.

23. Ibid., p. 365.

XXVI · Thomas in America

he year's option that the Bessemer Steel Company held on the Thomas process was due to expire in February of 1881, and it now seemed advisable for Thomas to make his long-delayed trip to the United States for final negotiations. Thomas arrived in March of 1881, and stayed with the Holleys on Joralemon Street in Brooklyn. The round of social events that followed amazed the modest young man. The first night he stayed aboard ship but spent the evening going to a concert and afterward walking up Broadway with Holley's assistant, Laureau. Thomas had heard and read about this famous street and was eager to see it. Next day he joined the Holleys in Brooklyn. During the day he was introduced to numerous people and wrote his mother in delight that he had ridden four times on the elevated railroad. That night he dined with Abram Hewitt. He was enchanted to meet the renowned Peter Cooper, then ninety-two, but still, as the visitor wrote home, "Bright, intelligent and active old boy." Later that evening Thomas attended the opera *La Favorita* as the guest of Holley and his daughter, Gertrude. "Fine house but overpowering amount of talking" was the visitor's comment.

The days were crammed with luncheons, dinners, and appointments with important people. On Sunday, Mrs. Holley and Gertrude persuaded their guest to accompany them to the well-known Plymouth Church to hear Henry Ward Beecher, the famous preacher. Thomas found the church "hideous but crammed. Beecher preached for one and a quarter hours—most eloquent, original and outré sermon or address." He was introduced to Rossiter W. Raymond, Holley's close friend and neighbor, and thought him "a clever fellow, who is engineer, poet, novelist, editor, man of business, composer, and Sunday school teacher, all at the same time."

Inspection tours to the various steel plants then began. Holley and Thomas went up the Hudson River by steamer to Albany where they were entertained by Erastus Corning, the son of Holley's first sponsor. Although he was eager to see the pioneer installation at Troy, Thomas first was obliged to visit the Corning country home, where he was shown the famous hothouses filled with many varieties of rare and valuable plants. He also was shown Corning's equally outstanding butterfly collection.[1] At last, to his great relief, he was at liberty to go with Holley to visit the Bessemer plant. There the two joined Robert Hunt and watched the converters go through their paces. Excited by the precision and efficiency of the operations, Thomas said to his friend, "I should like nothing better than to sit down on an ingot mold and watch the work all day." To this remark, Holley replied with a sly smile, "If you want an ingot mold cool enough to sit on you will have to send to England for it."[2]

Holley's remark was apropos. A scant fifteen years had elapsed since the beginning of the experiments at Troy. No longer was steel production measured ton by ton; nor could eight men and a boy operate a converter plant. Bessemer steel production had become a major industry. In 1880 the eleven works in production, using twenty-four converters, turned out 1,203,173 net tons of Bessemer steel ingots. The twenty-four converters produced nearly the same amount of steel as the seventy-eight that had operated in Great Britain during the same year.[3]

From Troy, Thomas went to Bethlehem, then on to Chattanooga, Cincinnati, Chicago, Joliet, Springfield, Buffalo, Harrisburg, Philadelphia, and Hartford. He was feted everywhere, and moved in a world of banquets, private Pullman cars, and adulatory speeches. After the Corning episode he had written home, "The people have, to a stranger, few deficiencies, except a too evident money-worship, and (whence the money-worship proceeds) a reckless way of spending. They are hospitality itself." Near the end of the stay he assured his mother that "all the men I meet are the most remarkable in America, are also 'gorgeous,' 'princely,' 'heroic,' etc. Have not been introduced to an ordinary mortal yet."[4]

Thomas was much amused by the activities of the American woman and repeated to his mother what the young wife of one of the men at the Joliet steel plant had told him. "She explained to me the social points of Springfield thus. She belongs to a French class, an Elocution class, a Shakespeare class, an Art Club and a Married Folks Club. I find all the married women here go to classes for language or literature or something." Despite all the excitement of the attention showered upon him, Thomas maintained an unassuming simplicity, which Americans found charming. After a two-month tour at the end of which the Bessemer group at last purchased the rights to his process, Thomas ended his visit.[5]

The purchase of the rights to the Thomas process in May of 1881 for $275,000 concluded several years of frustrating negotiations. Carnegie finally was rewarded for his efforts. Although only six of the eleven companies favored the purchase, Carnegie's Edgar Thomson Steel Works was not assessed any portion of the purchase price, which was divided among the other ten. Furthermore, Carnegie received a commission of

$50,000 from Thomas, for persuading the Bessemer Steel Company to buy the rights. Holley, as he had told Thomas earlier, expected and received nothing.

Although patent negotiations for Thomas had ended, continuing controversy over certain of the Reese patents would not be settled until 1888. It had been assumed that Reese had sold all of his patents to the Bessemer Steel Company in 1879. Subsequently it was discovered that he had withheld some of them. The final court decision directed that all of Reese's patents pertaining to the manufacture of steel into rails, ingots, and billets, should have been included in the 1879 transactions.[6]

For some years after the Bessemer Steel Company bought the Thomas rights, it declined to grant rights licenses for their use. As a consequence, numerous accusations appeared in the newspapers. Other writers have expressed their opinions about the motives of the Bessemer Steel Company at that time. Thomas's sister wrote that the chief source of the iron ore in the United States was relatively low in phosphorus. Consequently, the Bessemer companies had little immediate need for the Thomas process. She believed that by withholding the rights the Bessemer companies may have delayed the development of the steel industry in the southern United States, where there were large deposits of high phosphorus iron ore that could be used satisfactorily only in a basic converter.[7]

In his presidential address, given before the American Institute of Mining Engineers in 1890, Abram Hewitt said:

> It is remarkable . . . that the vast deposits of hematites and red fossiliferous ores with which the South is endowed are adapted to the "basic" process, while the ores of Lake Superior are suited to the "acid" process. Thus the two sections are practically the complements of each other in the work of supplying the needs of the country for steel. It will doubtless excite surprise in the minds of our visitors to find that the basic process had made no progress in this country. The delay has been due partly to the recent development of the Southern ores, and partly to the illiberal spirit in which the basic patents have been managed. But it will not longer be possible to arrest the manifold destiny of the South, which is now erecting a large number of furnaces, the product of which must find a market through the basic process.[8]

In 1888 an article in the *Bulletin of the American Iron and Steel Association* commented:

> Error had been zealously fostered in some sections of the country by Free Traders and others that production of basic steel in the United States has been prevented by the selfishness of the Bessemer Steel Company which owns the English patents and has supposed that it also owned Mr. Reese's patents. It is true that, immediately after the Bessemer Company acquired

the ownership of the English patents, it declined to grant licenses to use them and for a variety of reasons not now necessary to inquire or to justify. It is also true that when controversy with Reese commenced several years ago no persons wanted to accept licenses from the company unless they were guaranteed against an infringement lawsuit from the other side which could not be given. It has thus happened that down to the present time the Bessemer Company has not granted a single license although for several years it had been ready and willing to grant licenses to all persons who would agree to pay a royalty of one dollar per ton for every ton of melted metal which should be converted by the basic process. It has not advertised this fact because of the legal difficulties with Reese which prevented it from guaranteeing licenses against legal proceedings.[9]

In the United States, the first experimental heat of basic steel in a converter was made at the Pennsylvania Steel Company on 24 March 1884. The first in the South was produced more than seven years later on 24 August 1891 at the Southern Iron Company in Chattanooga. The basic Bessemer process never was adopted in the United States because of the simultaneous development of the basic open hearth. Samuel Wellman produced the first basic open hearth steel at the Otis Iron and Steel Company in Cleveland on 18 January 1886. Only 1,000 tons of steel were produced before the bottom was taken out and the furnace converted to the manufacture of acid steel. The next installation for the production of basic open hearth steel was built at the Homestead Works of Carnegie, Phipps and Company on 28 March 1888.[10]

Holley's comment to Thomas that it would be difficult to find a cool ingot mold in America was a cogent remark about a rivalry that had been growing for years. The English producers recognized that the Americans, late starters in a field that the English long had dominated, rapidly were outpacing them. Taken aback, the English argued that rapid production impaired the quality of the steel. Foreign metallurgists asserted, "You really turn out no more steel than we do. As a matter of personal pride, your manager gets up a tremendous spurt for a day, or even a week, crowds on men wastefully, rushes through great quantities of metal, and then for months afterwards has to recuperate. The machinery has been strained, the owner's money has been wasted; the men have been prostrated—all to gratify the manager's vanity. Your output is no larger than ours."[11] Holley replied that these attitudes were only a cover for an inadequate plant, and that steel was no better because it took five hours instead of one to set a vessel bottom and twice the time to handle the raw materials and the finished product.

The rivalry came to a head in 1881 when Captain Jones presented a paper before the British Iron and Steel Institute about the operations of the Edgar Thomson plant. The paper was of particular interest because it explained in detail why the Americans had forged ahead so rapidly.

His remarks rocked the British steel industry to its foundations and many of those listening were inclined to doubt his veracity.

After describing the size of the Edgar Thomson plant and its production for 1880, Jones continued, "In order to keep the works and machinery in proper order, great care and watchfulness must be exercised, the time for repairs being extremely limited." He mentioned the important part that the workmen played in the success of the operations. If work stopped for a few weeks, the men became rusty and awkward and it sometimes took as long as six weeks before they recovered their skill and energy. He admitted the difficulty of introducing changes and improvements in the mill. The men were not only loath to accept them, but also were capable of deliberately obstructing efforts to test their merits. He added, however, in fairness to the men, that once the value of changes had been proved, the men would accept them readily.

At the Edgar Thomson plant, he said, the condition of the converters was watched closely. Because the linings were kept at so high a temperature they were inclined to fuse and wear away very rapidly. Every effort was made to prevent the converter lining from burning through during the week, so that any repairs could start as soon as operations stopped at four o'clock on Saturday. All repairs were made Saturday night and Sunday, so that the crew reporting on the first shift Monday morning found everything in readiness.[12]

Jones discussed his so-called scrap-heap policy, which later became an accepted program in the steel industry. No matter how expensive or recently installed a machine was, he never hesitated to discard it if a better one became available. He assigned credit for the large production in the American steel plants principally to the group of young men, headed by Holley, who had applied their brilliant talents to the development of the American Bessemer mills. He related the strong rivalry that had arisen between Hunt at Troy and Fry at Cambria and how other Bessemer superintendents had entered the lists. He also cited the diversity of nationalities in the working force.[13]

Jones described entering the production contest himself when the Edgar Thomson plant started up. Bessemer production in America then assumed proportions that astounded the steel companies in Europe. "The output soon reached 1550 tons of ingots per week, then 1800 tons, then 2000, and ultimately increasing to 3000, 3100, 3200 and 3400," he said. The size of production eventually was being governed by the ability to get the ingots out of the way. The thrifty Englishmen heard him suggest that new equipment must be extra heavy and strong in order to function efficiently. To the often-repeated criticism that fast running made inferior steel, he responded, "It is impossible to attain great speed in working while making bad steel. The pig metal must be good, the machinery in proper order, the ladles in good condition, the pouring clean, and the heats regular, in fact, during fast running the whole plant must work in harmony, and the operations must be efficiently conducted." Jones, however, was a genius at running a steel plant, and his achievements were to become a legend in the steel industry.[14]

The Jones paper caused such a stir that a few months later, Holley was occupied in assembling data for a report requested by his friend, Lord Stafford concerning the equipment and production of certain American Bessemer plants. In it everything Jones had said was verified and even heightened. Holley boasted that the Bessemer process in the United States was the same as abroad but that the output in America from a pair of vessels was twice that of the English converters of the same nominal capacity due chiefly, he said, to superior arrangement and details of plant layout and to the greater "workfulness and enthusiasm" of the men and foremen. Holley could not resist adding that "when one old-fashioned converter wouldn't make enough steel, it seemed to be the practice in England to add another old-fashioned converter."[15]

The English producers could not believe that these were the real reasons for the American success in producing steel, but Henry Howe put it succinctly when he said that the greater output did not necessarily mean that any of the machinery moved faster, it simply meant that there were shorter periods of rest and disuse. The important point was the more nearly continuous running, rather than swifter movement. Machinery was always the better for continuous operations, he said. In idle machines, grit accumulated on the bearings, foundations settled, rust could appear, and stresses set up by heating and cooling of machinery caused expansion and contraction each time the engine started and stopped.

> Personal pride has, doubtless, played its part; without it we would still be savages, or apes or mollusks, if you prefer. These baser motives are the mudsills of progress, the soiled roots of that growth of which civilization is the flower. Great outputs were made and published; generous rivalry sprang up. Each showed the other how he made his great output, each gained the experience bought by the other's toil, till the average output of today is far greater than the record-breaking output of the centennial year. Whatever may be the merits of our tariff system, we have pushed free-trade far in some important directions — free-trade in ideas, free-trade in land, free-trade in labour.[16]

NOTES

1. R. W. Burnie, ed., *Memoir and Letters of Sidney Gilchrist Thomas, Inventor* (London: John Murray, 1891), pp. 146-48.

2. *Memorial of Alexander Lyman Holley* (New York: American Institute of Mining Engineers, 1884), p. 73.

3. *Annual Report of the American Iron and Steel Association* (Philadelphia: AISA, 1881), p. 25.

4. Burnie, *Memoir and Letters of Sidney Gilchrist Thomas*, pp. 151, 155.

5. Lilian Gilchrist Thompson, *Sidney Gilchrist Thomas* (London: Faber and Faber, 1940), pp. 158-66 passim.

6. James M. Swank, *History of the Manufacture of Iron in All Ages*, 2d ed. (Philadelphia: American Iron and Steel Association, 1892), p. 406.

7. Thompson, *Sidney Gilchrist Thomas*, p. 277.

8. Abram S. Hewitt, "Iron and Labor," Presidential Address, *Transactions of the American Institute of Mining Engineers* 19 (1891): 484.

9. *Bulletin of the American Iron and Steel Association* 82 (3 and 10 October 1888, combined issue): 300.

10. George W. Maynard, "Introduction of the Thomas Basic Steel Process in the United States," *Transactions of the American Institute of Mining Engineers* 13 (1885): 291.

11. Henry M. Howe, "Notes on the Bessemer Process," *Journal of the Iron and Steel Institute* (London) 37, no. 2 (1890): 97.

12. Captain W. R. Jones, "On the Manufacture of Bessemer Steel and Steel Rails in the United States," *Journal of the Iron and Steel Institute* (London) 18, no. 1 (1881): 130-32. Captain Jones converted the net ton of 2,000 lbs. used in the United States to the ton of 2,240 lbs. used in Great Britain.

13. Ibid., pp. 136-37. Jones had concluded that men could not work the then customary twelve-hour day, and he instituted the eight-hour day in the Bessemer works with the plant operating under three shifts. However, he stood alone in that opinion and was forced to return to the long-established twelve-hour shift.

14. Ibid., pp. 129, 133, 137. Jones in 1881 believed that the steel industry probably had reached its limits in turning out 123,303 tons of ingots yearly from two ten-ton converters. Less than ten years later, this tonnage had been increased two and one-half times to 310,635 tons, and by the end of his life in 1889, Bessemer production had reached 3,281,829 net tons, and open hearth production 419,488 tons (James M. Swank, *History of the Manufacture of Iron in All Ages*, 2d. ed. (Philadelphia: American Iron and Steel Association, 1892, pp. 414, 425).

15. ALH to Lord Stafford, 2 June 1881; Sidney Gilchrist Thomas and Percy Carlyle Gilchrist, "A Note on Current Dephosphorising Practice," *Journal of the Iron and Steel Institute* (London) 19, no. 2 (1881): 413.

16. Howe, "Notes on the Bessemer Process," pp. 107-8. Professor Tunner of Austria in his report on the Centennial Exhibition in Philadelphia wrote that the large output of the American Bessemer works was mostly due to the "superior intelligence of American workmen." This statement led Robert Forsyth of the North Chicago Rolling Mill to submit a breakdown by nationality of the ninety-three men and boys he employed. He wrote that Germans and Bohemians formed 43 percent of the working force, Irish 27.96 percent, English 16.13 percent, Norwegian 8.60 percent and Americans 4.30 percent (*Bulletin of the American Iron and Steel Association* 12 [17 and 24 April 1878, combined issue]: 91).

XXVII · The Curtain Falls

lthough the Bessemer Steel Company had bought the Thomas patent, the group was reluctant to take Holley's patent covering the removable shell. Robert Hunt had assumed they would, but in answer to his friend's optimism, Holley replied:

> The enclosed correspondence will show you that the managers who have obstructed the thing for four months intend to continue to obstruct and virtually threaten that none of the members shall enjoy the use of the invention without coming to their terms. Mr. Baldwin, one of the ablest lawyers in the U.S. has given me a written opinion that the interest is mine exclusively and many other persons who know about these things say that there isn't a shadow of doubt of it. As to the value of the invention the opinions of yourself, Bill Jones, Forsyth, Smith of Joliet and several others don't seem to strike Kennedy and Wharton as of authority. I have not answered Kennedy's last letter and don't intend to. I seem to stir up a contention whenever I mention the subject and I am going to let it alone.

> I happen to know however that several of the members are very anxious to limit the use of the invention and I am anxious to accommodate those who mean to be fair. If you have a mind to suggest to Mr. Corning and he has a mind to suggest to six or more of the others to control this inaction without obstruction from Mr. Kennedy and Company I shall be happy to transfer it to them and I don't want but fifty percent of the money now—the rest in twelve months—only I want it settled before I sail on July 9th. There is enough time to do all this by telegraph if it is desirable.[1]

Holley's relations with the Bessemer group were steadily worsening. When August Wendel, the chemist at Troy, asked Holley for letters of introduction to foreign plants that he wished to visit, Holley was obliged to tell him that the president of the Bessemer Steel Company had requested that he not give any more such letters. Holley endeavored to take some of the sting out of the refusal by saying that certain foreign companies had made a similar request.[2]

As the time neared for what had become his annual trip to Europe, Holley decided to combine business and a vacation journey with his wife and daughters. They would spend several weeks visiting the English and Scottish lakes while he went about his business, and all would meet in London about the first of August, spend the month in Switzerland, travel through Lorraine, Belgium, and Westphalia, and end the journey in London late in the fall.[3]

In his efforts to make up for the time his serious illness of the year before had cost as well as to make up for the time he was to be away, Holley spent scarcely a day at home as he traveled his well-known circuit. The schedule proved to be so onerous that, as before, he was taken ill some time before sailing. In spite of his indisposition, he insisted on going through with all his plans. When the family arrived in Europe, Holley started out on the tight schedule that he had planned. By the time he reached Belgium, he suffered renewed attacks of the chills and fever that he had experienced the year before. Not wishing to interfere with his family's pleasure, he made light of his condition for fear that his wife would insist on abandoning the trip. His illness persisted, but Holley with great fortitude followed his schedule.

When the Holleys gathered in London, Gertrude had begun to show symptoms of typhoid fever and soon was desperately ill. Frantic with worry about her condition, Holley refused to consider his own health. After some harrowing weeks, Gertrude began to improve, and Holley, realizing that he could no longer ignore his own health, sought medical aid. The London doctors were puzzled about the cause of his ailment and recommended that he leave the unhealthy winter weather of London and return to America. Sir Polydor de Keyser, Lord Mayor of London, and Lady de Keyser, who were devoted friends, promised Holley that they would look after his family if only he would take steps on his own behalf. With these assurances, Holley agreed to return home. The family would follow as soon as Gertrude had recovered enough to travel.[4]

The New York atmosphere and familiar surroundings made little difference to the sick Holley. He remained continually optimistic but was unable to throw off the illness. Conducting as many of his affairs as he could from the house in Joralemon Street, he sent frequent cheerful letters to the family in London. On 21 January 1882, he wrote his half-sister Maria, who was in Connecticut nursing his father who was seriously ill with typhoid fever, "You will understand that my answer must be short as I have a great deal of social and business correspondence that must be recognized. I am much encouraged about Father to whom give my best love. I am confining myself to my room and have half a day or more in bed to choke off by powerful medicines under every advantage of uniform temperature the deep seated chills and fever. Yesterday was the first day I escaped a chill but I did not escape the fever. The jaundice is bad still but the doctor assures me that in every respect I am improving slowly and shall get well."[5]

By this time his family had started for home. Just before sailing, they had received from Holley's doctor a cable that assured them that there was every hope for Holley's recovery. With Gertrude on the road to good health and with the cheerful news from home, the were able to depart in high spirits.

On January 24 Holley suddenly developed peritonitis, and his friends knew then that he was indeed fatally ill. It now became a matter of much concern whether or not the family could arrive home while he still lived. Because the *Germanic* was delayed in making her pier in New York, the family did not reach the house in Brooklyn until ten minutes before eight, the evening of January 29. Holley had died twenty minutes before.[6]

Inasmuch as the cause of Holley's death was still unknown an autopsy was performed. It was found that he had suffered from an internal tumor that obstructed the gall duct and had produced the jaundice, chills, and fever.[7]

A host of Holley's friends attended the funeral services and followed the cortège to Green-Wood Cemetery in Brooklyn. There were the Bessemer superintendents headed by John Fritz, Captain Jones, and Robert Hunt; Chester Griswold, son of Holley's first partner at Troy; officers of the technical societies; editors of the professional journals; and representatives from Stevens Institute, Columbia School of Mines and Rensselaer

Alexander Lyman Holley residence at 89 Joralemon Street, Brooklyn, New York. This is where Holley lived in later years and where he died. (Courtesy of John Rudd.)

Polytechnic Institute, the schools that Holley admired so greatly. As they stood in the new fallen snow at Green-Wood in the clear and still day, their minds must have been full of thoughts of the man who had shown so many of them the way.[8]

Holley would have been fifty years old in the year of his death. During his comparatively short life he not only had witnessed the birth of the steel age, but also had been one of the assiduous attendants at the event. While he had made it possible for many men to accumulate vast fortunes, he was able to leave only a modest estate. The size of the bank account of his business of less than $1,000 was evidence of the extent to which Holley had shared in the profits of steel. Although he had invested whatever he was able to, at the time of his death his personal estate was not considered large.[9]

Holley never so much as implied that he may have received less than his due, but friends with most knowledge of his affairs believed that his rewards should have been larger than they were. Hunt pointed out that, of the various patents issued to Kelly, Mushet, Bessemer, and Holley covering the many facets of the pneumatic process, all save those of Holley had lapsed. Thus only on his patents could the Bessemer Steel Company still collect royalties.[10] The removable shell patent that Holley so anxiously had hoped the Bessemer group would take had not been purchased, although virtually all the European steel plants were using his changeable bottom, and three new works under construction had adopted the changeable shell.

Abram Hewitt, mourning Holley's passing, said, "Of all the men at this day, in this line, I know of none surviving him who has contributed as much to the growth of material wealth and the advancement of industrial progress in this country, as Holley did."[11]

The editor of *Iron Age,* James Bayles, devoted an entire first page of the journal to an account of Holley's career. Bayles wrote bitterly, "He asked for himself far less than his services were worth. His fortune was small and his family are left in circumstances only comfortable. If the members of the Bessemer Association do not double the estate he has left, they will show but a poor appreciation of his invaluable services to them while living."[12]

In an address at a memorial meeting held by the American Society of Mechanical Engineers on 19 April 1882, in Philadelphia, Bayles said:

> In the position of Consulting Engineer of the Bessemer Association, he worked with tireless industry, but with probably less advantage to his reputation and fortune than had he worked in another capacity. His responsibilities were great, and although his work cannot be said to have been profitable in proportion to its value to those for whom it was performed, he was too conscientious to consider his own interests to the sacrifice of theirs. It was not until a

manifestation of their feeling of absolute ownership stung Holley's proud and sensitive nature, that he sought to sever this connection by resigning his position. But they were in no mood to part with one on whom they had learned to depend so absolutely. He remained to the last trammelled by a professional engagement he could not resign and was loath to sever, but which gave him an opportunity in the line of his best abilities.[13]

It was Bayles who recalled a remark Holley made when he knew he could not recover. "I should like to live ten or fifteen years longer to aid in realizing the possibilities of the open hearth process. This would have rounded and completed my professional career; but I am satisfied."[14]

It had been common knowledge among the steel companies that much of the negotiation between Holley and the companies involved some controversy and even misunderstanding. Many members of the Bessemer Steel Company had opposed the purchase of the shell patent outright or had demanded terms more favorable than those that Holley had set. Although the Bessemer Steel Company did not double the value of Holley's estate as Bayles had advocated, it did appropriate $50,000 at its March 1882 meeting as compensation for Holley's last patent. Since the group probably was aware that the basic Bessemer process would not be used actively for some time they may have considered the sum to be more a contribution to the family than an investment.[15]

Three months after his death, Holley was awarded the Bessemer Medal, the most coveted award in the field in which he had worked, and one that many friends believed had not been given him earlier because of petty jealousies.[16] Each of the three professional societies of which Holley had been a member held memorial meetings within a few months of his death. By November 1883, almost two years later, the three associations had collected $10,000 for a memorial to Holley. They expected to have it placed in Central Park in New York City, but the memorial was placed instead in Washington Square in that city.

Holley carried on his activities among many companies and in many countries, and no individual society or company could claim him as its own. The book in which he kept a record of the companies to which he had sent drawings illustrated the wide scope of both his national and international work. Not only were all the Bessemer plants in the United States represented but here were also the names of many of the leading steel companies in Europe, such as Bolckow, Vaughan Company, West Cumberland Iron and Steel Company, and many others in England; De Wendel and Creusôt in France; Krupp in Germany; and Witkowitz in Austria, to mention only a few.

During the final years of his life Holley had been working to organize an international meeting of engineers. Initially it was planned that European and American engineers would meet in the United States in

Alexander Lyman Holley bust in Washington Square, New York City, (Courtesy of Francis Olmsted.)

1880. However, the meeting that Holley long had hoped for did not take place until 1890 when members of the Iron and Steel Institute, the Verein Deutscher Eisenhuttenleute of Germany, as well as the French iron and steel group, gathered in New York with members of the American Institute of Mining Engineers, the American Society of Mechanical Engineers, the Society of Civil Engineers, and the American Iron and Steel Association. Although Holley had been dead for eight years, many of the group remembered. On 2 October 1890, the group gathered for a memorial session at Chickering Hall to hear James Dredge speak at length about Holley's life and work, and afterward went to Washington Square for the unveiling and dedication of their monument to him.[17]

Much of what Dredge said in his address the members already knew but he felt that some of his thoughts could bear repeating. He described the early installation at Troy.

This was the beginning of the Bessemer steel industry in the United States. ... I think I echo the opinion of those who feel an interest in this subject when I ascribe all the credit to Alexander Lyman Holley. In doing so I in no way forget his great friends and contemporaries and their successors who cooperated with him from the commencement and afterwards. But it was to his genius and energy and patience that the Bessemer process in America was a success from the casting of the first at the Troy works. It was Holley, who from the beginning and during a number of years, adapted the English details of the process to suit American requirements, and who initiated that system of rapid production, combined with a high standard of excellence, which aroused at first the incredulity, and afterwards the admiration of European manufacturers.[18]

Many distinguished officers of large steel companies and metallurgists in the United States and abroad who could not attend the meeting sent letters to Dredge expressing the admiration and affection that they still felt for their American associate.

Percy Gilchrist's message was typical of others: "I have the greatest admiration for the work Holley achieved, as I think that the present proud position of the American steel trade is very largely due to his genius. A lesser man would have copied the European steel-works, but instead of this, he picked out the weak places, and in the American works that were fortunately subject to his advice, one sees how their defects were avoided by him, how breakdowns were eliminated, labour saved, and output consequently increased."[19]

As the years passed, Holley still was remembered for his achievements. In 1893, at the Palace of Mechanic Arts at the World's Columbian Exposition, the names of the four most outstanding American engineers were selected to be engraved on the entablature of the east facade, and Holley's name was among them.[20]

In 1912, thirty years after Holley's death, his friend and colleague Robert Hunt received the John Fritz medal for notable scientific or industrial achievement, which was awarded by the four national engineering societies. Hunt included some reminiscences in his acceptance speech. He recalled that the knowledge of chemistry and metallurgy was exceedingly limited in those pioneering days. Engines had not been developed to blow air at high pressure or in large volumes. The use of electricity as power was unknown and there was little development in the field of labor saving devices. "Things today easy were then hard," he recalled, "but we had the benefit of an element which under our then environment, exercised a greater power than I would believe would be possible today and that was personal friendship." Few companies then existed and thus the engineers were drawn together by their common difficulties and the fascinating problems they all encountered in trying to overcome them. Hunt was picturing a way of life long gone. Many of the men present found it difficult to imagine the relationships that had

existed in the group made up of Holley, George Fritz, Bill Jones, John
Fry, Robert Forsyth, and Robert Hunt with Uncle John Fritz serving as
their advisor and as monitor of the organization that Hunt fondly remembered as the "Order of the Bessemer Boys."

"This spirit of comradeship," he recalled, "was largely inaugurated
and fostered by the commanding intellect and more than all, by the
love-compelling personality of the man who became the consulting engineer to most of the works, Alexander Lyman Holley."[21]

NOTES

1. ALH to Robert Hunt, 28 June 1881, Holley Family Papers (hereafter cited as HF).
2. ALH to Dr. August Wendel, 20 June 1881, HF. He assured Wendel, however, that
   with the cards that he was enclosing with his letter and with credentials from his
   company, Wendel should have no difficulty in gaining admission to any plant he chose
   to visit.
3. ALH to James Dredge, 20 June 1881, HF.
4. James Dredge, "Holley Memorial Address," *Transactions of the American Institute of
   Mining Engineers* 20 (1891): 1i.
5. ALH to Maria Holley, 21 January 1882, HF.
6. *Iron Age* 29 (2 February 1882): 1.
7. *Engineering and Mining Journal* 33 (4 February 1882): 62.
8. Ibid., p. 63. Holley's father, seriously ill at his home in Lakeville, could not be told
   of his son's death. Knowing of his son's illness, he was depressed by the lack of news.
   Finally, because he was anxious and worried, the family considered it best that he be
   told. The old man was crushed by the loss of his last son and could only cover his
   face with his hands and murmur over and over, "My son, my son" (*Alexander Hamilton
   Holley, Memorial Volume* [Privately circulated, 1888]).
9. *Iron Age*, 2 February 1882, p. 1.
10. Chester Griswold, Secretary, Pneumatic Steel Association, *Patents Owned by the
    Pneumatic Steel Association and Still in Force* (New York: Pneumatic Steel Association,
    1876), 3 pp., Author's Collection; *Memorial of Alexander Lyman Holley* (New York:
    American Institute of Mining Engineers, 1884), p. 31.
11. *Memorial of Alexander Lyman Holley*, p. 24.
12. *Iron Age* 29 (2 February 1882): 1.
13. *Memorial of Alexander Lyman Holley*, p. 68.
14. Ibid., pp. 71–72, 74. James Bayles, as a journalist, had established a close relation-
    ship with Holley many years before when, as a young and inexperienced writer, Bayles
    first had approached the already established and busy engineer. "It is the experience
    of the journalist that truth is sometimes difficult of access," he said. "The statements
    made to him are often colored by a regard for real or supposed self-interest, and
    misrepresentation masquerades in the garb of frankness and confidence. But when
    I met Holley and sat with him for half an hour in conference, I would have pinned
    my faith to his deliberate statement, otherwise unsupported, against the oaths of all
    the world.
15. *Engineering and Mining Journal* 33 (4 March 1882): 118.
16. *Memorial of Alexander Lyman Holley*, p. 73.

17. The bronze bust of Holley was executed by J.Q.A. Ward and was mounted on a stone pedestal designed by Carrière and Hastings on which was inscribed:

HOLLEY

Born in Lakeville, Conn., July 20, 1832

Died in Brooklyn, N.Y., January 29, 1882

IN HONOR OF

ALEXANDER LYMAN HOLLEY

FOREMOST AMONG THOSE

WHOSE GENIUS AND ENERGY

ESTABLISHED IN AMERICA

AND IMPROVED THROUGHOUT THE WORLD

THE MANUFACTURE

OF BESSEMER STEEL

THIS MEMORIAL IS ERECTED

BY ENGINEERS

OF TWO HEMISPHERES

The Memorial was unveiled by Holley's six-year-old grandson, Alexander Holley Olmsted.

18. James Dredge, "Henry Bessemer, 1813–1898," *Transactions of the American Society of Mechanical Engineers* 19 (1898): 938.

19. Dredge, "Holley Memorial Address," pp. xxxi–xxxv passim.

20. William M. Henderson, "Four Distinguished Names," *Cassier's Magazine* 5 (November 1893–April 1894): 233–36.

21. Robert W. Hunt, "Starting Bessemer Steel Making in America," *Iron Age* 90 (12 December 1912): 1371.

Appendix 1 · Later Years of Holley's Friends and Associates

HENRY BESSEMER retired from active business in 1872 and spent the remainder of his life pursuing his hobbies, scientific and otherwise. He was more fortunate than many other inventors in the returns he received from his patents, particularly that for the steel process. It has been estimated that he received much more than £ 1,000,000 from this patent alone by the time it expired in 1870.

The interests he pursued after his retirement were the ship with the swinging saloon and an observatory, which he had built at his residence. He constructed a telescope for the observatory and later worked on the mechanisms necessary for the grinding and polishing of lenses. He also became interested in the subject of solar heat and its application in producing high temperatures in a solar furnace. Bessemer had established a grandson in business, and for it he planned a diamond cutting and polishing plant.

Perhaps his greatest pleasures were derived from the house and grounds located on a forty-acre site at Denmark Hill, then a suburb of London. The residence stood on high ground with a view of the famous Crystal Palace.

Several writers have commented that while Bessemer was financially successful he was less so in the number of honors and awards given him. The writer, Ernest Lange, thought that the rewards were "insignificant in view of his great services to the industrial world." Bessemer did receive awards from the technical societies to which he belonged, but a knighthood was delayed.

Bessemer never recovered from the death in 1897 of his wife, to whom he had been married for sixty years, and he died the following year at the age of eighty-five.

ANDREW CARNEGIE remained active in the steel industry until 1901, when he sold the Carnegie Steel Company to J. P. Morgan for $480,000,000.

Having written in his famous article "Wealth" which appeared in the *North American Review* of June 1889, that the man who died rich died disgraced, Carnegie then turned his attention to large-scale philanthropy.

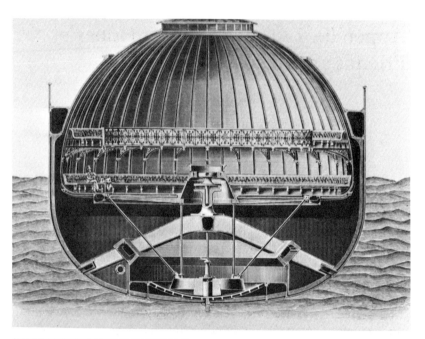

Two sectional views of early form of Bessemer saloon, one in still water and the other with vessel rolling. From *Sir Henry Bessemer: An Autobiography* (London: *Office of Engineering*, 1905).

The first of a host of projects was the Carnegie Institute in Pittsburgh. This was followed by the famous library program, under which about twenty-eight hundred free public library buildings were constructed. Other projects included the Scottish Universities Trust, the Dunfermline Trust, the Carnegie Institution in Washington, Support of a Simplified Spelling Program, the Hero Fund, and the Carnegie Endowment for International Peace, to mention a few.

But no matter how much Carnegie gave away, his fortune continued to mount, and it appeared certain that Carnegie would die a rich man. Finally, in 1911, he created the Carnegie Corporation of New York and transferred the larger portion of his fortune to this organization to be administered as a sacred trust for the "good of his fellow men." It was a source of relief to Carnegie to know that his wealth would benefit mankind long after his death. His wife and daughter had been provided for. Carnegie lived until 11 August 1919. He was buried in Sleepy Hollow Cemetery, North Tarrytown, New York.

WILLIAM DURFEE held various positions after leaving Wyandotte, the last being with the U.S. Mitis Company, which was engaged in promoting a new process for the production of iron and steel castings. He wrote many articles for magazines and also presented papers before various technical societies. He died in 1899 at the age of sixty-six.

JOHN FRITZ remained at the Bethlehem Iron Company until he retired at the age of seventy in 1892. One of the most loved and respected men in the iron and steel industry, he was the recipient of many honors and testimonials of esteem.

In 1893, he served as president of the American Institute of Mining Engineers and, in 1894, as president of the American Society of Mechanical Engineers. The prestigious John Fritz Medal awarded by the five engineering societies was initiated in his honor. He was awarded the Bessemer Medal in 1892.

Fritz became one of the trustees of Lehigh University when it was founded in 1866. At the age of eighty-seven he not only gave Lehigh $50,000 to build Fritz Engineering Laboratory, but he also took an active interest in its construction, selected much of its equipment, and bequeathed a further $150,000 to Lehigh University for its maintenance. Fritz died in 1913 at the age of ninety-one.

GÖRAN GÖRANSSON, to whose aid Bessemer owed so much of his first success, continued as chairman of the board of directors of the famous Sandvik Iron and Steel Works, which he had founded. In 1865, he received the gold medal of the Swedish Jernkontor. He was a Knight Grand Cross of the Royal Wasa Order and a Knight of the Royal Order of the Polar Star. He was a man of the broadest culture and of a most lovable personality. In 1898 members of the British Iron and Steel Institute visited Sweden.

Although he was very infirm, Mr. Göransson, welcomed them to the Sand-vik works. He died at Sandviken in 1900 at the age of eighty-one.

ABRAM HEWITT was involved in politics during much of his life. In 1871 he had played a major role in the reorganization of Tammany after the fall of the Tweed Ring. From 1875 until 1886, with the exception of the two years between 1879 and 1881, he served as a representative in Congress. He was elected as a reform mayor of New York City in 1886, winning over both Henry George and Theodore Roosevelt. The New York *World* said, after Hewitt had been in office for six months, that no other mayor ever had introduced so many important reforms in so short a time. All his life Hewitt was a man to be reckoned with; he had a quick temper, which was aroused by delays in the programs he wished to have adopted.

Hewitt was a staunch supporter of the gold standard and an ardent foe of William Jennings Bryan. He spoke often and boldly about the relation-ship between labor and capital. His thoughts on the subject frequently were a source of dismay to the industrialists, who were never quite certain what position he might take.

The Trenton Iron Company, of which Hewitt was a partner along with Edward Cooper, never had a strike. From 1873 to 1878 the mill operated at a yearly loss of $100,000, much of which Hewitt paid out of his own pocket. Allan Nevins writes that Hewitt was not a rich man and made sacrifices to maintain the men's positions at Trenton. He gave so gener-ously that he was frequently short of funds.

Probably most of his energy was devoted to placing Cooper Union, which his father-in-law Peter Cooper has started, on a more solid founda-tion. He was successful in securing donations of $600,000 from Andrew Carnegie, $250,000 from H. H. Rogers, and large amounts from many others.

He was elected president of the American Institute of Mechanical Engineers in 1890 and received the Bessemer Medal in the same year. Hewitt remained active until the end of his life, and took keen interest in his properties and in public affairs. His health never had been good and with the years it had deteriorated steadily. By January 1902, he was being kept alive only by oxygen and it became obvious that he was fatally ill. With his family at his bedside, he suddenly removed the tube from his mouth, whispering "I am officially dead." And in one second he was, at the age of eighty-one.

ROBERT W. HUNT remained in charge of the Albany & Rensselaer Iron & Steel Company, later the Troy Steel & Iron Company, until he resigned in April of 1888. During his long tenure at Troy he almost completely rebuilt the various works of the company and was also responsible for the large blast furnace built there. He greatly enlarged the number of products that the company could manufacture and developed grades of steel that previously had not been made in a Bessemer converter. In conjunction with Captain Jones and Max M. Suppes, he developed rail mill feed tables that were used in the majority of rail mills in the United States.

He was elected president of the American Institute of Mining Engineers in 1883 and held the same office in 1891 in the American Society of Mechanical Engineers. Through the years he had been interested especially in the field of testing materials and held important posts with organizations working in this area.

In April of 1888, he left Troy and moved to Chicago where he established the firm of Robert W. Hunt & Company, Inspection and Consulting Engineers. The company grew to be one of the largest of its kind in the world, with offices and laboratories in the principal cities of the United States as well as in London, Mexico City, and Canada.

Mr. Hunt was awarded the John Fritz Medal in 1912 for his contributions to the early development of the Bessemer process. He had the unique distinction of having been elected honorary member of every American engineering society. Hunt died in 1923 at the age of eighty-five.

CAPTAIN WILLIAM JONES remained with the Edgar Thomson Works until his death on 26 September 1889. Under his direction, the plant had become the most productive in the world.

He was responsible for many patents, among them washes for ingot molds, hot beds for bending rails, a feeding appliance for rolling mills, roll housings, and apparatus for removing ingots from molds. However, his last and perhaps his greatest invention was the Jones mixer, which was an ingenious method for mixing the molten pig iron taken from blast furnaces before charging it into a converter.

Jones died a terrible death. One of the Braddock furnaces had not been working satisfactorily. Cinder had lodged in the furnace, and, while the men were making efforts to release it, the furnace burst, throwing flame and gas upon the men who were standing before it. Almost immediately there was a deluge of hot ore, coke, and limestone, which was incandescent but not yet fused.

About twenty men were burned badly, among them Bill Jones, who was directing the work. Trying to escape the cascading inferno, he sprang back and fell ino a pit, striking his head upon the iron edge of a car. One of his workmen fell beside him and was killed instantly. Jones died the next day from burns and the fall, never having regained consciousness. He was only fifty years old.

Casson reports, "The five thousand workmen at Braddock were frantic with grief. Never before or since has the iron and steel world had so great a sorrow. Carnegie, looking upon poor Jones as he lay in the hospital, sobbed like a child. Ten thousand wet-eyed men marched with him to his grave, and today the veteran steel-maker's most precious memory is 'I worked with Bill Jones.'"[1]

WILLIAM KELLY did little more than look after his increasing royalties after his patent was renewed in 1871. He died in 1888 at the age of seventy-nine in Louisville, Kentucky.

PIERRE MARTIN was awarded the Bessemer Medal by the British Iron and Steel Institutue in 1915. In awarding the medal the presiding chairman described Martin's contributions as follows.

"Mr. Martin's first patent was taken out in July of 1865, and, as was well known, consisted of melting a mixture of pig iron with steel scrap or with oxide of iron or with both. The actual discovery that steel could be made in that way was not new. Many metallurgists before Martin's time had demonstrated the possibility of the method but all their efforts to carry the results into practical effect had failed mainly for the reason that they lacked the means of maintaining continuously a sufficiently high temperature in the hearth. In 1861 a great step in advance was made when Sir William Siemens invented the regenerative furnace by means of which the difficulty of temperature was overcome.

Pierre Martin carried out his experiments in a Siemens furnace of one ton capacity, and after many disappointments and trials he at length succeeded in applying the process to steel-making on an industrial scale. At the Paris Exhibition of 1867 Mr. Martin showed a number of commercial products obtained by his process, for which he was awarded the Gold Medal at the Exhibition and the process was shortly afterwards adopted at several French steelworks. The invention of Pierre Martin was incontestably a most important one, in that he obtained a class of steel of regular quality and composition by a new method. He had, however, attained his end by the use of a combination of appliances and processes most of which were already known. It could be no matter of surprise, therefore, that immediately the success of Martin seemed assured, the validity of his patent was attacked on all sides. Costly lawsuits followed which he unfortunately had not the means successfully to defend, and he was eventually compelled to abandon the struggle and retire into private life, without reaping the just reward of his industry and perseverance, although it had never been denied that the merit of the practical solution of the problem belonged to him alone. The obstacle to the extensive adoption of the Martin process in its early days was that by the time the invention had reached the practical stage, the Bessemer process already occupied the field, and on account of the greater cost of the Martin process, it was unable for many years to make headway against its powerful rival. But the excellence of the product told, and the Martin process gradually won its way. The introduction of the basic process by Thomas and Gilchrist in 1879 enormously increased its application until today it has far outdistanced the Bessemer in quantity of production. In 1913 upwards of 44,000,000 tons of acid and basic steel were produced by the Martin process whilst the production of the Bessemer process scarcely reached 30,000,000 tons. Those facts alone afforded to the world some indication of the vast importance of the invention with which Mr. Martin's name was honorably associated."[2]

Toward the end of 1909 Martin was discovered to be living in straitened circumstances near Nevers. When this fact became known the Comité des Forges de France instituted a fund with a gift of £4,000 as a form of reparation to one who had been so neglected. In England the Iron and Steel Institute took the initiative in contributing to the fund and was followed by the steelmaking countries—Germany, Belgium, Italy, and Austria-Hungary, among others.

A few days after he received the Bessemer Medal in 1915, Martin died at the age of ninety-one.

SIR WILLIAM SIEMENS pursued many interests and developed other inventions after the years during which the open hearth process was making its way.

The Siemens brothers' plant in England prospered along with the one in Berlin and, despite several setbacks, the company became famous for cable laying and work in telegraphy.

The years between 1870 and 1880 were probably the most productive of Siemens's life. In 1874 he designed a steamship, the *Faraday*, especially for cable laying. In addition to his work in telegraphy and in the field of electric power, he worked on electric lighting, the development of the electric pyrometer, and the bathometer and attraction meter.

He also wrote numerous papers describing his work and appeared often before technical societies. He regularly attended the meetings of the Royal Society and of the Institution of Civil Engineers.

He found time to lavish care on an estate of about one hundred sixty acres near Tunbridge Wells. In one of the buildings on the grounds he built a steam engine for operating a dynamo-electric machine that produced the electric current used in the house and gardens. Waste steam was utilized for heating.

Later years again were devoted to his first interest, the utilization of heat and electric power. Siemens was active in the efforts of the smoke abatement movememt dedicated to eliminating the London fog. At this time he developed an improved gas grate, which produced great heat but made little smoke. As one contemporary commented, it was part of the regenerative furnace brought into the drawing room.

Siemens also investigated growing plants under electric light, worked on the use of electric power for railways, and developed an electric furnace to melt steel. Siemens was interested in education for civil engineering and, as a member of the board of visitors of the Indian Engineering College, devoted considerable time to revising and improving the school's curriculum.

An associate who saw Siemens daily described a typical day in his life:

His secretary was with him at nine o'clock nearly every working day of the year; there was work for one society or another to be done, proofs of the abstracts of the Institution of Civil Engineers to be examined, letters and opinions on scientific subjects to be dictated, frequently also specifications of new inventions already schemed out. Then followed the walk across the Parks almost at racing speed, to Westminster; the business of the Landore Siemens Steel Company, of Messrs. Siemens Brothers and Co. (of both which large undertakings he was chairman), the work in connexion with the furnaces and metallurgical operations of which he was the inventor; visitors and enquirers to be seen, and in the afternoon attendance at council meetings of the learned societies, or directors' meetings of various companies. The evenings, again, were spent at one or other of the learned societies. This gives a faint idea only of the way Sir William Siemens passed his weeks, months and years. When a man has to address himself within the space of an hour to subjects so different as those for instance of telegraphy and metallurgy in their scientific aspects; when he has to consider workmen and wages one moment, licenses and specifications of inventions the next; when, as was constantly the case, half-a-dozen people were awaiting at the same time to see him, each thinking his own business the most important, and to each of whom, Sir William gave his attention, the wonder is that he has been able to work so long. [3]

Siemens and his wife, who was English, were a gregarious couple. They entertained frequently and lavishly in their town house. Perhaps the most spectacular of their parties was the reception given in honor of the emperor of Brazil. The Siemens's firm had laid a submarine cable between Rio de Janeiro and the coast of Uruguay for which Siemens had been decorated several times by the emperor. A few years later, in 1877, when he visited England, the emperor asked especially to spend an evening with the Siemens, and this was the occasion for the party in his honor.

In 1883, William Siemens was knighted, another honor added to the many given him by other countries. Shortly thereafter, as Siemens was walking home from a meeting at the Royal Institution he tripped on a curbstone and fell heavily on his left arm. Insisting that he was all right, he continued his walk home and continued his activities. Ten days later he awoke with a sharp pain in the region of his heart. He steadily weakened and died on 19 November 1883. An autopsy indicated that the presence of a serious heart ailment and a slight rupture from the fall had aggravated the condition.

The funeral services, which were held in Westminster Abbey, were attended by noted scientists and other distinguished people. The line of carriages in the drive to Kensal Green, where Siemens was buried, was so long that at no point in the route could the whole procession be seen. As a memorial, members of the five engineering societies with which Siemens was connected most intimately had installed a stained glass window in Westminster Abbey in the north side of the nave next to the memorial of Robert Stephenson.

The Siemens had no children. His personal estate amounted to a little over £380,000, far less than that of Bessemer. His biographer writes that the amount left by Siemens "does not represent nearly the money he earned, seeing the heavy losses that at different times he had to sustain, the liberal gifts he made, and the large sums he spent in other ways. It was said by one of his friends, that he had made three fortunes; of which he lost one, spent one, and kept one."[4]

SIDNEY GILCHRIST THOMAS, despite recurring bouts with a respiratory ailment, in February 1881 became active in the establishment of a steel plant especially designed for the production of basic steel. As its first and most active director, he was the moving spirit in its organization. Thomas also secured most of the necessary capital. Later the company became a part of Dorman Long & Company which, in turn, became a part of the British Steel Corporation.

In March of 1881, Thomas made his celebrated trip to the United States. For a man in his state of health the trip was overly strenuous even though it excited and delighted him.

In July of 1881 Thomas returned home and soon he was traveling frequently to the steelworks abroad. By fall it was evident that he was in no condition to spend the winter in London and his doctor advised him to go to the south of France. But Thomas believed that he could not spare the time and compromised by going to Torquay. No matter where Thomas went, however, he was followed by large amounts of correspondence, much of which related to the basic process, but a large amount to the numerous social causes that Thomas's growing fortune enabled him to support.

His health steadily deteriorated. He sought relief by taking a trip around the world and making a lengthy stay in Algeria, but to no avail. In Paris, where he had sought further treatment, Thomas died on 1 February 1885. He was not yet thirty-six years old. He had asked to be buried in Passy Cemetery in order to spare his family the ordeal of a channel crossing with his body in bad weather.

Thomas appointed his sister Lilian as his sole executrix and legatee. At the age of twenty-five she was entrusted with a difficult task. In his will, Thomas wrote that she never must spend more that £300 a year (at the most £350) on herself, in addition to the expenses of administering the requirements of the will. The balance she was to spend doing good with discretion. In her old age, Lilian Thomas gave an accounting of how the money had been spent. Thomas indeed would have been proud of the varied paths his money had traveled in "doing good." He might also have been surprised by the bold ventures his sister undertook.

For her first projects she investigated the so-called home industries. The first of these was the handling of matches that caused lead poisoning, the next was fur pulling, which entailed pulling out the long hair of rabbit skins in order to leave the close fur for the furrier. The workers labored in the rooms in which they lived, and the air was full of dust and fur, which caused a high death rate from lung and stomach troubles, particularly among infants. In both instances, Miss Thomas won. The match companies put an end to the dreadful conditions. The matter of fur pulling went directly to the House of Commons, and a private bill was introduced listing fur pulling as a dangerous trade and prohibiting it as a home industry. Aid was provided for those who were thrown out of work as a consequence.

Families who wished to emigrate to Canada and Australia were helped. Hostels were set up in Canada to give aid to the new emigrants. Cottages were built to replace hovels in various sections of England.

The Thomas contributions to the National Trust were to be used to save beautiful old examples of English craftsmanship from demolition. Money was spent on the creation and maintenance of parks and the building of parish and village halls. Scholarships were granted. Much money was spent in the successful fight to prohibit the opening of saloons at the pit mouths of collieries. There were special projects to help improve the working conditions of factory girls and shop assistants.

Five thousand pounds was given to the Thames Police Court to establish the Sidney Gilchrist Thomas Trust. While he worked at the Thames Police Court, with little money, he often had paid for investigations of the home lives of prisoners so that justice could be done and innocent victims of circumstances assisted. The establishment of the trust made it possible to engage a person for such full-time activity who could be helpful to the persons brought into the court, and thus carry out the intentions of Thomas.

For help and advice, Miss Thomas enlisted the aid of such women as Mrs. Ramsay MacDonald, Margaret Bonfield, Edith Hogg, and the Women's Industrial Council.

The responsibilities were heavy. She presented her accounts twice a year to four businessmen and corresponded with foreign licensees on points that arose from time to time. There was also the responsibility of investing and accounting for the "fighting fund" of £50,000 provided for her use in the event any defense of patent rights might be necessary.

CAPTAIN EBER WARD bought the Wyandotte experimental steel works from the holding group in 1865 and abandoned it in 1869. The converter built by William Durfee was only one of Ward's great number of enterprises. His holdings were one of the early examples of the conglomerate organization. He was probably the richest man in the Middle West. Although occupied with business affairs, he still had time to be active in politics both nationally and locally.

Ward never had been robust but he chose to ignore ill health and persisted in enlarging the scope of his activities even after a severe heart attack in 1869. After having been an invalid for a year, he recovered, but later collapsed on a Detroit street and died in January of 1875 at the age of sixty-four.

SAMUEL WELLMAN remained with the Otis Steel Company until 1 January 1889, and then became consulting engineer to the Illinois Steel Company. In 1890 he became president of the Wellman Iron and Steel Company, formerly the Chester Rolling Mills Company, in Chester, Pennsylvania. After four years of operation the company failed.

Wellman then returned to Cleveland where, along with his brother Charles and John W. Seaver, he organized the Wellman Seaver Engineering Company, later the Wellman-Seaver-Morgan Company, which specialized in the construction of open hearth installations. While at Otis, Wellman invented the open hearth charging machine and the electro-magnet for handling pig iron and scrap steel. The savings brought about by these two inventions ran into millions of dollars, and the inventions consequently were used wherever there was an open hearth plant of any size in the world.

Wellman died suddenly at the age of seventy-two on 11 July 1919. He had been on his way to a camping trip.

NOTES

1. Herbert N. Casson, *The Romance of Steel* (New York: A. S. Barnes & Co., 1907), p. 33.
2. *Journal of the Iron and Steel Institute* (London) 91, no. 1 (1915): 5-7.
3. William Pole, *The Life of Sir William Siemens* (London: John Murray, 1888), p. 395.
4. Pole, *Siemens*, p. 397.

Appendix 2 · Catalog of the Books, Addresses, and Professional Papers Written by Alexander Lyman Holley

Prepared by Rossiter W. Raymond, the list appears as Part VII, pages 142–152, in the *Memorial of Alexander Lyman Holley* (New York: American Institute of Mining Engineers, 1884).

I. BOOKS.

THE PERMANENT WAY AND COAL-BURNING LOCOMOTIVES OF EUROPEAN RAILWAYS; *with a Comparison of the Working Economy of European and American Lines, and the Principles upon which Improvement must Proceed.* By Zerah Colburn and Alexander L. Holley. With fifty-one plates by J. Bien. New York: Holley and Colburn, 1858. Folio, 168 pp. of text.

AMERICAN AND EUROPEAN RAILWAY PRACTICE *in the Economical Generation of Steam, including the Materials and Construction of Coal-burning Boilers, Combustion, the Variable Blast, Vaporization, Circulation, Superheating, Supplying and Heating Feed-water, etc., and the Adaptation of Wood and Coke-burning Engines to Coal-burning; and in Permanent Way, including Road-bed, Sleepers, Rails, Joint-fastenings, Street Railways, etc.* By Alexander L. Holley, B.P. With seventy-seven plates, engraved by J. Bien. New York: D. Van Nostrand; London: Sampson Low, Son & Co., 1860. (Second Edition 1867.) Folio, 192 pp. of text.

A TREATISE ON ORDNANCE AND ARMOR: *Embracing Descriptions, Discussions and Professional Opinions concerning the Material, Fabrication, Requirements, Capabilities and Endurance of European and American Guns for Naval, Sea-coast and Ironclad Warfare, and their Rifling, Projectiles and Breech-loading. Also, Results of Experiments against Armor, from Official Records. With an Appendix,*

Mr. Holley's articles in the Railway Advocate, the American Engineer, and the American Railway Review, and his editorial paragraphs in Van Nostrand's Eclectic Engineering Magazine are not included in this list. Their number, and the difficulty of identifying them with certainty, forbids the attempt to catalogue them. But the files of the Railway Advocate for 1855, 1856 and 1857, the American Engineer for a brief period in 1857, and the mechanical department of the American Railway Review for 1860 and the first half of 1861, abound in his work. He was editor of Van Nostrand's from its foundation in January 1869, for one year.—[R W. R.]

Referring to Gun-Cotton, Hooped Guns, etc. With 493 Illustrations. New York: D. Van Nostrand; London: Trübner & Company, 1865. Octavo, 900 pp.

II. ADDRESSES AND PROFESSIONAL PAPERS.

An Essay on Pen and Pocket Cutlery, Embracing a Detailed Description of the Mechanical, Chemical and Manual Operations Performed on Certain Raw Materials, to Convert them into the Means, Implements and Materials for Manufacturing Pen and Pocket Knives. By A. L. Holley. Published in Poor's *American Railroad Journal,* New York, during May, June and July, 1850.

The Natural Motors. Graduating Oration (unpublished), delivered September 7th, 1853, at Brown University, Providence, R.I.

Water Considered as a Carrier: the Properties upon which its Qualification for the Office Depends. By Dr. Hugag. Published in the *Litchfield Enquirer,* January 26th, 1854.

Corliss's Stationary Steam Engine. Published in the *Polytechnic Journal,* New York City, 1854. 10 pp., with diagrams.

The Dominion of Mind. An Oration, Delivered before the Theta Delta Chi Fraternity at their Annual Convention, by Alexander L. Holley, June 1st, 1855. Providence, B. T. Albro, Printer, 1852. Octavo pamphlet, 18 pp.

Ironclad War Vessels. By A. L. Holley. Published in the *National Almanac* for 1863. About 6 pp., small octavo.

Ironclad Ships and Heavy Ordance. Published in the *Atlantic Monthly* for January 1863. 10 pp., octavo.

Steel, and the Bessemer Process. A paper read before the Polytechnic Association of the American Institute, New York City, October 12th, 1865. Published in the *American Artisan.*

The Bessemer Process and Works in the United States. (Originally an Article in the *Troy Daily Times,* July 27th, 1868.) New York: D. Van Nostrand, 1868. Octavo pamphlet, 39 pp.

Rolling versus Hammering Ingots. A paper read before the American Institute of Mining Engineers, February 21st, 1872. Published in the *Transactions,* Vol. I., p. 203. Octavo, 3 pp.

Three-high Rolls. A paper read before the American Institute of Mining Engineers, October, 17th, 1872. Published in the *Transactions,* Vol. I., p. 287. Octavo, 6 pp., with plate.

Tests of Steel. A paper read before the American Institute of Mining Engineers, October 22d, 1872. Published in the *Transactions,* Vol. II., p. 116. Octavo, 6 pp.

Bessemer Machinery. A lecture delivered before the students of the Stevens Institute of Technology, Hoboken, N.J., in 1873. Published in the *Journal of the Franklin Institute,* Vol. LXIV. Octavo, 30 pp., with plates.

Recent Improvements in Bessemer Machinery. A paper read before the American Institute of Mining Engineers, February 24th, 1874. Published in the *Transactions,* Vol. II., p. 263. Octavo, 10 pp., with plates.

On American Rolling Mills. A paper read before the Iron and Steel Institute of Great Britain. Published in the *Journal* of the Institute, No. II. for 1874. Octavo, 19 pp., with plates.

Setting Bessemer Converter-Bottoms. A paper read before the Iron and Steel Institute of Great Britain. Published in the *Journal* of the Institute, No. II. for 1874. Octavo 6 pp., with plates.

The Porter-Allen Engine: Report on its Practical Performance, Economy, Durability, etc., with a View to Establishing its Manufacture. By A. L. Holley, C.E. New York: D. Van Nostrand, 1875. Octavo pamphlet, 11 pp.

On the Use of Natural Gas for Puddling and Heating, at Leechburg, Pennsylvania. A paper read before the American Institute of Mining Engineers, May 27th, 1875. Published in the *Transactions,* Vol. IV., p. 32. Octavo, 3 pp.

The Form, Weight, Manufacture and Life of Rails. Remarks before the American Society of Civil Engineers, June, 1875. Published in the *Transactions,* Vol. IV., p. 233. Octavo, 3 pp.

Tests and Testing-Machines. Remarks before the American Society of Civil Engineers, June, 1875. Published in the *Transactions,* Vol. IV., p. 265. Octavo, 44 pp.

Some Pressing Needs of our Iron and Steel Manufactures. Presidential Address before the American Institute of Mining Engineers, October 26th, 1875. Published in the *Transactions,* Vol. IV., p. 77. Octavo, 23 pp.

What is Steel? A paper read before the American Institute of Mining Engineers, October 29th, 1875. Published in the *Transactions,* Vol. IV., p. 138. Octavo, 12 pp.

The Inadequate Union of Engineering Science and Art. Presidential Address before the American Institute of Mining Engineers, February 22d, 1876. Published in the *Transactions,* Vol. IV., p. 191. Octavo, 17 pp.

Iron and Steel at Philadelphia. A series of articles by A. L. Holley and Lenox Smith, published in London *Engineering* as follows:
I. Iron and Steel at Philadelphia, November 10th, 1876.
II. Swedish Exhibits, December 6th, 1876.
III. Belgian and French Exhibits, January 5th, 1877.
IV. German Exhibits, January 19th, 1877.
V. The Exhibits of Great Britain, February 9th, 1877.
VI. American Exhibits, March 2d, 1877.

American Iron and Steel Works. A series of illustrated articles, by A. L. Holley and Lenox Smith, published in London *Engineering,* as follows:
I. An Analysis of the American Bessemer Plant, Illustrated by the Vulcan Works, St Louis, March 9th, 1877.
II. An Analysis of the American Bessemer Plant, Illustrated by the Vulcan Works, St Louis, March 16th, 1877.
III. The Midvale Steel Works, Philadelphia, March 30th, 1877.
IV. The Works of Park, Brothers & Co., May 4th, 1877.
V. The Union Iron Works, Buffalo, N.Y., June 22d, 1877.
VI. The Otis Iron and Steel Works, Cleveland, O., July 27th, 1877.
VII. The Works of the Bethlehem Iron Co., Bethlehem, Pa., August 24th, 1877.
VIII. The Works of the Bethlehem Iron Co., Bethlehem, Pa., August 31st, 1877.
IX. The Works of the Bethlehem Iron Co., Bethlehem, Pa., September 14th, 1877.
X. The Works of the Bethlehem Iron Co., Bethlehem, Pa., October 19th, 1877.
XI. The Works of the Bethlehem Iron Co., Bethlehem, Pa., October 26th, 1877.
XII. The Crescent Steel Works, Pittsburg, Pa., November 23d, 1877.
XIII. The Cedar Point Iron Company's Works, December 21st, 1877.
XIV. The North Chicago Rolling Mill Company's Works, January, 18th, 1877.
XV. The North Chicago Rolling Mill Company's Works, February, 8th, 1878.
XVI. Rail Mill of the Philadelphia and Reading Railroad, Reading, Pa., February 22d, 1878.
XVII. Mines and Works of the Crown Point Iron Company, N.Y., March 15th, 1878.

XVIII. Mines and Works of the Crown Point Iron Company, N.Y., March 22d, 1878.

XIX. The Diamond Furnace — Ferro-manganese Manufacture in the United States, March 29th, 1878.

XX. Works of the Cleveland Rolling Mill Company, O., April 5th, 1878.

XXI. Works of the Edgar Thomson Steel Company, Limited, Pittsburg, Pa., April 9th, 1878.

XXII. Works of the Edgar Thomson Steel Company, Limited, Pittsburg, Pa., April 26th, 1878.

XXIII. Works of the Edgar Thomson Steel Company, Limited, Pittsburg, Pa., May 17th, 1878.

XXIV. Plate Mill of the Bay State Iron Company, Boston, Mass., May 24th, 1878.

XXV. The Cambria Iron and Steel Works, Johnstown, Pa., May 31st, 1878.

XXVI. The Cambria Iron and Steel Works, Johnstown, Pa., June 21st, 1878.

XXVII. The Cambria Iron and Steel Works, Johnstown, Pa., July 12th, 1878.

XXVIII. The Cambria Iron and Steel Works, Johnstown, Pa., July 19th, 1878.

XXIX. The Cambria Iron and Steel Works, Johnstown, Pa., August 23d, 1878.

XXX. The Cambria Iron and Steel Works, Johnstown, Pa., September 30th, 1878.

XXXI. The American Iron Works of Jones & Laughlin, Pittsburg, Pa., November 1st, 1878.

XXXII. The Lucy Furnaces, Pittsburg, Pa., November 22d, 1878.

XXXIII. The Union Iron Mills, Pittsburg, Pa., January 10th, 1879.

XXXIV. The Works of Cooper, Hewitt & Company, Trenton, N.J., and Riegelsville, Pa., January 31st, 1879.

XXXV. The Works of Cooper, Hewitt & Co., Trenton, N.J., and Riegelsville, Pa., February 21st, 1879.

XXXVI. The Works of the Meier Iron Company, April 18th, 1879.

XXXVII. Salisbury Iron; and the Works of the Barnum Richardson Company, May 30th, 1879.

XXXVIII. The Works of the Phoenix Iron Company, Phoenixville, Pa., February 6th, 1880.

XXXIX. The Springfield Iron Works, Springfield, Ill., May 14th, 1880.

XL. The Albany and Rensselaer Iron and Steel Works, Troy, N.Y., December 24th, 1880.

XLI. The Albany and Rensselaer Iron and Steel Works, Troy, N.Y., December 31st, 1880.

Notes on the Salisbury, Conn., Iron Mines and Works. A paper read before the American Institute of Mining Engineers, October 23d, 1877. Published in the *Transactions,* Vol. VI., p. 220. Octavo, 4 pp.

Notes on the Iron Ore and Anthracite Coal of Rhode Island and Massachusetts. A paper read before the American Institute of Mining Engineers, October 24th, 1877. Published in the *Transactions,* Vol. VI., p. 224. Octavo, 3 pp.

The Strength of Wrought Iron, as Affected by its Composition, and by its Reduction in Rolling. A paper read before the American Institute of Mining Engineers, February 27th, 1878. Published in the *Transactions,* Vol. VI., p. 101. Octavo, 24 pp.

Solid Steel Castings for Ordnance, Structures and General Machinery by the Terrenoire Process. Published in the *Metallurgical Review,* New York, May, June and July, 1878. Octavo, 48 pp.

Chemical and Physical Analyses of Phosphoric Steel. A paper read before the Institution of Civil Engineers (England). Published in the *Proceedings,* Vol. VIII., Session 1877-78. Part III. London 1878. Octavo, 28 pp.

The United States Testing Machine at Watertown Arsenal. A paper read before the American Institute of Mining Engineers, February 18th, 1879. Published in the *Transactions,* Vol. VII., p. 256. Octavo, 9 pp.

The Pernot Furnace. A paper read before the American Institute of Mining Engineers, February 19th, 1879. Published in the *Transactions,* Vol. VII., p. 241. Octavo, 14 pp., with plates.

The Tessié Gas Producer. A paper read before the American Institute of Mining Engineers, May 13th, 1879. Published in the *Transactions,* Vol. VIII., p. 27. Octavo, 7 pp., with plates.

Washing Phosphoric Pig-Iron for the Open-hearth and Puddling Processes. A paper read before the American Institute of Mining Engineers, September 17th, 1879. Octavo, 9 pp., with plates.

The Field of Mechanical Engineering. An address delivered at the preliminary meeting of the American Society of Mechanical Engineers, February 16th, 1880. Published in the *Transactions,* Vol. I., p. 7 (in the earliest copies it is at p. 28). Octavo, 5 pp.

Notes on the Siemens Direct Process. A paper read before the American Institute of Mining Engineers, February 19th, 1880. Published in the *Transactions,* Vol. VIII., p. 321. Octavo, 4 pp., with plates.

Engineering, the Intermediate Power between Nature and Civilization. An address delivered at the banquet in honor of Count Ferdinand de Lesseps, New York, March 1st, 1880. Published, with other addresses on the same occasion, by D. Appleton & Company, New York, 1880.

An Adaptation of Bessemer Plant to the Basic Process. A paper read before the American Society of Mechanical Engineers, November, 1880. Published in the *Transactions,* Vol. I., p. 124. Octavo, 8 pp., with plates.

Steel. An article in *Appleton's Cyclopoedia of Mechanics,* New York, 1880. Octavo, 15 pp.

On Rail Patterns. A paper read before the American Institute of Mining Engineers, February 17th, 1881. Published in the *Transactions,* Vol. IX., p. 360. Octavo, 16 pp., with plates and tables.

The Bethlehem Iron and Steel Works. An article in London *Engineering,* October 28th, 1881.

Index

Note: Italic numbers refer to diagrams and illustrations.